ELEMENTS
OF SIMULATION

ELEMENTS
OF SIMULATION

Byron J. T. Morgan

MATHEMATICAL INSTITUTE
UNIVERSITY OF KENT, CANTERBURY, UK

LONDON   NEW YORK
CHAPMAN AND HALL

First published 1984 by
Chapman and Hall Ltd
11 New Fetter Lane, London EC4P 4EE
Published in the USA by
Chapman and Hall
733 Third Avenue, New York NY 10017

© *1984 Byron J. T. Morgan*

Printed in Great Britain at
The University Press, Cambridge

ISBN 0 412 24580 9 (cased)
ISBN 0 412 24590 6 (Science Paperback)

British Library Cataloguing in Publication Data

Morgan, Byron J. T.
    Elements of simulation
    1. Simulation methods
    I. Title
    001.4'24   T57.62

    ISBN 0-412-24580-9
    ISBN 0-412-24590-6 Pbk

Library of Congress Cataloging in Publication Data

Morgan, Byron J. T., 1946-
    Elements of simulation.

    Bibliography: p.
    Includes indexes.
    1. Mathematical statistics—Data processing.
2. Digital computer simulation.   I. Title.
QA276.4.M666   1984   519.5'0724   83-25238
ISBN 0-412-24580-9
ISBN 0-412-24590-6 (pbk.)

CONTENTS

v

PREFACE

The use of simulation in statistics dates from the start of the 20th century, coinciding with the beginnings of radio broadcasting and the invention of television. Just as radio and television are now commonplace in our everyday lives, simulation methods are now widely used throughout the many branches of statistics, as can be readily appreciated from reading Chapters 1 and 9. The rapid development of simulation during this century has come about because of the computer revolution, so that now, at one extreme, simulation provides powerful operational research tools for industry, while at the other extreme, it is also taught in schools. It is employed by a wide variety of scientists and research workers, in industry and education, all of whom it is hoped will find this book of use. The aim is to provide a guide to the subject, which will include recent developments and which may be used either as a teaching text or as a source of reference.

The book has grown out of a fifteen-hour lecture course given to third-year mathematics undergraduates at the University of Kent, and it could be used either as an undergraduate or a postgraduate text. For elementary teaching the starred sections and Exercises may be omitted, but most of the material should be accessible to those possessing a knowledge of statistics at the level of *A Basic Course in Statistics* by Clarke and Cooke (1983). This book, henceforth referred to as '*ABC*', is widely used for teaching statistics in schools. Most of the standard statistical theory encountered that is not in the revision Chapter 2 is given a page reference in *ABC*.

Simulation may either be taught as an operational research tool in its own right, or as a mathematical method which cements together different parts of statistics and which may be used in a variety of lecture courses. In the last three chapters indications are made of the varied uses of simulation throughout statistics. Alternatively, simulation may be used to motivate subjects such as the teaching of distribution theory and the manipulation of random variables, and Chapters 4 and 5 especially will hopefully be useful in this respect. To take a very simple illustration, a student who finds subjects such as random variables and probability density functions difficult is greatly helped if the random variables can be simulated and histograms drawn that may be compared with density functions, as is done in Chapter 2.

Inevitably simulation requires algorithms, and computer programs for the operation of those algorithms. A seemingly complicated algorithm can often be programmed in a simple way, and so the approach has been adopted here of occasionally presenting illustrative computer programs in the BASIC computer language. BASIC is chosen because of its current wide use, from home microcomputers upwards, and the programs are written in accordance with the American Standard for what is called 'Minimal BASIC', so that they need no modification for running on a variety of different implementations of BASIC. Graduate students and research workers in industry and elsewhere will frequently use pre-programmed computerized algorithms in their simulation work. Ready supplies of these algorithms are provided by the NAG and IMSL libraries, and Appendix 1 describes the relevant routines from these libraries, indicating also which methods are being used. Computer simulation languages are discussed in Chapter 8.

There are 232 Exercises and complements, some of which are designed to be of interest to advanced readers and contain more recent and more difficult material. These, and more advanced sections are marked with a *. In each chapter a small number of straightforward exercises are marked with a † symbol, and it is recommended that all readers should attempt these exercises.

ACKNOWLEDGEMENTS

This book was originally planned as a joint venture with Barrie Wetherill, but because of his involvement in several other books, it became mine alone. I am therefore particularly grateful to Professor Wetherill for his early motivation, advice and enthusiasm. Many others have been extremely helpful at various stages, and I am especially indebted to Mike Bremner, Tim Hopkins, Keith Lenden–Hitchcock, Sophia Liow and my wife, Janet Morgan, for discussions and advice. Tim Hopkins's computing expertise proved invaluable. Christopher Chatfield, Ian Jolliffe and Brian Ripley were kind enough to comment in detail on an earlier draft, and I have benefited greatly from their perceptive comments. Finally, it is a pleasure to record my appreciation for all the hard work of Mrs Mavis Swain who typed the manuscript.

Canterbury
May, 1983.

LIST OF NOTATION

	arrival times and exponentially distributed service times (p. 49)
$M_{\mathbf{x}}(\theta)$	multivariate moment generating function (p. 44)
$N(\mu, \sigma^2)$	normal distribution with mean μ and variance σ^2 (p. 23)
$N(0, 1)$	standard normal distribution
$N(\boldsymbol{\mu}, \boldsymbol{\Sigma})$	multivariate normal distribution, with mean vector $\boldsymbol{\mu}$ and variance/covariance matrix $\boldsymbol{\Sigma}$ (p. 40)
p.d.f.	probability density function (p. 13)
p.g.f.	probability generating function (p. 42)
$\phi(x)$	probability density function of a standard normal random variable (p. 23)
$\Phi(x)$	cumulative distribution function of a standard normal random variable (p. 23)
$\Pr(A)$	probability of the event A (p. 12)
r.v.	random variable
$U(a, b)$	uniform distribution over the range $[a, b]$ (p. 20)
U, U_i	$U(0, 1)$ random variables
$\text{Var}(X)$	variance of a random variable X (p. 15)

1

INTRODUCTION

1.1 Simulation

'Simulation' is a word which is in common use today. If we seek its definition in the *Concise Oxford Dictionary*, we find:

'**simulate**, *verb transitive*. Feign, pretend to have or feel, put on; pretend to be, act like, resemble, wear the guise of, mimic, ... So **simulation**, *noun*.'

Three examples are as follows:

(a) Following the crash of a DC-10 jet after it had lost an engine, the *Observer* newspaper, published in Britain on 10 June 1979, reported that, 'A DC-10 flight training simulator is being programmed to investigate whether the aircraft would be controllable with one engine detached.'
(b) Model cars or aeroplanes in wind-tunnels can be said to simulate the behaviour of a full-scale car, or aeroplane, respectively.
(c) A British television 'Horizon' programme, presented in 1977, discussed, with the aid of a simulated *Stegosaurus*, whether or not certain dinosaurs were hot-blooded. In this case the simulating model was very simply a tube conducting hot water; the tube was also equipped with cooling vanes, the shape of which could be changed. Different shapes then gave rise to differing amounts of cooling, and the vane shapes of the *Stegosaurus* itself were shown to provide efficient cooling. Thus these shapes *could* have developed, by natural selection, to cool a hot-blooded creature.

For statisticians and operational-research workers, the term 'simulation' describes a wealth of varied and useful techniques, all connected with the mimicking of the rules of a model of some kind.

1.2   What do we mean by a model?

It frequently occurs that we find processes in the real world far too complicated to understand. In such cases it is a good idea to strip the processes of some of their features, to leave us with *models* of the original processes. If we can understand the model, then that may provide us with some insight into the process itself. Thus in examples (a) and (b) above, it proves much cheaper and easier to investigate real systems through simulated models. In the case of (b) a physical scale model is used, while, in the case of (a), the model would most likely have been a computer simulation. In example (c) one can only employ simulated models because dinosaurs are extinct!

Subjects such as physics, biology, chemistry and economics use models to greater and lesser extents, and the same is true of mathematics, statistics and probability theory. Differential equations and laws of mechanics, for instance, can be viewed as resulting from models, and whenever, in probability theory, we set up a sample-space (see for example *ABC*[†], p. 62, who use the term 'possibility space') and assign probabilities to its elements, we are building a model of reality. Some particular models are described later. When the models are given a mathematical formulation, but analytic predictions are not possible, then quite often simulation can prove to be a useful tool, not only for describing the model itself, but also for investigating how the behaviour of the model may change following a change in the model; compare the situation leading to example (a) above. Abstract discussion of models is best accompanied by examples, and several now follow.

1.3   Examples of models which may be investigated by simulation

(a) Forest management

A problem facing foresters is how to manage the felling of trees. One possible approach is to 'clear fell' the forest in sections, which involves choosing a section of forest, and then felling all of the trees in it before moving on to a new section. An alternative approach is to select only mature, healthy trees for felling before moving on to the next section. A disadvantage of the former approach is that sometimes the trees felled will not all be of the same age and size, so that some will only be useful for turning into pulp, for the manufacture of paper, say, while others will be of much better quality, and may be used for construction purposes. A disadvantage of the latter approach that has been encountered in the Eucalypt forests of Victoria, in Australia, is that the resulting tree stumps can act as a food supply for a fungal disease called *Armilleria* root-rot. Spores of this fungus can be transmitted by the wind,

---

[†] *ABC: A Basic Course in Statistics*, by Clarke and Cooke (1983).

alight on stumps and then develop in the stumps, finally even proceeding into the root-system of the stump, which may result in the transmission of the infection to healthy neighbouring trees from root-to-root contact.

While one can experiment with different management procedures, trees grow slowly, and it could take a lifetime before proper comparisons could be made. One can, however, build a model of the forest, which would include a description of the possible transmission of fungal disease by air and root contact, and then simulate the model under different management policies. One would hope that simulation would proceed very much faster than tree growth. For a related example, see Mitchell (1975).

(b) Epidemics

Diseases can spread through animal, as well as plant populations, and in recent years there has been much interest in mathematical models of the spread of infection (see Bailey, 1975). Although these models are often extremely simple, their mathematical solution is not so simple, and simulation of the models has frequently been employed (see Bailey, 1967).

(c) Congestion

Queues are a common feature of everyday life. They may be readily apparent, as when we wait to pay for goods in a shop, cash a cheque at a bank, or wait for a bus, or less immediately obvious, as, for example, when planes circle airports in holding stacks, waiting for runway clearance; time-sharing computers process computer jobs; or, in industry, when manufacturers of composite items (such as cars) await the supply of component parts (such as batteries and lights) from other manufacturers. The behaviour of individuals in the more apparent type of queue can vary from country to country: Mikes (1946) wrote that the British, for instance, are capable of forming orderly queues consisting of just single individuals!

Congestion in queues can result in personal aggravation for individuals, and costly delays, as when planes or industries are involved. Modifying systems with the aim of reducing congestion can result in unforeseen secondary effects, and be difficult and costly. Here again, models of the systems may be built, and the effect of modifications can then be readily appreciated by modifying the model, rather than the system itself. An example we shall encounter later models the arrival of cars at a toll; the question to be answered here was how to decide on the ratio of manned toll-booths to automatic toll-booths.

Another example to be considered later deals with the queues that form in a doctor's waiting-room. Some doctors use appointment systems, while others do not, and a simulation model may be used to investigate which system may be better, in terms of reducing average waiting times, for example. (See also Exercise 1.6.)

(d) Animal populations

Simulation models have been used to mimic the behaviour of animal populations. Saunders and Tweedie (1976) used a simulation model to mimic the development of groups of individuals assumed to be colonizing Polynesia from small canoe-loads of migrants, while Pennycuick (1969) used a computer model to investigate the future development of a population of Great Tits. Gibbs (1980) used simulation to compare two different strategies for young male gorillas. The background to this last example is the interesting sociology of the African mountain gorilla, which lives in groups headed by a single male, who is the only male allowed to mate with the group females. In such a society young males who do not head groups have to choose between either biding their time in a hierarchy until the head male and other more dominant males all die, or attempting to set-up a group themselves, by breaking away and then trying to attract females from established groups. Both of these strategies involve risks, and the question to be answered was 'which strategy is more likely to succeed?'

(e) Ageing

The age-structure of the human populations of many societies is changing, currently with the fraction of individuals that are elderly increasing. The prospect for the future is thus one of progressively more elderly individuals, many of whom will be in need of social services of some kind. Efficient distribution of such services depends upon an understanding of the way in which the population of elderly is composed, and how this structure might change with time.

   Harris *et al.* (1974) developed categories which they called 'social independence states' for the elderly, the aim being to simulate the changes of state of elderly under various assumptions, and thereby to evaluate the effect of those assumptions. See also Wright (1978), and Jolliffe *et al.* (1982), who describe a study resulting in a simple classification of elderly into a small number of distinct groups.

1.4   The tools required

Models contain parameters, such as the birth and death rates of gorillas, the rate of arrival of cars at a toll-booth, the likelihood of a car-driver to possess the correct change for an automatic toll-booth, and so forth. In order for the models to be realistic, these parameters must be chosen to correspond as closely as possible to reality, and typically this means that before a model can be simulated, the real-life process must be observed, and so sampled that parameter estimates can be obtained; see, for example, *ABC*, chapter 10. In the case of modelling forests, the forests must first of all be surveyed to assess the extent of fungal infection. In the case of a doctor's waiting room, variables such as consultation times and inter-arrival times must be observed; see Exercises 1.6 and 1.7. With epidemics, one needs reports of the spread of disease, possibly within households on the one hand, or on a larger scale from field observations, as occurred with the recent spread of rabies in Europe, or of myxomatosis in Australia in the early 1950s. For useful comments on data collection see *ABC*, chapter 11, and Barnett (1974).

The models of Section 1.3 all involve random elements, so that the predictions of the models are not known or calculable with certainty. In some cases one encounters models which do not involve random elements, and then simulation is simply a case of enumerating the certain predictions of the model. Such models can occur in the area of manpower planning, for example; see Bartholomew and Forbes (1979). Models with random elements are the concern of this book, and the simulation of such models then necessitates the generation of such random variates as waiting times, service times and life times. In some cases longer times will be more likely than shorter times, while in others the converse will be true, and in general statistical terms what we need to be able to do is simulate random variables from a whole range of different statistical distributions; see for example *ABC*, chapters 7, 9, 13, 14 and 19. This, then, is the motivation for the material of Chapters 3, 4 and 5 of this book, while Chapter 2 provides a résumé of some of the results regarding random variables and distributions that we shall need in the remainder of this book.

An additional feature of the simulation of models is how to assess the performance of the model. Measures such as expected waiting times have already been mentioned, and this is a topic to which we shall return later. (See also Exercise 1.9).

1.5   Two further uses of simulation

As we shall see later, not only may we use simulation to mimic explicitly the behaviour of models, but also we can use simulation to evaluate the behaviour of complicated random variables whose precise distribution we are unable to evaluate mathematically. An early example of this also provides us with an interesting footnote to the history of statistics. Let us suppose that $(x_1, x_2, \ldots, x_n)$ form a random sample from a normal distribution of mean μ and variance σ^2 (details are given in Section 2.9). We can estimate the parameter σ^2 by

$$s^2 = \frac{1}{(n-1)} \sum_{i=1}^{n} (x_i - \bar{x})^2$$

(1.1)

where
$$\bar{x} = \sum_{i=1}^{n} x_i/n.$$

The quantity $z = \sqrt{n}(\bar{x} - \mu)/s$ is itself a realization of a random variable with a t-distribution on $(n - 1)$ degrees of freedom.

These t-distributions were first studied by W. S. Gossett, who published his work under the pen-name of 'Student'. In 1908, he wrote:

'Before I had succeeded in solving my problem analytically, I had endeavoured to do so empirically [i.e. by simulation]. The material used was a . . . table containing the height and left middle finger measurements of 3000 criminals . . . . The measurements were written out on 3000 pieces of cardboard, which were then very thoroughly shuffled and drawn at random . . . each consecutive set of 4 was taken as a sample . . . [i.e., $n = 4$ above] . . . and the mean [and] standard deviation of each sample determined. . . . This provides us with two sets of . . . 750 z's on which to test the theoretical results arrived at. The height and left middle finger . . . table was chosen because the distribution of both was approximately normal . . .'

This use of simulation has increased over the years, and is remarkably common today. We have seen that simulation can be used to assess the behaviour of models, and also of certain random variables. Similarly, it can be used to gauge the performance of various techniques.

Examples can be provided from the area of multivariate analysis (see Everitt, 1980; Jolliffe, 1972), time-series analysis (see Anderson, 1976), and also ecology (see Jeffers, 1972). In the area of cluster analysis, many different techniques have been devised for detecting groups in populations of individuals. One way of assessing the performance of these techniques, and of comparing different techniques, is to simulate populations of known structure and then to see to what extent the techniques identify that structure. This is done, for example, by Everitt (1980, p. 77) and Gardner (1978). In the ecological case, sampling techniques exist for assessing the size of mobile animal populations, such as sperm whales, for example. Frequently these techniques involve marking captured animals, releasing them, and then considering the proportions of marked individuals in further samples. One can examine the performance of such techniques by simulating the behaviour of animal populations of known size, and then considering how well the techniques recover the known population size. This is another application of simulation in common use today, and, like the investigation of sampling distributions, it also requires the material of Chapters 3, 4 and 5.

Both of the uses considered here are still, fundamentally, simulating models, though perhaps not quite so obviously as before, and in the second use above the model is only of secondary interest. We shall return to the examples of this section in Chapter 9.

1.6 Computers, mathematics and efficiency

Most simulation in statistics is carried out today with the aid of a computer. Interesting exceptions do exist, however. One example is provided by Fine (1977), in which we find a description of a mechanical model which mimics the rules of an epidemic model. Another is provided by Moran and Fazekas de St Groth (1962), who sprayed table-tennis balls with paint in a particular way, to mimic the shielding of virus surfaces by antibodies.

As computers become more powerful and more widely available, one might even ask whether simulation might render mathematical analysis of models redundant. The fact is that, useful though simulation is, an analytical solution is always preferable, as we can appreciate from a simple example. Exercise 1.4 presents a simple model for a queue. If $p < \frac{1}{2}$ in this model then, after a long period of time since the start of the queueing system, we can write the probability that the queue contains $k \geq 0$ customers as

$$p_k = \left(\frac{q-p}{2pq}\right)\left(\frac{p}{q}\right)^k \qquad \text{for } k \geq 1$$

with

$$p_0 = \left(\frac{q-p}{2q}\right), \qquad \text{where } q = 1 - p.$$

Mathematical analysis has, therefore, provided us with a detailed prediction for the distribution of queue size, and while we might, for any value of $p < \frac{1}{2}$, use simulations to estimate the $\{p_k, k \geq 0\}$ distribution, we should have to start again from scratch for every different value of p.

Analytical solutions could be mathematically complex, in which case simulation may provide a reassuring check, as was true of Gossett's simulations mentioned earlier. In some cases, approximate analytical solutions are sought for models, and then simulations of the original model can provide a check of the accuracy of the approximations made; examples of this are provided by Lewis (1975) and Morgan and Leventhal (1977).

The relative power of an analytical solution, as compared with a simulation approach to a model, is such that even if a full analytical solution is impossible, such a solution to part of the model, with the remainder investigated by simulation, is preferable by far to a simulation solution to the whole model. We find comments to this effect in, for example, Estes (1975), Brown and Holgate (1974), Mitchell (1969) and Morgan and Robertson (1980). Mitchell (1969) puts the point as follows:

'It is well to be clear from the start that simulation is something of a sledgehammer technique. It has neither elegance nor power as these terms would be understood in a mathematical context. In a practical context it can, however, be an extremely powerful way of understanding and modelling a system. More importantly from the practical point of view, it can be the only way of deriving an adequate model of a system.'

Mitchell goes on to emphasize a major disadvantage of simulation, which is that it could well be time-consuming and expensive. Even with powerful computers, poor attention to efficiency could make a simulation impractical. Efficiency in simulating is therefore very important, and is a common theme underlying a number of the following chapters, especially Chapter 7.

1.7 Discussion and further reading

We have seen from this chapter that simulation can be a powerful technique of wide-ranging applicability. Of course, the examples given here comprise only a very small part of a much larger and more varied picture. The current interest in simulation is reflected by the papers in journals such as *Mathematics and Computers in Simulation*, *Communications in Statistics*, Part B: *Simulation and Computation*, and the *Journal of Statistical Computation and Simulation*.

Further examples of models and their simulation can be found in Barrodale, Roberts and Ehle (1971), Gross and Harris (1974), Hollingdale (1967), Kendall (1974), Smith (1968) and Tocher (1975). In addition, many books on operational research contain useful chapters dealing with simulation, as do also Bailey (1964), Cox and Miller (1972) and Cox and Smith (1967).

1.8 Exercises and complements

†**1.1**   Toss a fair coin 500 times and draw a graph of
  (i) r/n vs. n, for $n = 1, 2, \ldots, 500$, where n is the number of tosses and r is the number of heads in those n tosses;
  and
  (ii) $(2r - n)$ vs. n, i.e. the difference between the number of heads and the number of tails.
  Here you may either use a coin (supposed fair), the table of random digits from Appendix 2 (how?) or a suitable hand-calculator or computer if one is available. A possible alternative to the table of random digits is provided by the digits in the decimal expansion of π, provided in Exercise 3.10.

  Comment   (a) on the behaviour of r/n
  and   (b) on the behaviour of $(2r - n)$.

1.2   In 1733, George Louis Leclerc, later Comte de Buffon, considered the following problem:

> 'If a thin, straight needle of length l is thrown at random onto the middle of a horizontal table ruled with parallel lines a distance $d \geq l$ apart, so that the needle lies entirely on the table, what is the probability that no line will be crossed by the needle?'

You may like to try to prove that the solution is $1 - 2l/\pi d$, but this is not entirely straightforward.

Dr E. E. Bassett of the University of Kent simulated this process using an ordinary sewing needle, and obtained the following results from two classes of students:

(i) 390 trials, 254 crossings;   (ii) 960 trials, 638 crossings.

Explain how you can use these data to estimate π; in each of these simulations, d was taken equal to l, as this gives the greatest precision for estimating π, a result you may care to investigate. We shall return to this topic in Chapter 7.

†1.3   Give five examples of processes which may be investigated using simulation.

*1.4   The following is a *flow diagram* for a very simple model of a queue with a single server; time is measured in integer units and is denoted by $T \geq 0$. T changes only when the queue size changes, by either a departure or a new arrival. $Q \geq 0$ denotes queue size. Explain how you would use the table of random digits in Appendix 2 to simulate this model

(i) when $p = 0.25$ and   (ii) when $p = 0.75$,

in each case drawing a graph of Q vs. T for $0 \leq T \leq 50$. For the case (i) estimate p_0, the proportion of time the queue is empty over a long period, and compare your estimate with the known theoretical value for p_0 of $p_0 = (q - p)/(2q)$.

Modify the flow diagram so that (a) there is limited waiting room of r individuals; (b) there are two servers and customers join the smaller queue when they arrive.

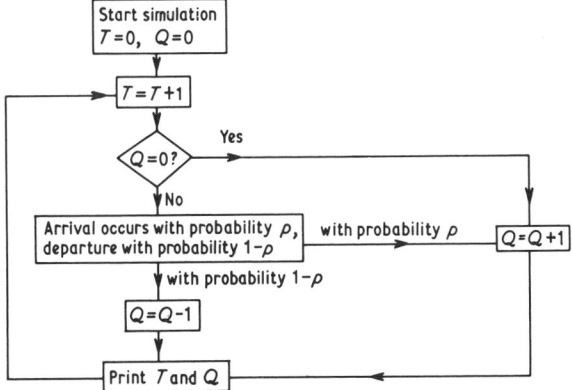

1.5   A canoe containing three men and three women lands on a previously uninhabited island. Discuss the information you require in order to model the society of these individuals, and how their population size changes with time.

†1.6   Consider how you would simulate the following model of a doctor's surgery operating an appointment system:
    (i) Patients are scheduled to arrive every 5 units of time.
    (ii) Independently of all other patients, each patient fails to turn up with probability 0.1.
    (iii) Independently of all other patients, each patient has arrival times with the following distribution:

	Early		On time	Late	
	2 units	1 unit		1 unit	2 units
probability:	$\frac{1}{10}$	$\frac{1}{5}$	$\frac{2}{5}$	$\frac{1}{5}$	$\frac{1}{10}$

    (iv) Consultation times have the following distribution:

Time units	2	3	4	5	6	7	8	9
Probability	$\frac{1}{10}$	$\frac{1}{10}$	$\frac{1}{10}$	$\frac{1}{5}$	$\frac{1}{5}$	$\frac{1}{10}$	$\frac{1}{10}$	$\frac{1}{10}$

    (v) Patients are seen in the order in which they arrive.

    Discuss how you might use this model to investigate the efficiency of the appointment system.

1.7   Criticize the assumptions made in the last exercise. It may help to consider the following data collected from a British medical practice by Keith Gibbs as part of his mathematics undergraduate dissertation at the University of Kent:

		Patient			
		Male		Female	
	Consulting time	Adult	Child	Adult	Child
Morning	≤ 5 minutes	8	2	24	11
	> 5 minutes	8	2	12	8
Afternoon	≤ 5 minutes	21	9	11	6
	> 5 minutes	6	4	6	2

1.8   Construct suitable histograms (*ABC*, p. 8) of $z = (\bar{x}\sqrt{n}/s)$, using samples of size $n = 2$ from the table of standard normal variables given in Appendix 2, and, separately, using samples of size $n = 4$ from the same table. Repeat this, but now for $z = (\bar{x} - 1)\sqrt{n}/s$, using the table of exponential random variables, and comment on the results.

1.9 One way of comparing different queueing systems is to compare mean waiting times. Can you envisage a situation in which other features of the entire distributions of waiting times might also be important?

1.10 Simulate English text, using the word frequencies of Dewey (1923). Examples can be found in Shannon and Weaver (1964).

2

SOME PROBABILITY
AND STATISTICS
REVISION

We have seen from Chapter 1 that in many uses of simulation, statisticians need to simulate discrete and continuous random variables of different kinds, and techniques for doing this are provided in Chapters 4 and 5. The aim of this chapter is to specify what we mean by random variables, and generally to provide revision of the material to be used later.

It will be assumed that the reader is familiar with the axioms of probability and the ideas of independence, and conditional probability. The material assumed is covered in, for example, chapters 1–6 of *ABC*. In the following we shall write $\Pr(A)$ for the probability of any event A. We shall begin with general definitions, and then proceed to consider particular important cases.

2.1 Random variables

Underlying all statistical investigations is the concept of a random experiment, such as the tossing of a coin k times. The set of all possible outcomes to such an experiment is called the *sample-space*, introduced by von Mises in 1921 (he called it a *Merkmalraum*), and we can formally define random variables as functions over the sample-space, about which it is possible to make probability statements. In the simple model of a queue given in Exercise 1.4, the change in queue size forms a random experiment, with just two possible outcomes, namely, an arrival or a departure. With this random experiment we can associate a random variable, X, say, such that $X = +1$ if we have an arrival, and $X = -1$ if we have a departure. The model is such that $\Pr(X = +1) = p$, and $\Pr(X = -1) = 1 - p$. In detail here, the sample-space contains just two outcomes, say ω_1 and ω_2, corresponding to arrival and departure, respectively, and we can write the random variable X as $X(\omega)$, so that $X(\omega_1) = +1$ and $X(\omega_2) = -1$. However, we find it more convenient to suppress the argument of $X(\omega)$, and simply write the random variable as X in this example. We shall adopt the now standard practice of using capital letters for random variables, and small letters for values they may take. When a random variable X is simulated n times then we obtain a succession of values: $\{x_1, x_2, \ldots, x_n\}$, each of which provides us with a *realization* of X.

Another random experiment results if we record the queue size in the model of Exercise 1.4. From Section 1.6 we see that if $p < \frac{1}{2}$, then after a long period of time since the start of the queueing system we can denote the queue size by a random variable, Y, say, such that Y may take any non-negative integral value, and $\Pr(Y = k)$ for $k \geq 0$ is as given in Section 1.6.

2.2 The cumulative distribution function (c.d.f.)

For any random variable X, the function F, given by $F(x) = \Pr(X \leq x)$ is called the *cumulative distribution function of X*. We have

$$\lim_{x \to \infty} F(x) = 1; \qquad \lim_{x \to -\infty} F(x) = 0$$

$F(x)$ is a nondecreasing function of x, and $F(x)$ is continuous from the right

(i.e. if $x > x_0$, $\lim_{x \to x_0} F(x) = F(x_0)$).

The nature of $F(x)$ determines the type of random variable in question, and we shall normally specify random variables by defining their *distribution*, which in turn provides us with $F(x)$. If $F(x)$ is a step function we say that X is a *discrete* random variable, while if $F(x)$ is a continuous function of x then we say that X is a *continuous* random variable. Certain variables, called *mixed* random variables, may be expressed in terms of both discrete and continuous random variables, as is the case of the waiting-time experienced by cars approaching traffic lights; with a certain probability the lights are green, and the waiting-time may then be zero, but otherwise if the lights are red the waiting-time may be described by a continuous random variable. Mixed random variables are easily dealt with and we shall not consider them further here. Examples of many common c.d.f.'s are given later.

2.3 The probability density function (p.d.f.)

When $F(x)$ is a continuous function of x, with a continuous first derivative, then $f(x) = dF(x)/dx$ is called the *probability density function* of the (continuous) random variable X. If $F(x)$ is continuous but has a first derivative that is not continuous at a finite number of points, then we can still define the probability density function as above, but for uniqueness we set $f(x) = 0$, for instance, when $dF(x)/dx$ does not exist; an example of this is provided by the c.d.f. of the random variable Y of Exercise 2.25.

The p.d.f. has the following properties:

(i) $f(x) \geq 0$

(ii) $\displaystyle\int_{-\infty}^{\infty} f(x)\,dx = 1$

(iii) $\Pr(a < X < b) = \Pr(a \leq X < b) = \Pr(a < X \leq b) = \Pr(a \leq X \leq b)$

$$= \int_{a}^{b} f(t)\,dt$$

EXAMPLE 2.1

Under what conditions on the constants α, β, γ can the following functions be a p.d.f.?

$$g(x) = \begin{cases} e^{-\alpha x}(\beta + \gamma x) & \text{for} \quad x \geq 0 \\ 0 & \text{for} \quad x < 0 \end{cases}$$

We must verify that $g(x)$ is non-negative, and that $\int_{-\infty}^{\infty} g(x)\,dx = 1$. If $\alpha \leq 0$, this integral cannot be finite, and so we must have $\alpha > 0$.

$$\int_{-\infty}^{\infty} g(x)\,dx = \int_{-\infty}^{\infty} e^{-\alpha x}(\beta + \gamma x)\,dx = \frac{\beta}{\alpha} + \frac{\gamma}{\alpha^2}$$

Thus we must have $\gamma = \alpha^2 - \alpha\beta = \alpha(\alpha - \beta)$ resulting in the p.d.f.

$$g(x) = e^{-\alpha x}(\beta + \alpha(\alpha - \beta)x) \qquad \text{for} \quad x \geq 0$$

In order that $g(x) \geq 0$ for $x \geq 0$, we must have $\beta \geq 0$, and $\alpha \geq \beta$. Hence set $\beta = \theta\alpha$ and $\gamma = \alpha^2(1 - \theta)$, for $\alpha > 0$ and $0 \leq \theta \leq 1$.

We sometimes abbreviate 'probability density function' to just 'density'.

2.4   Joint, marginal and conditional distributions

In the case of two random variables X and Y we can define the *joint* c.d.f. by $F(x, y) = \Pr(X \leq x \text{ and } Y \leq y)$, and then the univariate distributions of X and Y are referred to as the *marginal* distributions. If

$$f(x, y) = \frac{\partial^2 F(x, y)}{\partial x\, \partial y}$$

is a continuous function, except possibly at a finite number of points, then $f(x, y)$ is called the *joint* p.d.f. of X and Y, and in this case the marginal p.d.f.'s are given by:

$$f_X(x) = \int_{-\infty}^{\infty} f(x, y)\,dy$$

and

$$f_Y(y) = \int_{-\infty}^{\infty} f(x, y) \, dx$$

Here we have adopted a notation we shall employ regularly, of subscripting the p.d.f. of a random variable with the random variable itself, so that there should be no confusion as to which random variable is being described. The same approach is adopted for c.d.f.'s, and also, sometimes, for joint distributions.

The *conditional* p.d.f. of the random variable Y, given the random variable X, may be written as $f_{Y|X}(y|x)$, and is defined by

$$f_{Y|X}(y|x) = f_{X,Y}(x, y)/f_X(x) \qquad \text{if} \qquad f_X(x) > 0$$

For two *independent* continuous random variables X and Y, with joint p.d.f. $f_{X,Y}(x, y)$, we have

$$f_{X,Y}(x, y) = f_X(x) f_Y(y) \qquad \text{for any } x \text{ and } y$$

The above definitions of independence and joint, marginal and conditional p.d.f.'s have straightforward analogues for discrete random variables. For example, the marginal distribution of a discrete random variable X may be given by:

$$\Pr(X = x) = \sum_y \Pr(X = x, Y = y)$$

Furthermore, while we have only discussed the bivariate case, these definitions may be extended in a natural way to the case of more than two random variables.

2.5 Expectation

The *expectation* of a random variable X exists only if the defining sum or integral converges absolutely. If X is a continuous random variable we define the expectation of X as:

$$\mathscr{E}[X] = \int_{-\infty}^{\infty} x f(x) \, dx \qquad \text{if} \qquad \int_{-\infty}^{\infty} |x| f(x) \, dx < \infty$$

where $f(x)$ is the p.d.f. of X. Similarly, if X is a discrete random variable which may take the values $\{x_i\}$, then

$$\mathscr{E}[X] = \sum_i x_i \Pr(X = x_i) \qquad \text{if} \qquad \sum_i |x_i| \Pr(X = x_i) < \infty$$

The *variance* of a random variable X is defined as

$$\text{Var}(X) = \mathscr{E}[(X - \mathscr{E}[X])^2]$$

and the *covariance* between random variables X and Y is defined as

$$\text{Cov}(X, Y) = \mathscr{E}[(X - \mathscr{E}[X])(Y - \mathscr{E}[Y])]$$

The expectation of a random variable X is frequently used as a measure of location of the distribution of X, while the variance provides a measure of spread of the distribution. Independent random variables have zero covariance, but in general the converse is not true.

The *correlation* between random variables X and Y is defined as

$$\text{Corr}(X, Y) = \frac{\text{Cov}(X, Y)}{\sqrt{[\text{Var}(X)\,\text{Var}(Y)]}}$$

2.6 The geometric, binomial and negative-binomial distributions

Consider a succession of independent experiments, such as tosses of a coin, at each of which either 'success' (which we could identify with 'heads' in the case of the coin) or 'failure' ('tails' for the coin) occurs. This rudimentary succession of experiments, or trials, provides the framework for three important discrete distributions.

$$\text{Let } p = \text{Pr(success)} \quad \text{and} \quad q = 1 - p = \text{Pr(failure)}$$

The simplest is the *geometric* distribution, which results if we let X be the discrete random variable measuring the number of trials until the first success. We have

> Geometric distribution:
> $$\Pr(X = i) = q^{i-1}p \quad \text{for} \quad 1 \le i \le \infty$$

$$\mathscr{E}[X] = 1/p \quad \text{and} \quad \text{Var}(X) = 1/p^2$$

Figure 2.1(a) gives a bar-chart illustrating the geometric distribution for the case $p = 0.5$. Figure 2.1(b) demonstrates the result of simulating such a geometric random variable 100 times.

The *binomial* distribution results if we fix a number of trials at $n \ge 1$, say, and

---

Figure 2.1   (a) Bar-chart illustrating the geometric distribution with $p = 0.5$. (b) Bar-chart illustrating the results from simulating a random variable with the distribution of (a) 100 times. Here i is observed n_i times, $i \ge 1$.

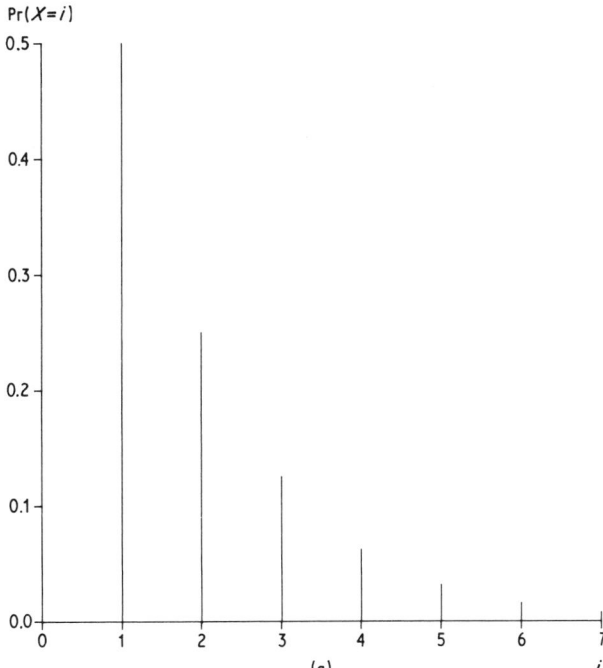

(a)

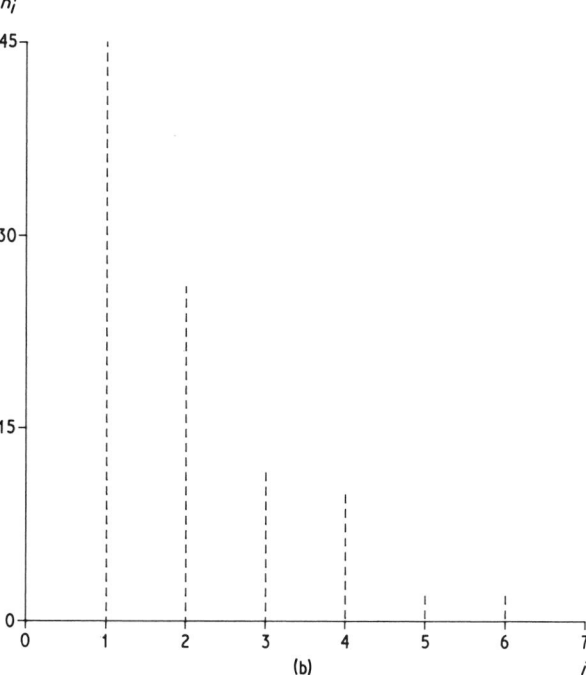

(b)

count the number, X, of successes. This gives

Binomial distribution:

$$\Pr(X = i) = \binom{n}{i} p^i q^{n-i} \qquad \text{for} \qquad 0 \le i \le n$$

$$\mathscr{E}[X] = np \qquad \text{and} \qquad \text{Var}(X) = npq$$

Figure 2.2(a) gives a bar-chart illustrating the binomial distribution for the case $p = 0.5$ and $n = 5$. Figure 2.2(b) demonstrates the result of simulating such a binomial random variable 100 times. We shall refer to such a random variable as possessing a $B(n, p)$ distribution, thus specifying the two parameters, n and p.

A geometric random variable provides the *waiting-time* measured by the number of trials until the first success. The random variable X which measures the waiting-time until the nth success has a *negative-binomial* distribution. When $X = n + i$, for $i \ge 0$, then the $(n + i)$th trial results in success, and the remaining $(n - 1)$ successes occur during the first $(n + i - 1)$ trials, and we can write

---
Negative-binomial distribution:

$$\Pr(X = n + i) = \binom{n + i - 1}{i} p^i q^n \qquad \text{for} \quad 0 \le i \le \infty$$

$$\mathscr{E}[X] = n/p \quad \text{and} \quad \text{Var}(X) = n/p^2$$

As is shown in Exercise 2.18, there is a simple relationship between the binomial and the negative-binomial distributions.

2.7 The Poisson distribution

A random variable X with a *Poisson* distribution of parameter λ is described as follows:

Poisson distribution:

$$\Pr(X = i) = \frac{e^{-\lambda} \lambda^i}{i!} \qquad \text{for} \qquad 0 \le i \le \infty$$

$$\mathscr{E}[X] = \lambda \quad \text{and} \quad \text{Var}(X) = \lambda$$

Figure 2.2 (a) Bar-chart illustrating the binomial distribution for the case $n = 5$, $p = 0.5$. (b) Bar-chart illustrating the results from simulating a random variable with the distribution of (a) 100 times. Here i is observed n_i times, $0 \le i \le 5$.

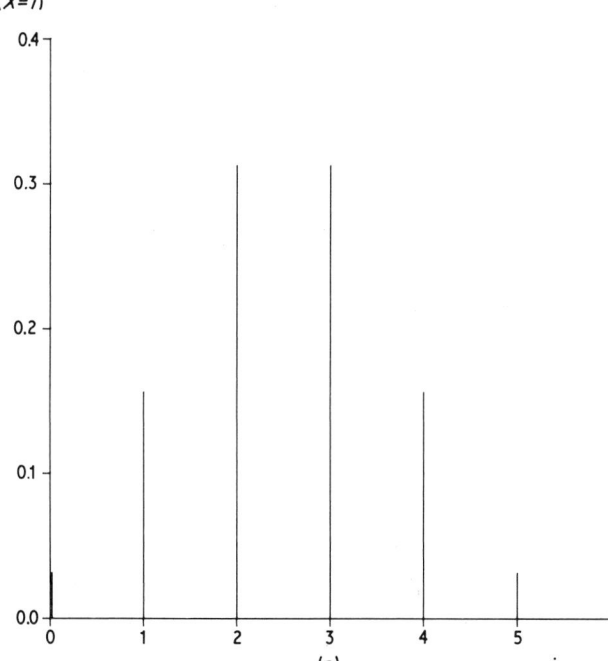

(a)

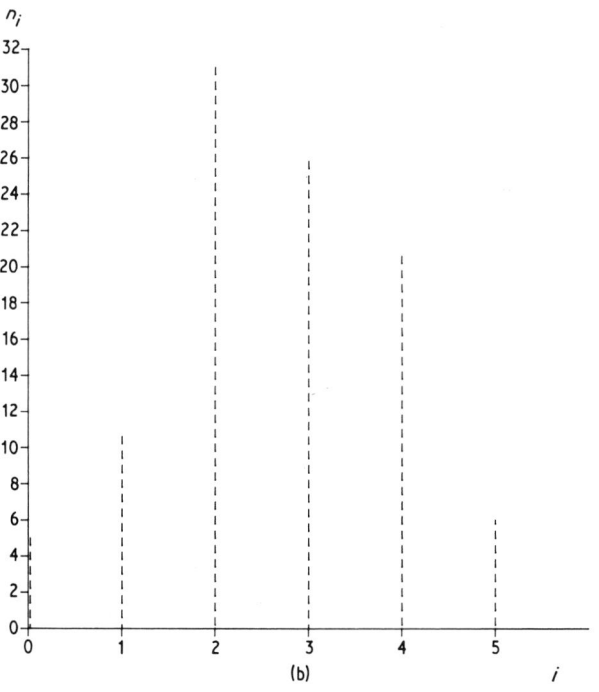

(b)

Named after the French mathematician, S. D. Poisson, who derived the distribution in 1837, the distribution had been obtained earlier by De Moivre. The Poisson distribution is often useful as a description of data that result when counts are made of the occurrence of events, such as the occurrence of telephone calls in fixed intervals of time, or the numbers of plants within areas of a fixed size. This is because the real-life processes giving rise to the data approximate to a model called a *Poisson process*, which predicts a Poisson distribution for the data. We shall discuss the Poisson process in detail in Section 4.4.2.

Figure 2.3(a) gives a bar-chart illustrating the Poisson distribution for $\lambda = 5$, and Fig. 2.3(b) describes the results of simulating such a Poisson random variable 100 times.

2.8   The uniform distribution

The simplest continuous random variables have *uniform* (sometimes called rectangular) distributions. As we shall see later, uniform random variables form the basis of most simulation investigations. A uniform random variable over the range $[a, b]$ has the p.d.f.

> Uniform p.d.f. over $[a, b]$:
>
> $$f(x) = \frac{1}{(b-a)} \qquad \text{for } a < x < b$$
>
> $$f(x) = 0 \qquad \text{for } x < a \text{ and } x > b$$

We shall frequently refer to this as the $U(a, b)$ p.d.f., the most important case being when $a = 0$ and $b = 1$.

The c.d.f. of a $U(0, 1)$ random variable X is given by

$$F(u) = \int_0^u 1 \, dx = u \qquad \text{for } 0 \le u \le 1$$

and so for any $0 \le \alpha \le \beta \le 1$,

$$\Pr(\alpha \le X \le \beta) = \Pr(0 \le X \le \beta) - \Pr(0 \le X \le \alpha)$$

$$= F(\beta) - F(\alpha) = (\beta - \alpha)$$

Figure 2.3   (a) Bar-chart illustrating the Poisson distribution for $\lambda = 5$. (b) Bar-chart illustrating the results from simulating a random variable with the distribution of (a) 100 times. Here i is observed n_i times, for $i \ge 0$.

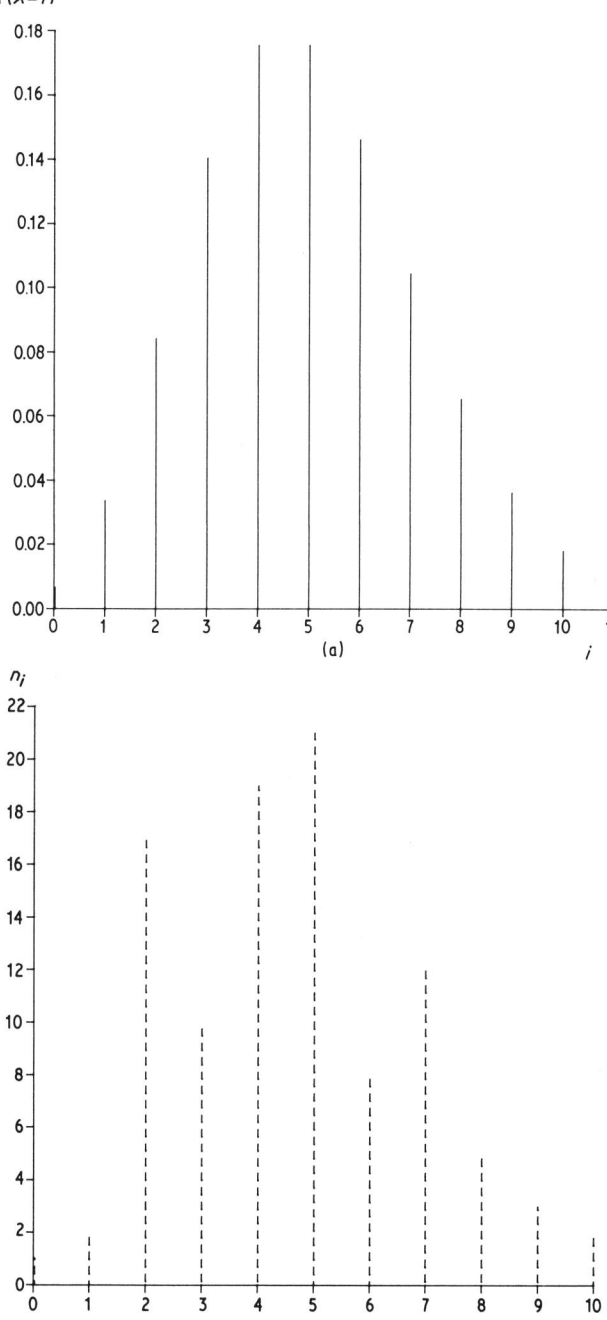

(a)

(b)

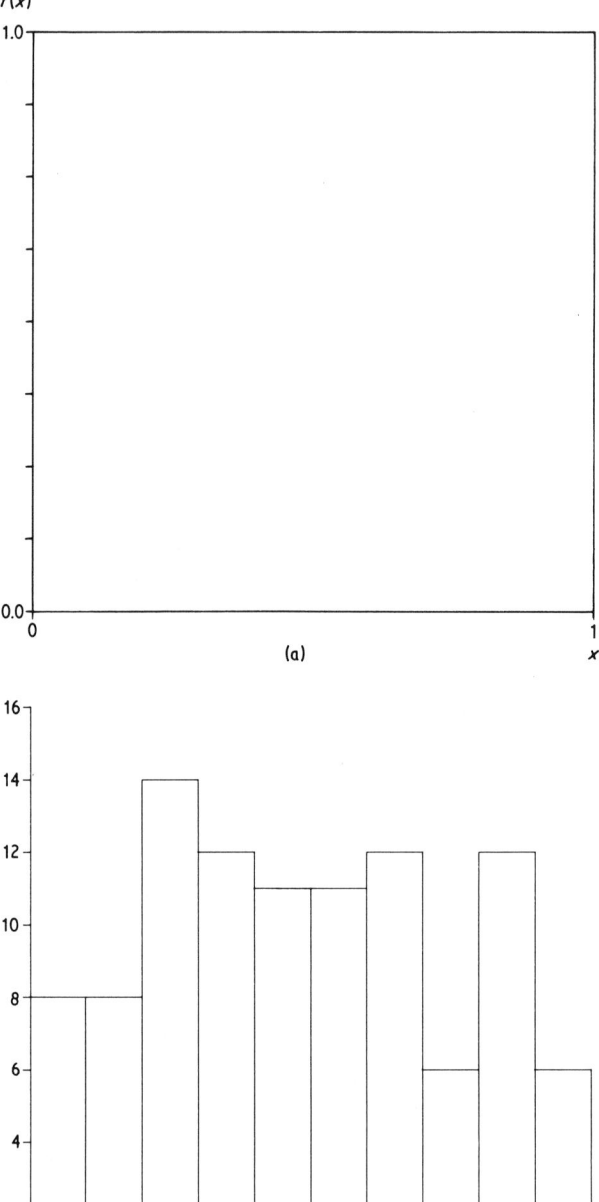

(a)

(b)

a result which is much used in later chapters. For a $U(0, 1)$ random variable X, $\mathscr{E}[X] = \frac{1}{2}$; $\text{Var}(X) = 1/12$. Figure 2.4 illustrates the $U(0, 1)$ probability density function, and also a histogram resulting from a random sample of size 100 from this density.

2.9 The normal distribution and a central limit theorem

A continuous random variable with a *normal* distribution, and mean μ and variance σ^2 has the p.d.f.

Normal probability density function:

$$f(x) = \frac{1}{\sigma \sqrt{(2\pi)}} \exp\left[-\frac{1}{2}\left(\frac{x-\mu}{\sigma}\right)^2\right] \quad \text{for } -\infty \leq x \leq \infty$$

Early work on this distribution was by such pioneers as De Moivre, Laplace and Gauss, towards the end of the 18th century and at the start of the 19th century. The normal distribution is so called because of its common occurrence in nature, which is due to 'central limit theorems', which state that, under appropriate conditions, when one adds a large number of random variables, which may well not be normal, the resulting sum has an approximately normal distribution. A formal statement of the commonest central limit theorem is that:

if $X_1, X_2, \ldots, X_n$ are independent, identically distributed random variables, with $\mathscr{E}[X_i] = \mu$ and $\text{Var}(X_i) = \sigma^2$, then for any real x,

$$\lim_{n \to \infty} \Pr\left\{\frac{1}{\sqrt{n}} \sum_{i=1}^{n} \left(\frac{X_i - \mu}{\sigma}\right) \leq x\right\} = \Phi(x)$$

where $\Phi(x)$ is the c.d.f. of a normal random variable with zero mean and unit variance.

For a more general central limit theorem, and historical background, see Grimmett and Stirzaker (1982, p. 110).

We shall use the notation $N(\mu, \sigma^2)$, to denote the distribution of a normal random variable with mean μ and variance σ^2. The $N(0, 1)$ case is frequently called the *standard* normal, when the p.d.f. is denoted by $\phi(x)$. Figure 2.5 illustrates $\phi(x)$ and also presents a histogram resulting from a random sample of size 100 from this density.

Figure 2.4 (a) The $U(0, 1)$ probability density function. (b) Histogram summarizing a random sample of size 100 from the density function of (a).

2.10   Exponential, gamma, chi-square and Laplace distributions

We say that a continuous random variable X has an *exponential* distribution with parameter λ when we can write the p.d.f. as

> Exponential probability density function:
>
> $$f(x) = \lambda e^{-\lambda x} \qquad \text{for } 0 \le x \le \infty$$

$$\mathscr{E}[X] = 1/\lambda \qquad \text{and} \qquad \text{Var}(X) = 1/\lambda^2$$

Some authors (see for example Barnett, 1965) call this the 'negative exponential' p.d.f.

The Poisson process, mentioned in Section 2.7, is often used to model the occurrence of events in time. It predicts that

$$\text{Pr}(k \text{ events in a time interval of length } t) = \frac{e^{-\lambda t}(\lambda t)^k}{k!} \qquad \text{for} \quad 0 \le k \le \infty$$

where $\lambda > 0$ is the *rate* parameter for the model, and is equal to the average number of events per unit time.

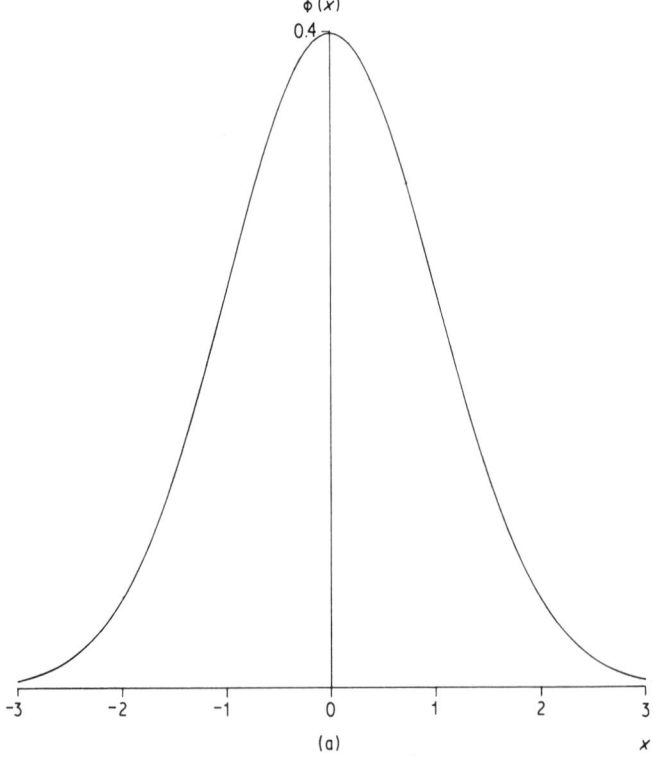

(a)

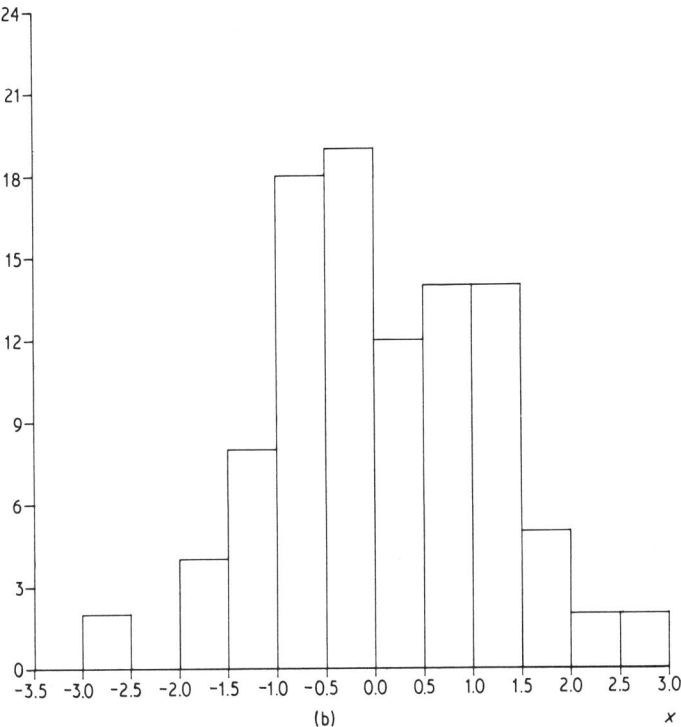

Figure 2.5 (a) The standard normal probability density function,

$$\phi(x) = \frac{1}{\sqrt{(2\pi)}} \exp\left(\frac{-x^2}{2}\right)$$ over the range $-3 \le x \le 3$. (b) Histogram summarizing a random sample of size 100 from the density function of (a).

If T is a random variable denoting the time to the next event in the Poisson process, measuring time from some arbitrary time origin, then

$$\text{Pr}\,(T \ge t) = \text{Pr}\,(\text{no events in the time interval } (0, t))$$

$$= e^{-\lambda t}$$

i.e. $$f(t) = \lambda e^{-\lambda t}$$

and so times between events in a Poisson process have an exponential distribution.

If we form the sum

$$S = \sum_{i=1}^{n} X_i$$

in which the X_i are independent random variables, each with the above exponential distribution, then (see Exercise 2.6 and Example 2.6) S has a *gamma* distribution with the p.d.f.

Gamma probability density function:

$$f(x) = \frac{e^{-\lambda x}\lambda^n x^{n-1}}{\Gamma(n)} \quad \text{for} \quad 0 \le x \le \infty$$

$$\mathscr{E}[X] = n/\lambda \quad \text{and} \quad \text{Var}(X) = n/\lambda^2$$

We shall refer to such a gamma distribution by means of the notation $\Gamma(n, \lambda)$. In this derivation, n is a positive integer, but in general gamma random variables have the above p.d.f. in which the only restriction on n is $n > 0$.

Figure 2.6 presents an exponential p.d.f., and two gamma p.d.f.'s, and a histogram summarizing a random sample of size 100 from the exponential p.d.f.

A random variable with a $\Gamma(v/2, \frac{1}{2})$ distribution is said to have a *chi-square* distribution with parameter v. For reasons which we shall not discuss here, the parameter v is usually referred to as the 'degrees-of-freedom' of the distribution. A random variable X with a $\Gamma(v/2, \frac{1}{2})$ distribution is also said to have a χ_v^2 distribution, with the p.d.f.

Chi-square probability density function with v degrees of freedom:

$$f(x) = \frac{e^{-x/2}x^{v/2-1}}{\Gamma(v/2)2^{v/2}} \quad \text{for } x \ge 0$$

$$\mathscr{E}[X] = v \quad \text{and} \quad \text{Var}(X) = 2v$$

The exponential and gamma distributions describe only non-negative random variables, but the exponential distribution forms the basis of the *Laplace* distribution, discovered by Laplace in 1774 and given below:

Laplace probability density function:

$$f(x) = \frac{\lambda}{2}e^{-\lambda|x|} \quad \text{for} \quad -\infty \le x \le \infty$$

$$\mathscr{E}[X] = 0 \quad \text{and} \quad \text{Var}(X) = 2/\lambda^2$$

Figure 2.6 (a) The $\Gamma(1, 1)$ (i.e., exponential), $\Gamma(2, 1)$ and $\Gamma(5, 1)$ probability density functions. (b) Histogram summarizing a random sample of size 100 from the exponential density function of (a).

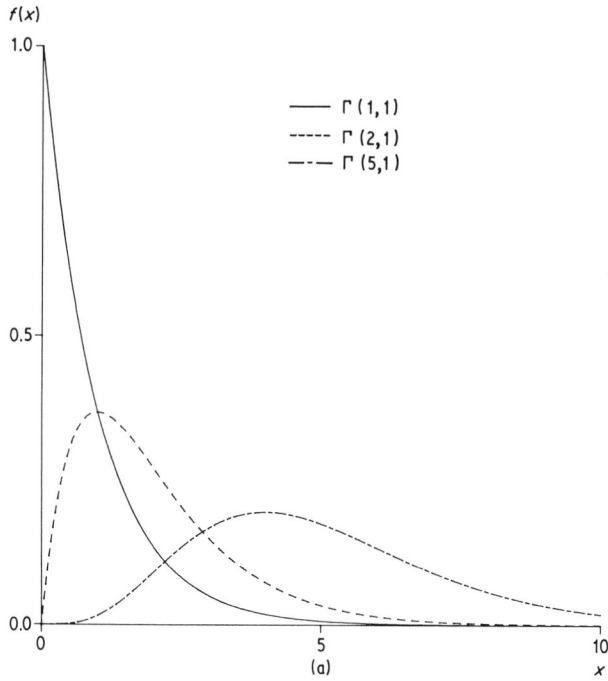

(a)

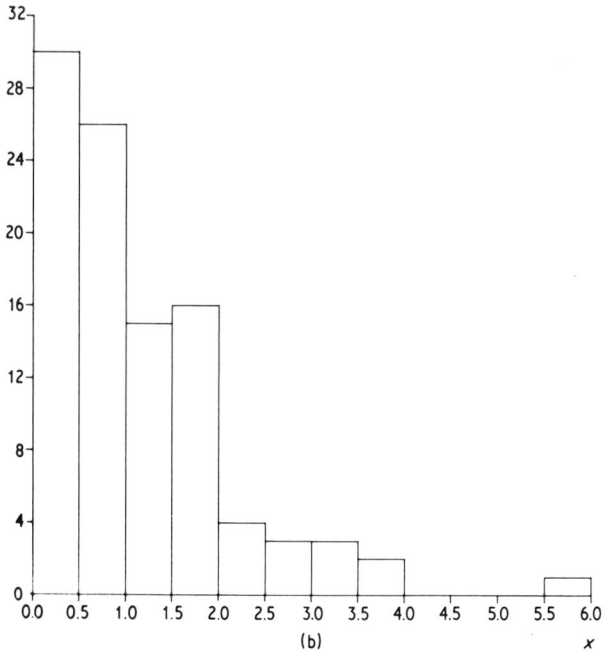

(b)

Just as the geometric and negative-binomial distributions describe waiting times when time is measured in integer units, the exponential and gamma distributions describe waiting times (in a Poisson process) when time is a continuous quantity, and we shall return to this point again later.

2.11   Distributions of other continuous random variables

Here we shall simply list the standard forms of the p.d.f. of a number of other common continuous random variables to which we shall refer later.

A probability density function that has the same qualitative shape as the normal p.d.f. is the *logistic* p.d.f., given in standard form below:

> The standard logistic probability density function:
>
> $$f(x) = \frac{e^{-x}}{(1 + e^{-x})^2} \qquad \text{for } -\infty \leq x \leq \infty$$

$$\mathscr{E}[X] = 0 \qquad \text{and} \qquad \text{Var}(X) = \pi^2/3$$

We shall later make use of the logistic c.d.f., which in standard form is

$$F(x) = (1 + e^{-x})^{-1} \qquad \text{for} \qquad -\infty \leq x \leq \infty$$

A unimodal symmetric p.d.f. with more weight in the tails than either the normal or logistic is the *Cauchy* p.d.f., so called because of its appearance in a paper by Cauchy in 1853. In its standard form the Cauchy p.d.f. is as follows:

> The standard Cauchy probability density function:
>
> $$f(x) = \frac{1}{\pi(1 + x^2)} \qquad \text{for } -\infty \leq x \leq \infty$$

Because of the large weight in the tails of this p.d.f., a random variable with this distribution does not possess a finite mean or variance.

Finally, we give below the p.d.f. of a random variable with a *beta* distribution over $[0, 1]$.

> The beta probability density function over $[0, 1]$:
>
> $$f(x) = \begin{cases} \dfrac{x^{\alpha-1}(1-x)^{\beta-1}\Gamma(\alpha+\beta)}{\Gamma(\alpha)\Gamma(\beta)} & \text{for } 0 < x < 1 \\ 0 & \text{for } x < 0 \text{ and for } x > 1 \end{cases}$$

$$\mathscr{E}[X] = \alpha/(\alpha+\beta) \quad \text{and} \quad \text{Var}(X) = \frac{\alpha\beta}{(\alpha+\beta)^2(\alpha+\beta+1)}$$

The beta distribution contains the uniform distribution as a special case: $\alpha = \beta = 1$. If the random variable X has this distribution, then such a beta random variable will be said to have a $B_e(\alpha, \beta)$ distribution.

Figures 2.7–2.9 provide examples of these p.d.f.'s together with histograms summarizing random samples of size 100 from the respective p.d.f.'s

The logistic and Cauchy p.d.f.'s are given in standard, parameter-free form, but we can simply introduce location and scale parameters β and α respectively by means of the transformation $Y = \alpha X + \beta$. This is an example of transforming one random variable to give a new random variable, and we shall now consider such transformations in a general setting.

2.12   New random variables for old

Transforming random variables is a common statistical practice, and one which is often utilized in simulation. The simplest transformation is the linear transformation, $Y = \alpha X + \beta$. In the case of certain random variables, such as uniform, logistic, normal and Cauchy, this transformation does not change the distributional form, and merely changes the distribution parameters, while in other cases the effect of this transformation is a little more complicated.

In the case of single random variables, a general transformation is $Y = g(X)$, for some function g. In such a case, if X is a discrete random variable then the distribution of Y may be obtained by simple enumeration, using the distribution of X and the form of g. Thus, for example, if $Y = X^2$, $\Pr(Y = i) = \Pr(X = -\sqrt{i}) + \Pr(X = \sqrt{i})$. Such enumeration is greatly simplified if g is a strictly monotonic function, so that in the last example, if X were a non-negative random variable then we simply have

$$\Pr(Y = i) = \Pr(X = \sqrt{i})$$

The simplification of the case when g is a strictly monotonic function applies also to the case of the continuous random variables X. Two possible examples are shown in Fig. 2.10.

For the case (a) illustrated in Fig. 2.10, the events $\{Y \le y\}$ and $\{X \le x\}$ are clearly equivalent, while for case (b) it is the events $\{Y \ge y\}$ and $\{X \le x\}$ that are equivalent, so that

$$\left. \begin{array}{ll} \text{for case (a),} & F(y) = \Pr(Y \le y) = \Pr(X \le x) = F(x) \\ \text{and for case (b),} & 1 - F(y) = \Pr(Y \ge y) = \Pr(X \le x) = F(x) \end{array} \right\} \quad (2.1)$$

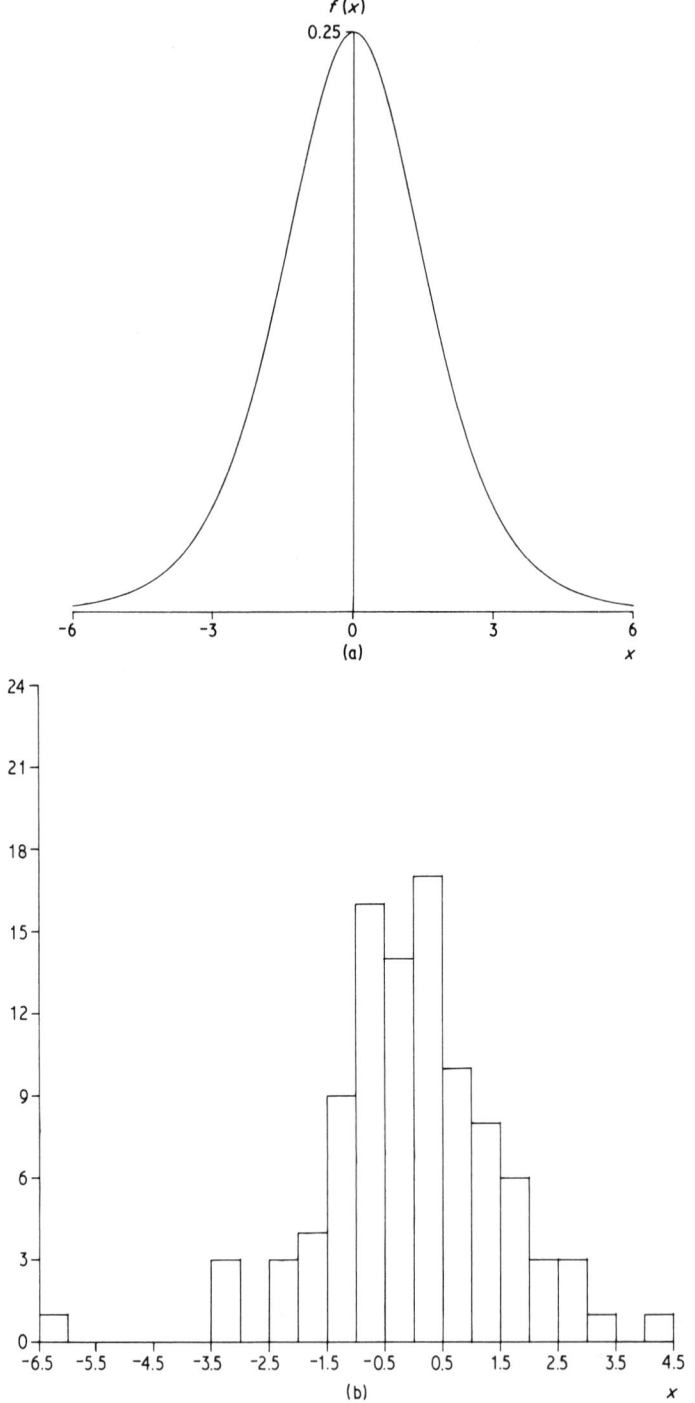

Figure 2.7 (a) The standard logistic density function for $|x| \le 6$. (b) Histogram summarizing a random sample of size 100 from the density function of (a).

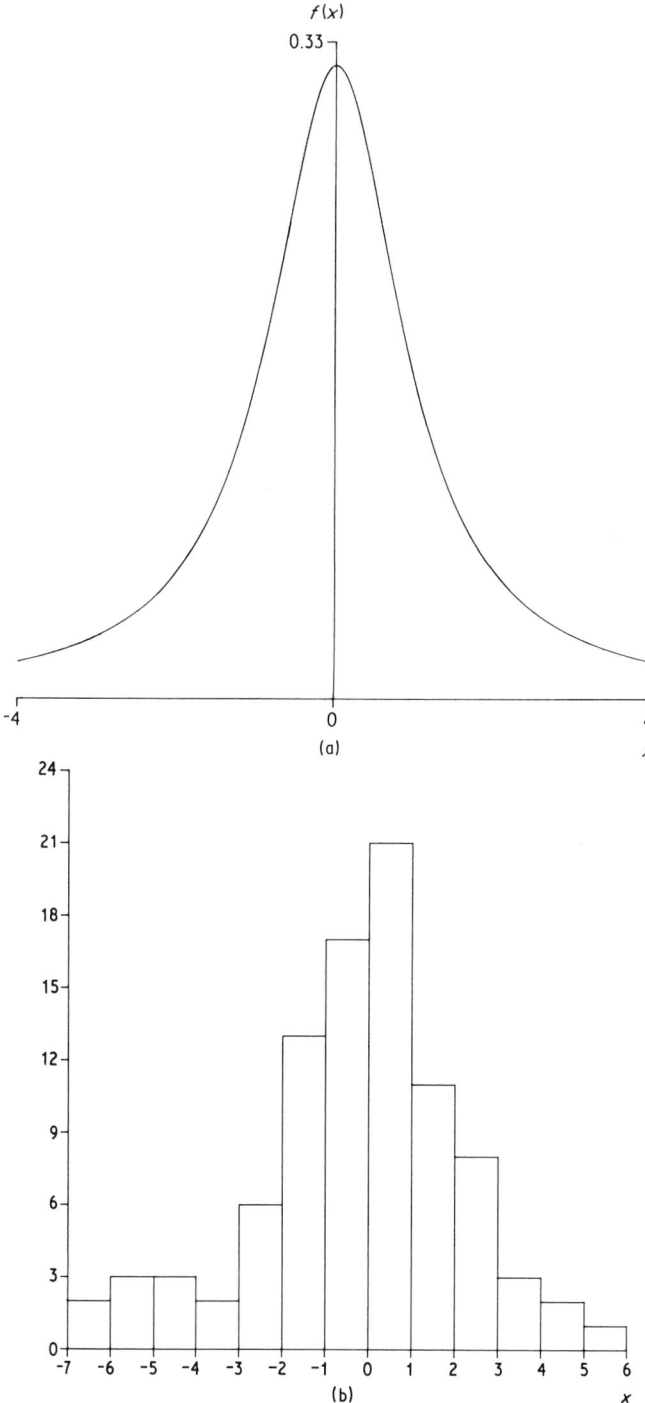

Figure 2.8   (a) The standard Cauchy density function for $|x| \leq 4$. (b) Histogram
summarizing a random sample of size 100 from the density function of (a).

31

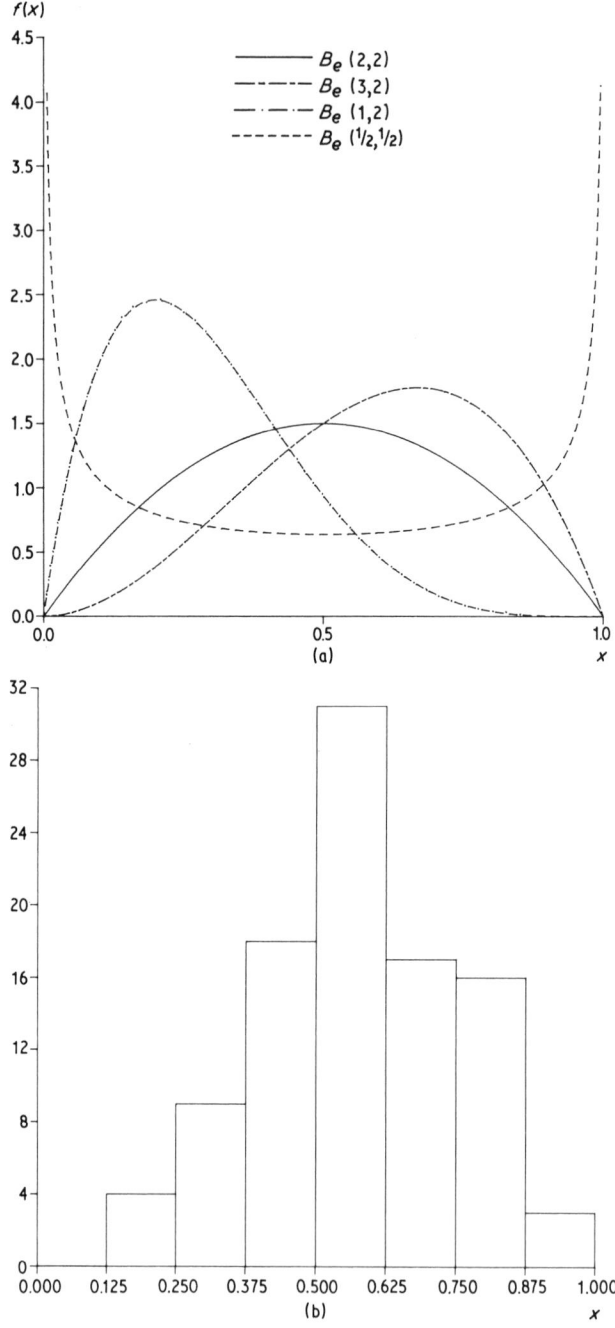

Figure 2.9 (a) The $B_e(2, 2)$, $B_e(3, 2)$, $B_e(1, 2)$ and $B_e(\frac{1}{2}, \frac{1}{2})$ density functions. (b) Histogram summarizing a random sample of size 100 from the $B_e(3, 2)$ density function.

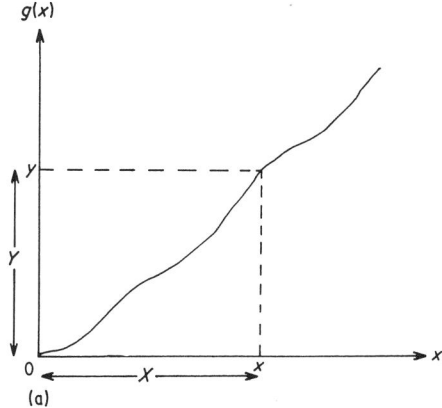

(a)

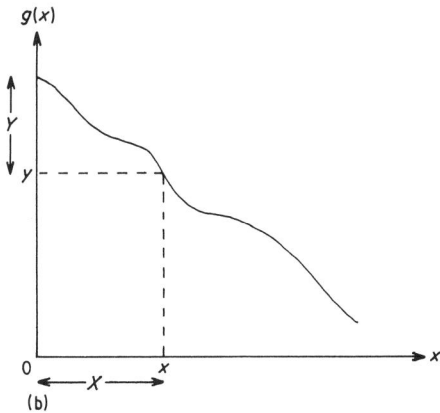

(b)

Figure 2.10   Illustrations of $y = g(x)$ where g is strictly monotonic, (a) increasing and (b) decreasing.

leading naturally to:

$$f(y) = f(x)\frac{dx}{dy} \qquad \text{for case (a)}$$

and $$f(y) = -f(x)\frac{dx}{dy} \qquad \text{for case (b)}$$

(2.2)

Two examples of case (a) now follow.

EXAMPLE 2.2

$$y = x^2$$

$$f_X(x) = \lambda e^{-\lambda x} \quad \text{for } x \geq 0$$

$$f_Y(y) = \frac{\lambda e^{-\lambda x}}{2x} = \frac{\lambda e^{-\lambda \sqrt{y}}}{2\sqrt{y}}$$

See Fig. 2.11 for the case $\lambda = 1$.

EXAMPLE 2.3

$$y = \sqrt{x}$$

$$f_X(x) = \lambda e^{-\lambda x} \quad \text{for } x \geq 0$$

$$f_Y(y) = 2\lambda y e^{-\lambda y^2}.$$

See Fig. 2.12 for the case $\lambda = 1$.

We can see from Figs 2.11 and 2.12 how the two different transformations have put different emphases over the range of x, resulting in the two different forms for $f_Y(y)$ shown. Thus in Fig. 2.12, 'small' values of x are transformed into larger values of y (for $0 < x < 1$, $\sqrt{x} > x$), with the result that the mode of $f_Y(y)$ is to be found at $y = \sqrt{(1/2\lambda)} > 0$. However, in Fig. 2.11, for $0 < x < 1$, $x^2 < x$, and the mode of $f_Y(y)$ remains at 0.

The aim in the above has been to obtain $f(y)$ as a function of y alone, and to do this we have substituted $x = g^{-1}(y)$. Cases (a) and (b) in Equation (2.2) are both described by:

$$f_Y(y) = f_X(g^{-1}(y)) \left| \frac{dx}{dy} \right| \tag{2.3}$$

If g does not have a continuous derivative, then strictly (2.3) does not hold without a clear specification of what is meant by dx/dy. In practice, however, such cases are easily dealt with when they arise (see Exercise 2.25), since the appropriate result of Equation (2.1) always holds, giving $F(y)$.

The result of (2.3) is very useful in the simulation of random variables, as we shall see later. It may be generalized to the case of more than one random variable, when the derivative of (2.3) becomes a *Jacobian*. Thus, for example, if

$$w = g(x, y)$$

and

$$z = h(x, y)$$

provide us with a one-to-one transformation from (x, y) to (w, z), then the Jacobian of the transformation is given by the determinant

$$J = \begin{vmatrix} \dfrac{\partial w}{\partial x} & \dfrac{\partial w}{\partial y} \\[2mm] \dfrac{\partial z}{\partial x} & \dfrac{\partial z}{\partial y} \end{vmatrix}$$

and if $J \neq 0$ and all the partial derivatives involved are continuous, we can write the joint density of W and Z as:

$$f_{W,Z}(w, z) = f_{X,Y}(x, y) \left| J^{-1} \right| \tag{2.4}$$

As with the case of a single random variable, we express the right-hand side of (2.4) as a function of w and z only. It is sometimes useful to note that

$$J^{-1} = \begin{vmatrix} \dfrac{\partial x}{\partial w} & \dfrac{\partial x}{\partial z} \\[2mm] \dfrac{\partial y}{\partial w} & \dfrac{\partial y}{\partial z} \end{vmatrix}$$

It often occurs that we require the distribution of the random variable, $W = g(X, Y)$. Introduction of some suitable function, $Z = h(X, Y)$, may result in a one-to-one transformation, so that (2.4) will give the joint density function of W and Z, from which we may then derive the required density of W as the marginal density:

$$f_W(w) = \int f_{W,Z}(w, z)\, dz$$

(See Exercises 2.14 and 2.15 for examples.) We shall now consider an example of the use of (2.4).

EXAMPLE 2.4
Let N_1 and N_2 be independent $N(0, 1)$ normal random variables. The pair (N_1, N_2) defines a point in two dimensions, by Cartesian co-ordinates. The transformation, from Cartesian to polar co-ordinates given by

$$N_1 = R \cos \Theta$$

$$N_2 = R \sin \Theta$$

is one-to-one, and all the partial derivatives involved are continuous, so that we

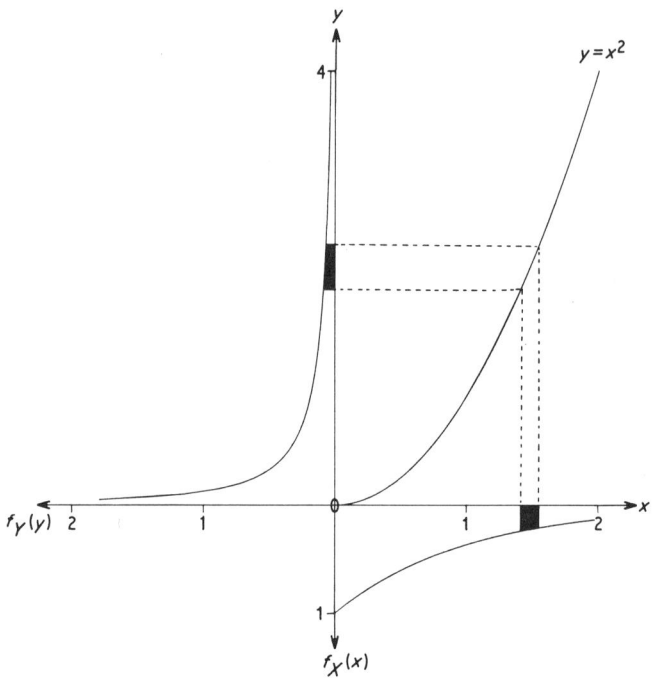

Figure 2.11 An illustration of the transformation $y = x^2$ and the densities $f_X(x) = e^{-x}, f_Y(y) = \dfrac{e^{-\sqrt{y}}}{2\sqrt{y}}$. The shaded regions have the same area.

may use (2.4) to derive the joint density function R and Θ as follows:

$$J^{-1} = \begin{vmatrix} \dfrac{\partial n_1}{\partial r} & \dfrac{\partial n_1}{\partial \theta} \\[2mm] \dfrac{\partial n_2}{\partial r} & \dfrac{\partial n_2}{\partial \theta} \end{vmatrix} = \begin{vmatrix} \cos\theta & -r\sin\theta \\ \sin\theta & r\cos\theta \end{vmatrix} = r$$

Thus $\quad f_{R\Theta}(r, \theta) = \dfrac{r}{2\pi} \exp\left[-\tfrac{1}{2}(n_1^2 + n_2^2) \right]$

$$= \dfrac{r}{2\pi} \exp\left[-r^2/2 \right] \qquad \text{for } 0 \le \theta \le 2\pi, \, 0 \le r \le \infty.$$

We thus see that R and Θ are independent random variables, with $f_\Theta(\theta) = 1/2\pi$, i.e. Θ is uniform over $[0, 2\pi]$, and $f_R(r) = r\exp\left[-r^2/2 \right]$, i.e. (see Example 2.3), R^2 has an exponential distribution of parameter $\tfrac{1}{2}$.

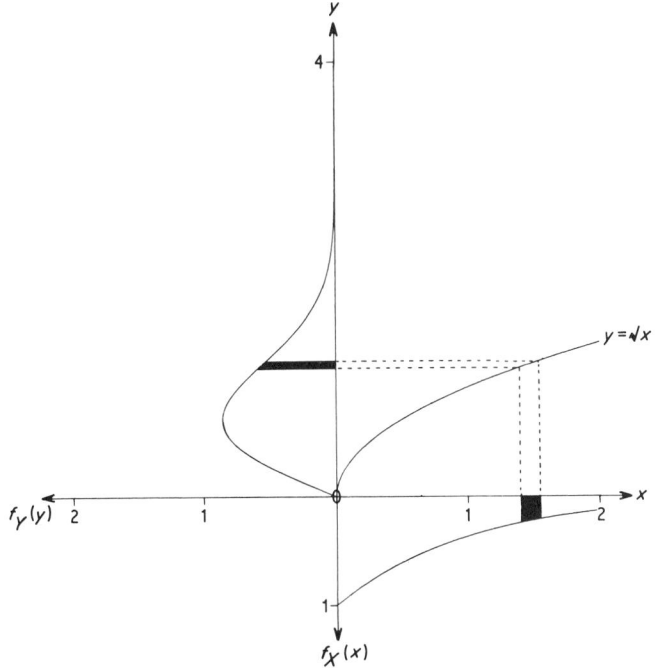

Figure 2.12 An illustration of the transformation $y = \sqrt{x}$ and the densities $f_X(x) = e^{-x}, f_Y(y) = 2ye^{-y^2}$. The shaded regions have the same area.

2.13 Convolutions

We have seen earlier that a further common transformation is a linear combination of a number of independent random variables. Again, in some cases the distributional form of the components of the sum is preserved, as occurs with Poisson, normal and Cauchy random variables, for example, while in other cases the distributional form changes, as when a sum of independent exponential random variables has a gamma distribution, as we have seen above.

The sum of mutually independent random variables is called a *convolution*. Its distribution may be evaluated by a convolution sum, or integral, as appropriate, as can be seen from the two examples that now follow.

EXAMPLE 2.5
Suppose X_1 has a $B(n_1, p)$ distribution, X_2 has a $B(n_2, p)$ distribution, and that X_1 and X_2 are independent.

Let $S = X_1 + X_2$

$$\Pr(S = k) = \sum_{i=0}^{\min(k, n_1)} \Pr(X_1 = i)\Pr(X_2 = k - i)$$

$$= \sum_{i=0}^{\min(k, n_1)} \binom{n_1}{i} p^i (1-p)^{n_1 - i} \binom{n_2}{k-i} p^{k-i} (1-p)^{n_2 - k + i}$$

$$= p^k (1-p)^{n_1 + n_2 - k} \sum_{i=0}^{\min(k, n_1)} \binom{n_1}{i}\binom{n_2}{k-i}$$

which can be shown to equal:

$$\binom{n_1 + n_2}{k} p^k (1-p)^{n_1 + n_2 - k} \qquad \text{for } 0 \leq k \leq n_1 + n_2$$

Thus S has a $B(n_1 + n_2, p)$ distribution.

EXAMPLE 2.6

Suppose X_1 and X_2 are independent exponential random variables, each with the p.d.f. $\lambda e^{-\lambda x}$ for $x \geq 0$.

Let $S = X_1 + X_2$

$$f_S(s) = \int f_{X_1}(x) f_{X_2}(s - x)\, dx$$

$$= \int_0^s \lambda^2 e^{-\lambda x} e^{-\lambda(s - x)}\, dx$$

$$= \lambda^2 e^{-\lambda s} s \qquad \text{for } s \geq 0$$

i.e. S has a $\Gamma(2, \lambda)$ distribution.

The result of this last example was anticipated in Section 2.10, and further examples of convolutions are given in Exercises 2.5–2.8. An important and often difficult feature in the evaluation of convolution sums and integrals is the correct determination of the admissible range for the convolution sum or integral.

2.14 The chi-square goodness-of-fit test

In the above figures illustrating distributions we can see the good qualitative match between the shapes of distributions and the corresponding shapes of histograms or bar-charts. For larger samples we would expect this match to improve. Whatever the sample size, however, we can ask whether the match between, say, probability density function and histogram is good enough. This is an important question when it comes to testing a procedure for simulating random variables of a specific type.

Special tests exist for special distributions, and we shall encounter some of these in Chapter 6; however, a test, due to K. Pearson, exists which may be applied in any situation. When this test was established by Pearson in 1900 it formed one of the cornerstones of modern statistics. The test refers to a situation in which, effectively, balls are being placed independently in one of m boxes. For any distribution we can divide up the range of the random variable into m disjoint intervals, observe how many of the simulated values (which now correspond to the balls) fall into each of the intervals (the boxes), and compare the observed numbers of values in each interval with the numbers we would expect. We then compute the statistic,

$$X^2 = \sum_{i=1}^{m} \frac{(O_i - E_i)^2}{E_i}$$

where we have used O_i and E_i to denote respectively the observed and expected numbers of values in the ith interval. If the random variables are indeed from the desired distribution then the X^2 statistic has, asymptotically, a chi-square distribution on an appropriate number of degrees of freedom. The rule for computing the degrees of freedom is

degrees of freedom = number of intervals $-1 -$ number of parameters, suitably estimated, if any

This test is useful because of its universal applicability, but simply because it may be applied in general it tends not to be very powerful at detecting departures from what one expects. A further problem with this test is that the chi-square result only holds for 'large' expected values. Although in many cases this may simply mean that we should ensure $E_i > 5$, for all i, we may well have to make a judicious choice of intervals for this to be the case. For further discussion, see Everitt (1977, p. 40), and Fienberg (1980, p. 172). The distribution of X^2 when cell values are small is discussed by Fienberg; this case may be investigated by simulation, and an illustration is given in Section 9.4.1. We shall use this test in Chapter 6 (see also Exercise 2.24).

*2.15   Multivariate distributions

In the last section we encountered the simplest of all multivariate distributions, the multinomial distribution, which results when we throw n balls independently into m boxes, with $p_i = \text{Pr}$ (ball lands in the ith box), for $1 \le i \le m$ and $\sum_{i=1}^{m} p_i = 1$.

Here we have a *family* of random variables, $\{X_i, 1 \leq i \leq m\}$, where X_i denotes the number of balls falling into the ith box, and so $\Sigma_{i=1}^{m} X_i = n$. The joint distribution of these random variables is given below.

Multinomial distribution:

$$\Pr(X_i = x_i, 1 \leq i \leq m) = \binom{n}{x_1, x_2, \ldots, x_m} \prod_{i=1}^{m} p_i^{x_i},$$

$$\text{where } \sum_{i=1}^{m} x_i = n \text{ and } \sum_{i=1}^{m} p_i = 1$$

Here

$$\binom{n}{x_1, x_2, \ldots, x_m} = \frac{n!}{x_1! x_2! \ldots x_m!},$$

the multinomial coefficient.

An important continuous multivariate distribution is the multivariate normal distribution, also called the multi-normal distribution. In its bivariate form the multivariate normal density function is

---

Bivariate normal probability density function:

$$\phi(x_1, x_2) = \frac{1}{2\pi\sigma_1\sigma_2(1-\rho^2)^{1/2}} \exp\left\{ -\frac{1}{2(1-\rho^2)} \left[\left(\frac{x_1-\mu_1}{\sigma_1}\right)^2 \right.\right.$$

$$\left.\left. -2\rho\left(\frac{x_1-\mu_1}{\sigma_1}\right)\left(\frac{x_2-\mu_2}{\sigma_2}\right) + \left(\frac{x_2-\mu_2}{\sigma_2}\right)^2 \right] \right\}$$

$$\text{for } -\infty < x_1, x_2 < \infty$$

Here ρ is the correlation between the two random variables; Fig. 2.13 illustrates two possible forms for $\phi(x_1, x_2)$. The p-variate density function has the following form:

---

p-variate multivariate normal probability density function:

$$\phi(\mathbf{x}) = (2\pi)^{-p/2} |\boldsymbol{\Sigma}|^{-1/2} \exp\left(-\tfrac{1}{2}(\mathbf{x}-\boldsymbol{\mu})' \boldsymbol{\Sigma}^{-1} (\mathbf{x}-\boldsymbol{\mu})\right)$$

$$\text{for } -\infty < x_i < \infty, 1 \leq i \leq p$$

notation used: $N(\boldsymbol{\mu}, \boldsymbol{\Sigma})$

Here $\boldsymbol{\mu}$ is the mean vector, $(\mathbf{x}-\boldsymbol{\mu})'$ is the transpose (row vector) of the column vector $(\mathbf{x} - \boldsymbol{\mu})$, and $\boldsymbol{\Sigma}$ is the variance/covariance matrix, i.e. $\boldsymbol{\Sigma} = \{\sigma_{ij}, 1 \leq i, j \leq p\}$, in which σ_{ij} is the covariance between the component random variables X_i and X_j. Thus σ_{ii} is the variance of X_i, for $1 \leq i \leq p$.

It can readily be shown (see, e.g., Morrison, 1976, p. 90, and cf. Exercise 2.16) that if $\mathbf{Y} = \mathbf{AX}$ and $\mathbf{X}$ has the $N(\boldsymbol{\mu}, \boldsymbol{\Sigma})$ distribution, where $\mathbf{A}$ is a nonsingular $p \times p$ matrix, then $\mathbf{Y}$ has the $N(\mathbf{A}\boldsymbol{\mu}, \mathbf{A}\boldsymbol{\Sigma}\mathbf{A}')$ distribution.

*2.16   Generating functions

The material of this section is not used extensively in the remainder of the book, and many readers may prefer to move on to Section 2.17.

It is often convenient to know the forms of *generating functions* of random variables. For any random variable X, we define the moment generating function (m.g.f.) as

$$M_X(\theta) = \mathscr{E}\left[e^{\theta X}\right]$$

for an appropriate range of the dummy variable, θ. Not all random variables have m.g.f.'s: the Cauchy distribution provides a well-known example. However, if $M_X(\theta)$ exists for a nontrivial interval for θ, then the m.g.f.

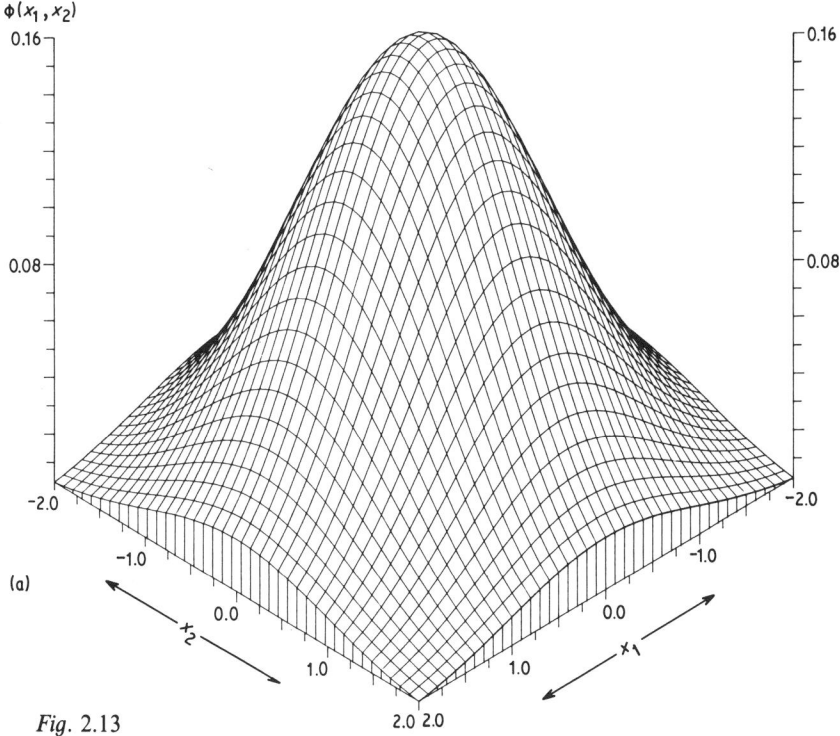

Fig. 2.13

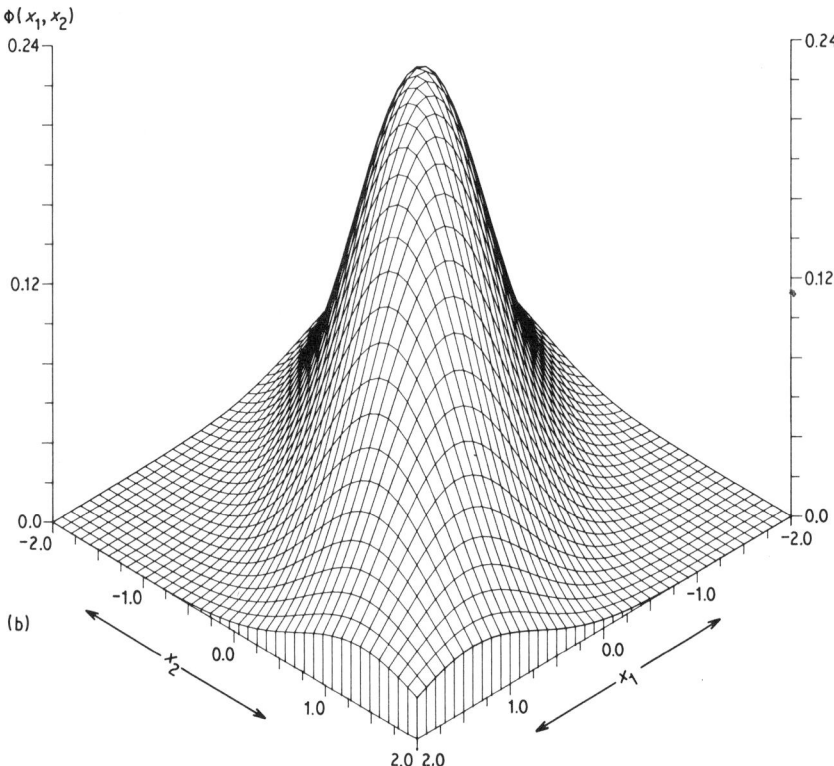

Figure 2.13   Illustration of the bivariate normal density function for the cases

(a) $\mu_1 = \mu_2 = 0,\quad \sigma_1 = \sigma_2 = 1,\quad \rho = 0$
(b) $\mu_1 = \mu_2 = 0,\quad \sigma_1 = \sigma_2 = 1,\quad \rho = 0.5$

characterizes the random variable. An alternative generating function is the probability generating function, defined by

$$G(z) = \mathscr{E}\left[z^X\right]$$

M.g.f.'s for some of the distributions considered earlier in this chapter are given in Table 2.1.

For the distributions of Table 2.1, the m.g.f. may be used to check the values of means and variances given earlier, since

$$M'(0) = \mathscr{E}[X], \quad \text{and} \quad M''(0) = \mathscr{E}[X^2],$$

illustrating why the m.g.f. is so named.

A glance at the m.g.f.'s of Table 2.1 shows that binomial, negative-binomial and gamma random variables can be expressed as convolutions of identically

Table 2.1   Common distributions and associated moment generating functions

Distribution	m.g.f.
geometric: $\Pr(X = i) = q^{i-1}p$	$pe^{\theta}(1 - qe^{\theta})^{-1}$, for $qe^{\theta} < 1$
binomial: $B(n, p)$: $\Pr(X = i) = \binom{n}{i}p^{i}q^{n-i}$	$(q + pe^{\theta})^{n}$
negative-binomial: $\Pr(X = n + i)$	$p^{n}e^{n\theta}(1 - qe^{\theta})^{-n}$, for $qe^{\theta} < 1$
$= \binom{n + i - 1}{i}p^{i}q^{n}$	
Poisson: $\Pr(X = i) = \dfrac{e^{-\lambda}\lambda^{i}}{i!}$	$e^{\lambda(e^{\theta} - 1)}$
normal: $N(0, 1)$: $f(x) = \dfrac{e^{-x^{2}/2}}{\sqrt{(2\pi)}}$	$e^{\theta^{2}/2}$
exponential: $f(x) = \lambda e^{-\lambda x}$	$\dfrac{\lambda}{\lambda - \theta}$ for $\theta < \lambda$
gamma: $\Gamma(n, \lambda)$: $f(x) = \dfrac{e^{-\lambda x}\lambda^{n}x^{n-1}}{\Gamma(n)}$	$\left(\dfrac{\lambda}{\lambda - \theta}\right)^{n}$ for $\theta < \lambda$

distributed random variables. We see why this is so as follows:

Let
$$S = \sum_{i=1}^{n} X_i$$

then
$$M_S(\theta) = \mathscr{E}\left[\exp\left(\theta \sum_{i=1}^{n} X_i\right)\right]$$

$$= \mathscr{E}\left[\prod_{i=1}^{n} \exp(\theta X_i)\right]$$

and if the $\{X_i\}$ are mutually independent, then

$$M_S(\theta) = \prod_{i=1}^{n} \mathscr{E}[\exp(\theta X_i)] = \prod_{i=1}^{n} M_{X_i}(\theta).$$

Furthermore, if the $\{X_i\}$ have the common m.g.f., $M_X(\theta)$, say, then

$$M_S(\theta) = (M_X(\theta))^{n} \tag{2.5}$$

Thus, for example, a random variable X with the $\Gamma(n, \lambda)$ distribution can be written as

$$X = \sum_{i=1}^{n} E_i$$

where the E_i are independent, identically distributed exponential random variables with parameter λ (cf. Exercise 2.6).

Moment generating functions may also be defined for m jointly distributed random variables $X_1, X_2, \ldots, X_m$, as follows:

$$M_{\mathbf{X}}(\boldsymbol{\theta}) = \mathscr{E}\left[\prod_{i=1}^{m} \exp\left(\theta_i X_i\right)\right]$$

Thus for the multinomial distribution of Section 2.15, we have the multivariate moment generating function

$$M_{\mathbf{X}}(\boldsymbol{\theta}) = \left(\sum_{i=1}^{m} p_i \exp \theta_i\right)^n,$$

while the multivariate normal distribution of Section 2.15 has the multivariate m.g.f.

$$M_{\mathbf{X}}(\boldsymbol{\theta}) = \exp\left(\boldsymbol{\theta}'\boldsymbol{\mu} + \tfrac{1}{2}\boldsymbol{\theta}'\boldsymbol{\Sigma}\boldsymbol{\theta}\right)$$

A bivariate Poisson distribution which we shall encounter later is simply defined by its m.g.f.:

$$M_{\mathbf{X}}(\boldsymbol{\theta}) = \exp\left[\lambda_1\left(e^{\theta_1} - 1\right) + \lambda_2\left(e^{\theta_2} - 1\right) + \lambda_3\left(e^{\theta_1 + \theta_2} - 1\right)\right] \qquad (2.6)$$

We shall conclude this section with two examples which complement work earlier in the chapter and illustrate further the utility of generating functions.

EXAMPLE 2.7   *Proof of a central limit theorem*
A $B(n, p)$ random variable W can be written as a convolution:

$$W = \sum_{i=1}^{n} X_i$$

where $\mathscr{E}[X_i] = p$ and $\mathrm{Var}(X_i) = pq$, where $q = 1 - p$.

Let $S_n = \dfrac{(W - np)}{\sqrt{(np)}}$

then

$$M_{S_n}(\theta) = \mathscr{E}\left[\exp\left(\frac{(W - np)\theta}{\sqrt{(npq)}}\right)\right]$$

$$= \mathscr{E}\left[\exp\left(\sum_{i=1}^{n}(X_i - p)\phi\right)\right]$$

where $\phi = \theta/\sqrt{(npq)}$

and so by (2.5), as the $\{X_i\}$ are independent,

$$M_{S_n}(\theta) = (M_Y(\phi))^n, \quad \text{where} \quad Y = (X_i - p)$$

From the above, $\mathcal{E}[Y] = 0$ and $\mathcal{E}[Y^2] = pq$, and so

$$M_{S_n}(\theta) = \left(1 + \frac{\phi^2 pq}{2} + \text{higher order terms in } \phi\right)^n$$

$$= \left(1 + \frac{\theta^2}{2n} + \text{higher order terms in } \left(\frac{\theta}{\sqrt{n}}\right)\right)^n$$

and by a result similar to that of Exercise 2.22,

$$M_{S_n}(\theta) \to \exp(\theta^2/2) \text{ as } n \to \infty.$$

Thus as $n \to \infty$ the m.g.f. of $S_n \to$ the m.g.f. of an $N(0, 1)$ random variable, and so the distribution of $S_n \to N(0, 1)$. A similar limiting operation applied to the multinomial distribution results in the multivariate normal distribution.

EXAMPLE 2.8 *Deriving the Poisson distribution from the binomial distribution*

If W has a $B(n, p)$ distribution, then

$$M_W(\theta) = (1 - p + pe^{\theta})^n$$

Now let us keep $np = \lambda$, say, fixed, while we let $n \to \infty$ (and consequently $p \to 0$).

Now
$$M_W(\theta) = \left(1 + \frac{\lambda(e^{\theta} - 1)}{n}\right)^n,$$

and as $n \to \infty$, $\quad M_W(\theta) \to \exp[\lambda(e^{\theta} - 1)], \quad$ (see Exercise 2.22)

i.e. the m.g.f. of a Poisson random variable with parameter λ. Hence under this limiting operation the distribution of W tends to this Poisson form.

It is possible to derive the exponential and gamma distributions by similar limiting processes applied, respectively, to the geometric and negative-binomial distributions (see Exercise 2.23). This approach may be used to provide an heuristic proof that the rules of the Poisson process result in a predicted Poisson distribution (see Parzen, 1960, p. 253).

2.17 Discussion and further reading

While we have dichotomized random variables as usually discrete or continuous, we have not mentioned, for instance, that most discrete random variables simply take integer values. Furthermore, the continuous random variables we have considered are, formally, absolutely continuous random variables. Such discussion is not necessary for the material to follow, but it may be found in books such as Blake (1979) and Parzen (1960). Additional discrete and continuous distributions will arise throughout the book.

In this chapter we have presented only the tip of a very large iceberg. Much more detail can be found in, for example, Haight (1967), Johnson and Kotz (1969, 1970a, 1970b and 1972), Kendall and Stuart (1961) and Ord (1972). Mardia (1970) considers families of bivariate distributions, while Douglas (1980) describes the interesting distributions which can result from special combinations of distributions such as the binomial and Poisson. Cramér (1954) discusses and proves different forms of central limit theorems, and Bailey (1964), Cox and Miller (1965) and Feller (1957) provide the necessary background to the Poisson process. Apostol (1963) is a good reference for the full transformation-of-variable theory, which is also well described by Blake (1979). Further discussion of the chi-square goodness-of-fit test is provided by Cochran (1952) and Craddock and Flood (1970), whose small-sample study is the subject of Section 9.4.1. A more leisurely introduction to some of the material of this chapter is provided by Folks (1981), and Cox and Smith (1967) provide a good introduction to the mathematical theory of queues, relevant to Exercises 2.26–2.28.

2.18   Exercises and complements

(a) Transforming random variables

†2.1   Derive the density function of the random variable

$$X = -\log_e U, \text{ where } U \text{ is } U(0, 1).$$

2.2   Consider the effect of the transformation $Y = aX$, where a is a fixed constant, and X is, e.g., an exponential, normal, gamma, or Poisson random variable.

2.3   Show that if X has the distribution of Exercise 2.1, and $W = \gamma X^{1/\beta}$ then W has a Weibull distribution with p.d.f.

$$f_W(\omega) = \frac{\beta}{\gamma^\beta}\, \omega^{\beta - 1} \exp\left[-(\omega/\gamma)^\beta\right]$$

for $0 \leq \omega < \infty$, $\beta > 0$, $\gamma > 0$.

2.4   Find the distribution of $Y = N^2$, where N is an $N(0, 1)$ random variable.

2.5   If $Y_1, Y_2, \ldots, Y_n$ are all mutually independent $N(0, 1)$ random variables, show, by induction or otherwise, that $\Sigma_{i=1}^{n} Y_i^2$ has the χ_n^2 distribution.

2.6   If $Y_1, Y_2, \ldots, Y_n$ are all mutually independent exponential random variables with p.d.f. $\lambda e^{-\lambda y}$ for $\lambda > 0$, $y \geq 0$, show, by induction and using the convolution integral, that $\Sigma_{i=1}^{n} Y_i$ has the $\Gamma(n, \lambda)$ distribution.

*2.7   If $Y_1, Y_2, \ldots, Y_n$ are all mutually independent Cauchy random variables with p.d.f., $(\pi(1 + y^2))^{-1}$, derive the distribution of $1/n \Sigma_{i=1}^{n} Y_i$.

†2.8 X, Y are independent random variables. Find the distribution of $X + Y$ when:

(a) X, Y are $N(\mu_1, \sigma_1^2)$, $N(\mu_2, \sigma_2^2)$ respectively
(b) X, Y are Poisson, with parameters λ, μ, respectively
(c) X, Y are exponential, with parameters λ, μ, respectively.

2.9   If X, Y are as in Exercise 2.8(b), find

$$\Pr(X = r \mid X + Y = n) \qquad 0 \leq r \leq n.$$

2.10   If X and Y are independent random variables, find the distribution of $Z = \max(X, Y)$ in terms of the c.d.f.'s of X and Y.

*2.11 $X_1, X_2, \ldots, X_n$ are independent random variables with the distribution of Exercise 2.1. Prove that the following random variables have the same distribution:

$$Y = \max(X_1, X_2, \ldots, X_n)$$

$$Z = X_1 + \frac{X_2}{2} + \ldots + \frac{X_n}{n}.$$

*2.12   Random variables Y_1 and Y_2 have the exponential p.d.f., e^{-x} for $x \geq 0$. Let $X_1 = Y_1 - Y_2$ and $X_2 = Y_1 + Y_2$. Find the joint distribution of (X_1, X_2).

*2.13   Let X_1, X_2 be two independent and identically distributed non-negative continuous random variables. Find the joint probability density function of $\min(X_1, X_2)$ and $|X_1 - X_2|$. Deduce that these two new random variables are independent if and only if X_1 and X_2 have an exponential distribution. In such a case, evaluate $\Pr(X_1 + X_2 \leq 3 \min(X_1, X_2) \leq 3b)$, where b is constant.

†2.14   Random variables X, Y are independently distributed as χ_{2a}^2 and χ_{2b}^2 respectively. Show that the new random variables, $S = X + Y$ and $T = X/(X + Y)$ are independent, and T has a beta, $B_e(a, b)$ distribution.

*2.15   If N_1, N_2, N_3, N_4 are independent $N(0, 1)$ random variables, show that:

(a) $X = |N_1 N_2 + N_3 N_4|$ has the exponential p.d.f., e^{-x} for $x \geq 0$.
(b) $C = N_1/N_2$ has the Cauchy distribution of Exercise 2.7.

*2.16 $\mathbf{X}$ is a p-dimensional column vector with the multivariate $N(\mathbf{0}, \mathbf{I})$

distribution, in which **0** denotes a p-variate zero vector and **I** is the $p \times p$ identity matrix. If $\mathbf{Z} = \mathbf{A}\mathbf{X} + \boldsymbol{\mu}$, where **A** is an arbitrary $p \times p$ matrix, and $\boldsymbol{\mu}$ is an arbitrary p-dimensional column vector, show that **Z** has the $N(\boldsymbol{\mu}, \mathbf{A}\mathbf{A}')$ distribution.

(b) Manipulation of random variables, and questions arising from the chapter

*2.17   Two independent Poisson processes have parameters λ_1 and λ_2. Find and identify the distribution of the number of events in the first process which occur before the first event in the second process.

*2.18   Random variables X and Y have the related distributions:

$$\Pr(Y = k) = \binom{n+m}{k} (1 - \theta)^k \, \theta^{n+m-k} \qquad \text{for } 0 \le k \le n+m$$

$$\Pr(X = k) = \binom{n+k-1}{k} \theta^k (1 - \theta)^n \qquad \text{for } k \ge 0$$

Here n, m are positive integers, and $0 < \theta < 1$, so that Y is binomial, and X is negative-binomial. By finding the coefficient of z^i in $(1 + z)^{n+m}/(1 + z)^{m+1-i}$, for $0 \le i \le m$, or otherwise, show that

$$\Pr(X \le m) = \Pr(Y \ge n).$$

*2.19   Use a central limit theorem approach to show that

$$\lim_{n \to \infty} e^{-n} \sum_{r=0}^{n} \frac{n^r}{r!} = \frac{1}{2}.$$

*2.20   Show that a random variable with the negative-binomial distribution has the moment generating function

$$M_X(\theta) = p^n e^{n\theta} (1 - qe^{\theta})^{-n}.$$

*2.21   Show that a random variable with the gamma $\Gamma(n, \lambda)$ distribution has the moment generating function

$$\left(\frac{\lambda}{\lambda - \theta} \right)^n \qquad \text{for } \theta < \lambda.$$

*2.22   Show that $\lim\limits_{n \to \infty} \left(1 + \dfrac{x}{n} \right)^n = e^x$.

*2.23   Suppose X is a random variable with a geometric distribution of parameter p. Let $Y = aX$. If $a \to 0$ and $p \to 0$ in such a way that $\lambda = a/p$ is a constant, show that the distribution of Y tends to that of a random variable with an exponential distribution with parameter λ^{-1}.

2.24 Use the chi-square goodness-of-fit test to compare the observed and expected values in the intervals: $(0, 0.1)$, $(0.1, 0.2)$, etc., for the example of Fig. 2.4, arising from the $U(0, 1)$ distribution. The grouped data frequencies are, in increasing order: 8, 8, 14, 12, 11, 11, 12, 6, 12, 6.

2.25 X is a random variable with the exponential p.d.f., e^{-x} for $x \geq 0$. We define Y as follows:

$$\text{for } 0 \leq X \leq 1, \qquad Y = X$$

$$\text{for } X \geq 1, \qquad Y = 2X - 1.$$

Obtain the distribution of Y.

(c) Questions on modelling, continuing Exercises 1.4, 1.6 and 1.7

†**2.26** The simple queue of Exercise 1.4 measured time in integral units. More realistically, times between arrivals, and service times, would be continuous quantities, sometimes modelled by random variables with exponential distributions. Observe a real-life queue, at a post-office, for example, make a record of inter-arrival and service times and illustrate these by means of histograms. What underlying distributions might seem appropriate?

2.27 (*continuation*)   The BASIC program given below simulates what is called an M/M/1 queue (see e.g., Gross and Harris, 1974, p. 8). In this queue, inter-arrival times are independent random variables with $\lambda e^{-\lambda x}$ exponential density function, and service times are independent random variables with $\mu e^{-\mu x}$ exponential density function. There is just one server and $\lambda/(\lambda + \mu)$ plays the rôle of p in Exercise 1.4. Run this program for cases: $\lambda = \mu$, $\lambda > \mu$ and $\lambda < \mu$, and comment on the results.

NOTE that the statements 100, 150 and 190 below simulate a $U(0, 1)$ random variable. The method used by the computer is described in the next chapter. The function of statements 110 and 160 should be clear from the solutions to Exercises 2.1 and 2.2. An explanation of why this program does in fact simulate an M/M/1 queue is given in Section 8.3.1.

```
10    REM THIS PROGRAM SIMULATES AN M/M/1/ QUEUE, STARTING EMPTY
20    REM AS INPUT YOU MUST PROVIDE ARRIVAL AND DEPARTURE RATES
30    REM NOTE THAT THERE IS NO TERMINATION RULE IN THIS PROGRAM
40    PRINT "TYPE LAMBDA AND MU, IN THAT ORDER "
50    INPUT L,M
60    LET S = L+M
70    LET I = L/S
80    PRINT "QUEUE SIZE.......AFTER TIME"
90    RANDOMIZE
100   LET U = RND
110   LET E = (-LOG(U))/L
120   REM E IS THE TIME TO FIRST ARRIVAL AT AN EMPTY QUEUE
```

```
130   LET Q = 1
140   PRINT Q,E
150   LET U = RND
160   LET E = (-LOG(U))/S
170   REM E IS TIME TO NEXT EVENT, IE., ARRIVAL OR DEPARTURE
180   REM WE MUST NOW FIND THE TYPE OF THAT EVENT
190   LET U = RND
200   IF U > I THEN 250
210   REM THUS WE HAVE AN ARRIVAL
220   LET Q = Q+1
230   PRINT Q,E
240   GOTO 150
250   REM THUS WE HAVE A DEPARTURE
260   LET Q = Q-1
270   PRINT Q,E
280   IF Q = 0 THEN 100
290   GOTO 150
300   END
```

*2.28 (*continuation*) We have seen that exponential distributions result from Poisson processes, and we can consider the parameters λ and μ of Exercise 2.27 to be rate parameters in Poisson processes for arrivals and departures, respectively. In some cases it may seem realistic for λ and μ each to be functions of the current queue size, n, say. For example, if $\lambda_n = 2/(n+1)$ and $\mu = 1$, we have simple 'discouragement' queue, with an arrival rate which decreases with increasing queue size. Modify the BASIC program of Exercise 2.27 in order to simulate this discouragement queue, and compare the behaviour of this queue with that of the M/M/1 queue with $\lambda = 2$, $\mu = 3$. We shall continue discussion of these queues in Chapters 7 and 8.

3

GENERATING UNIFORM RANDOM VARIABLES

3.1 Uses of uniform random numbers

Random digits are used widely in statistics, for example, in the generation of random samples (see Barnett, 1974, p. 22), or in the allocation of treatments in statistical experiments (see Cox, 1958, p. 72). More generally still, uniform random numbers and digits are needed for the conduct of lotteries, such as the national premium bond lottery of the United Kingdom (see Thompson, 1959).

A further use for random digits is given in the following example.

EXAMPLE 3.1
The randomized response technique

In conducting surveys of individuals' activities it may be of interest to ask a question which could be embarrassing to the interviewee; possible examples include questions relating to car driving offences, sex, tax-evasion and the use of drugs. Let us denote the embarrassing question by E, and suppose, for the population in question, we know the frequency, p, of positive response to some other, non-embarrassing, question, N, say. We can now proceed by presenting the interviewee with both questions N and E, and a random digit simulator, producing 0 with probability p_0, and producing 1 with probability $1 - p_0$. The interviewee is then instructed to answer N if the random digit is 0, say, and to answer E if the random digit is 1. The interviewer does not see the random digit. From elementary probability theory (see *ABC*, p. 85)

$$\text{Pr(response} = \text{Yes)} = \text{Pr(response} = \text{Yes}|\text{question is } N)p_0$$

$$+ \text{Pr(response} = \text{Yes}|\text{question is } E)(1 - p_0)$$

Knowing p_0 and $\text{Pr(response} = \text{Yes}|\text{question is } N)$, and estimating $\text{Pr(response} = \text{Yes)}$ from the survey, enables one to estimate $\text{Pr(response} = \text{Yes}|\text{question is } E)$. This illustration is an example of a randomized-response technique (RRT), and for further examples and discussion, see Campbell and Joiner (1973) and Exercises 3.1–3.4.

Uniform random numbers are clearly generally useful. Furthermore, in Chapters 4 and 5 we shall see that if we have a supply of $U(0, 1)$ random variables, we can simulate any random variable, discrete or continuous, by suitably manipulating these $U(0, 1)$ random variables.

Initially, therefore, we must consider how we can simulate uniform random variables, the *building-blocks* of simulation, and that is the subject of this chapter. We start by indicating the relationships between discrete and continuous uniform random variables.

3.2 Continuous and discrete uniform random variables

If U is a $U(0, 1)$ random variable, and we introduce a discrete random variable D such that

$$D = i \text{ if and only if } i \le 10\,U < i + 1, \quad \text{for } i = 0, 1, 2, \ldots, 9$$

then
$$\Pr(D = i) = \Pr(i \le 10\,U < i + 1)$$

$$= \frac{1}{10} \quad \text{for } i = 0, 1, 2, \ldots, 9$$

The random variable D thus provides equi-probable (uniform) random digits.

Conversely, if we write a $U(0, 1)$ random variable, U, in decimal form,

$$U = \sum_{k \ge 1} D(k)\,10^{-k}$$

Then intuitively we would expect $D(k)$ to be a uniform random digit, for each $k \ge 1$,

i.e.
$$\Pr(D(k) = i) = \frac{1}{10}, \quad \text{for } 0 \le i \le 10,$$

$$\text{and } k \ge 1$$

This and further results are proved by Yakowitz (1977, pp. 29–31). We see, therefore, that $U(0, 1)$ random variables can readily give us uniform random digits, while given a means of simulating random digits we can combine them to give $U(0, 1)$ variables to whatever accuracy is required.

3.3 Dice and machines

The simplest random number generators are coins, dice and bags of coloured balls, the very bread-and-butter of exercises in elementary probability theory.

Thus in the RRT example above, the interviewee could be given a well-shaken bag of balls, a proportion p_0 of which are white, with the remainder being black. Without looking, the interviewee then selects a ball from the bag, and answers question N if the ball chosen is white, and answers question E if the ball chosen is black. Similar physical devices are sometimes used in lotteries, and games of chance such as bingo and roulette. Certain countries such as Australia, Canada, France and West Germany televise, once a week, the operation of a complex physical device for selecting winning lottery numbers. West (1955) provides an analysis of the results of a lottery carried out in Rhodesia.

The random digits we usually need are uniform over the 0–9 range, and such digits can be obtained by suitably manipulating simple devices such as coins, as in the following example:

EXAMPLE 3.2

A fair coin is tossed four times. If we record a head as 0 and a tail as 1, then the result of the experiment is four digits, *abcd*, written in order, e.g., 0110. We can interpret *abcd* as the number, $(a \times 2^3) + (b \times 2^2) + (c \times 2) + d$, so that 0110 is interpreted as 6. If the resulting number is greater than 9 we reject it and start again. If the resulting number is in the 0–9 range then it is a realization of a uniformly distributed random digit over that range. (Based on part of an A-level question, Oxford, 1978.)

We can see this simply by enumerating the possible outcomes to the experiment:

Outcome	Resulting number
0000	0
1000	8
0100	4
0010	2
0001	1
1100	12
1010	10
1001	9
0110	6
0101	5
0011	3
1110	14
1101	13
1011	11
0111	7
1111	15

We are just using the coin to simulate the binary form of the digits 0–15. This method therefore does give rise to uniform random digits over 0–9, but it is rather wasteful, as resulting numbers are rejected 3/8 of the time.

Manipulations of this kind are avoided by the direct use of simple dice to produce 0–9 uniform random digits. Unfortunately, a regular 10-sided figure does not exist, but one can use icosahedral dice (giving regular 20-sided figures), each digit 0–9 appearing separately on two different faces. Further possibilities include rolling a regular 10-faced cylinder, or throwing a 10-faced di-pyramid, with each face being an isosceles triangle of some fixed size. These simple devices are illustrated in Fig. 3.1.

Figure 3.1 (a) Three icosahedral dice. Note the need to distinguish between 6 and 9 (b) A regular, 10-faced cylinder (c) Three 10-faced di-pyramids, with truncated isosceles triangles of the same size as faces. Note that the two pyramids are so attached that when the body is at rest a face is uppermost.

Because of the general demand for random digits, tables, such as those of Appendix 2, are now widely available. A sequence of random digits can be obtained by reading the table by rows, by columns, or by any other rule. The first table of this kind was produced by Tippett in 1927, and it was regarded as a 'godsend' by the statisticians of the day (Daniels, 1982).

In using physical devices such as dice to simulate random digits one is reversing the customary model/reality relationship. As described in Chapter 1, one usually takes a real-life situation, and builds a model of it. Here we start with a model, such as a uniform random digit, seek a real-life mechanism to correspond to that model, and then take observations from the real-life mechanism. There is always a discrepancy between model and reality – coins may not be fair (see, e.g., Kerrich, 1946), dice may be biased, and so on – therefore the numbers produced by physical devices are tested, to ensure that no drastic non-randomness is present. This is simply a form of quality control of random numbers, and one applies only a finite subset of the infinity of tests that are possible. We shall return to the subject of testing of numbers in Chapter 6.

Any process in nature that is thought to be random may be used to try to simulate uniform random numbers. Kendall and Babbington-Smith (1939a) used a rotating disk with ten uniform segments, which was stopped at random. Tippett (1925) used digits read from tables of logarithms. ERNIE, the computer used for selecting winning premium bonds in the British national lottery, uses the electronic 'noise' of neon tubes. The digits of Table 3.1 were obtained from reading the last three digits of successive numbers from the Canterbury region telephone directory. (Cf. Section 6.7 and Exercise 6.8. The relative frequencies of these digits are considered in Example 6.1.)

Student (1908a) drew samples from a set of physical measurements taken on criminals, as described in Section 1.5. In his case we have an illustration of sampling from a non-uniform population, (approximately normal in this case), and similarly exponential and Poisson random variables may be simulated directly if one can observe a process in nature which provides a good approximation to a Poisson process (see Section 4.4.2).

Dice and machines are impractical for all but the smallest simulations, which are now in any case likely to be conducted with the aid of readily available tables (see for instance, Neave, 1981, and Murdoch and Barnes, 1974). Large-scale simulations are usually conducted using computers, and early computers were equipped with built-in random-number generators of the physical kind, using random electronic features, as in ERNIE. Tocher (1975, chapter 5) provides many examples here, and even circuit diagrams. More recently, Isida (1982) presented a compact physical random-number generator based on the noise of a Zener diode. The modern equivalent of this can be found in certain hand-calculators, which have an RND button for simulating $U(0, 1)$ random variables. A problem with all physical devices is the danger that they may

Table 3.1   Digits from telephone numbers

	874	580	873	824	564	663
	478	658	540	561	360	082
	661	839	996	261	052	938
	334	420	356	571	081	866
	569	166	045	091	961	610
(a)	471	378	936	569	107	022
	916	865	961	838	303	826
	665	014	148	764	276	638
	504	776	237	682	634	207
	659	654	774	217	609	684
	423	213	423	002	960	273
	183	059	563	379	252	955
	202	410	451	887	467	427
(b)	207	483	809	265	117	891
	061	658	145	950	135	495
	716	232	955	771	747	699
	693	757	952	053	659	459
	991	876	091	431	316	283
	499	223	743	037	891	729
	611	998	650	527	073	665

become unreliable, through changes to the device in time; thus dice, for instance, could become unevenly worn, resulting in bias. Frequent checks of the generated numbers should therefore be carried out.

The modern approach to large-scale simulation is quite different from that of this section, and it avoids the need for such frequent checking by producing a sequence of numbers that can be shown mathematically to possess certain desirable features. This approach, which is also not without its drawbacks, is described in the next section.

3.4   Pseudo-random numbers

The digits of Table 3.2 superficially have the *appearance* of the digits of Table 3.1, but they have been generated in a blatantly non-random fashion, from the recursion formula

$$u_{n+1} = \text{fractional part of } (\pi + u_n)^5 \qquad \text{for } n \geq 0 \qquad (3.1)$$

where u_0 is some specified number in the range $0 < u_0 < 1$. u_0 is, rather graphically, termed the 'seed'. Knowledge of the formula of (3.1) provides one with complete knowledge of the sequence of numbers resulting in Table 3.2., but in many applications one may find these digits as suitable as those, say, of Table 3.1, and much more easily generated on a calculator or computer. Formula (3.1) can be likened to a 'black box' which takes the place of a physical black-box such as a die. Recursion formulae are most suitable for use on computers

Table 3.2   Digits from the recursion of Equation (3.1).

254	032	329	233	252	444
794	807	600	974	884	454
797	354	440	855	159	290
162	053	737	489	953	381
051	091	224	843	075	513
703	740	755	750	070	002
301	810	903	392	970	915
690	642	767	038	140	051
962	283	420	435	835	150
574	108	551	564	209	788
810	657	491	939	365	537
612	514	020	950	567	239
119	865	638	032	062	491
966	619	460	553	850	096
255	550	872	019	601	282
474	943	141	486	022	074
013	589	023	454	681	854
489	857	712	412	307	910
826	305	753	610	885	458
346	008	309	763	890	300

Each triple is obtained from the first three decimal places of the u_i, when (3.1) was operated using a 32-bit computer and floating-point arithmetic. Successive numbers were obtained moving from left to right across the rows, and down the table.

and calculators, and furthermore the properties of the numbers they produce can be investigated mathematically. If the resulting numbers satisfy a variety of tests, then because of the deterministic nature of a recursion formula, additional application of these tests at a later stage is not necessary, as there is no danger of bias creeping into the black-box, with the progress of time.

In some applications it may be required to re-run a simulation using the same random numbers as on a previous run. Such a requirement may seem unlikely, but we shall see in Chapter 7 that it can be very useful in certain methods for variance-reduction. Knowledge of u_0 for a formula such as (3.1) enables one to do this quite easily, whereas such a 're-run' facility is not possible with physical generators unless a possibly time-consuming record is made of the numbers used, Inoue *et al.* (1983) describe the generation and testing of random digits which may be supplied on magnetic tapes.

3.5   Congruential pseudo-random number generators

An alternative mathematical representation of formula (3.1) is:

$$u_{n+1} = (\pi + u_n)^5 \pmod 1 \qquad \text{for } n \geq 0$$

Currently the recursion formula that is most frequently adopted is:

$$x_{n+1} = ax_n + b \qquad (\text{mod } m) \qquad \text{for } n \geq 0 \qquad (3.2)$$

in which a, b and m are suitably chosen fixed *integer* constants, and the seed is an integer, x_0. Starting from x_0, the formula (3.2) gives rise to a sequence of integers, each of which lies in the 0 to $(m-1)$ range. Because the resulting numbers can be investigated by the theory of congruences, such generators are termed 'congruential'. Although terminology is not always uniform here, we shall call a generator with $b = 0$, 'multiplicative', and one with $b \neq 0$, 'mixed'. Approximations to $U(0, 1)$ variables can be obtained from setting $u_i = x_i/m$, as discussed in Exercise 3.15. For an example, see the solution to Exercise 3.21.

Formulae such as (3.1) are sometimes used to play games involving random elements on hand calculators. We can examine the numbers produced and we may find that they satisfy many criteria of random numbers. However, there is no guarantee, in general, that at some stage the sequence of numbers produced by such formulae may not seriously violate criteria of random numbers, and thus, in general, such formulae are of little use for scientific work. As we shall see, an advantage of the formula (3.2) is that certain guarantees *are* available for the resulting numbers.

The constants a, b and m are chosen with a number of aims in mind. For a start, one wants the arithmetic to be efficient. Human beings do arithmetic to base 10, and so if the formula (3.2) was being operated by hand, using pencil and paper, it would be sensible for m to be some positive integral power of 10. For example, if we have

$$x_0 = 89, \ a = 1573, \ b = 19, \ m = 10^3$$

then from (3.2),

$$x_1 = 140\,016 \quad (\text{mod } 10^3) = 16$$
$$x_2 = 25\,187 \quad (\text{mod } 10^3) = 187$$

etc.

Clearly, if one naturally does arithmetic to number base r, say, then the operation of division by m is most efficiently done if $m = r^k$ for some positive integer k. For most computers this entails setting $m = 2^k$, where k is selected so that m is 'large' (see below) and the numbers involved are within the accuracy of the machine.

A moment's thought shows that the generator of (3.2) can produce no more than m different numbers before the cycle repeats itself, again and again. Thus a second aim in choosing the constants a, b, m is that the cycle length, which could certainly be less than m, is reasonably large. It has been shown (see Hull and Dobell, 1962, and Knuth, 1981, pp. 16–18) that for the case $b > 0$, the maximum possible cycle length m is obtained if, and only if, the following relations hold:

(i) b and m have no common factors other than 1;

(ii) $(a - 1)$ is a multiple of every prime number that divides m;

(iii) $(a - 1)$ is a multiple of 4 if m is a multiple of 4.

If $m = 2^k$, relation (iii) will imply that $a = 4c + 1$ for positive integral c. Such an a then also satisfies relation (ii). When $m = 2^k$, relation (i) is easily obtained by setting $b =$ any odd positive constant. Proofs of results such as these are usually given in general number-theoretic terms; however, following Peach (1961), in Section 3.9 we provide a simple proof of the above result for the commonly used case: $m = 2^k$, $a = 4c + 1$ and b odd (c, b, and k positive integers).

Although multiplicative congruential generators involve less arithmetic than mixed congruential generators, it is not possible to obtain the full cycle length in the multiplicative case. Nevertheless, if $m = 2^k$ for a multiplicative generator, then a cycle length of 2^{k-2} may be obtained. This is achieved by setting $a = \pm 3$ (mod 8), and now also imposing a constraint on x_0, namely, choosing x_0 to be odd. A suitable choice for a is an odd power of 5, since, for positive, integral q,

$$5^{2q+1} = (1 + 4)^{2q+1} = (1 + 4(2q + 1)) \quad \text{mod (8)}$$
$$= -3 \quad (\text{mod } 8)$$

Five such generators that have been considered are:

a	k
5^{13}	36, 39
5^{17}	40, 42, 43

For further discussion of multiplicative congruential generators, see Exercise 3.31.

When one first encounters the idea of a sequence of 'random' numbers cycling, this is disturbing. However, it is put in perspective by Wichmann and Hill (1982a), who present a generator, which we shall discuss later, with a cycle length greater than 2.78×10^{13}. As they remark, if one used 1000 of these numbers a second, it would take more than 800 years for the sequence to repeat!

Large cycle lengths do not necessarily result in sequences of 'good' pseudo-random numbers, and a third aim in the choice of a, b, m is to try to produce a small correlation between successive numbers in the series; for truly random numbers, successive numbers are uncorrelated, but we can see that this is not likely to be the case for a generator such as (3.2). Greenberger (1961) has shown that an approximation to the correlation between x_n and x_{n+1} is given by:

$$\rho \approx \frac{1}{a} - \frac{6b}{am}\left(1 - \frac{b}{m}\right) \pm \frac{a}{m} \tag{3.3}$$

Greenberger gives the following two examples of sequences with the same full cycle length:

	a	b	m	ρ
(i)	$2^{34}+1$	1	2^{35}	0.25
(ii)	$2^{18}+1$	1	2^{35}	$\ll 2^{-18}$

Expressions such as (3.3) are obtained by averaging over one complete cycle of a full-period mixed generator (cf. Exercise 3.13) and exact formulae for ρ, involving Dedekind sums, are presented by Kennedy and Gentle (1980, p. 140). As is discussed by Kennedy and Gentle, and also by Knuth (1981, p. 84), choosing a, b and m to ensure small ρ can result in a poor generator in other respects. For instance, for sequences that are much shorter than the full cycle, the correlation between x_n and x_{n+1} may be appreciably higher than the value of ρ for the complete cycle. Also, higher-order correlations may be far too high; see Coveyou and MacPherson (1967) and Van Gelder (1967) for further discussion. It is sometimes recommended that one takes $a \approx \sqrt{m}$ (see e.g., Cooke, Craven and Clarke, 1982, p. 69). However, this approximate relationship holds for the RANDU generator, originally used by IBM and, as we can see from Exercise 3.25, this generator possesses a rather glaring Achilles heel. Unfortunately, as we shall see in Chapter 6, this is a defect which can be hard to detect using standard empirical tests. As a further example, $a \approx \sqrt{m}$ for the generator of Exercise 3.31 (ii), which passes the randomness tests of Downham and Roberts (1967) yet has since been shown to have undesirable properties by Atkinson (1980). Similar findings for this generator and that of Exercise 3.31 (i) are given by Grafton (1981).

The choice of the constants a, b and m is clearly a difficult one, but the convenience of pseudo-random number generators has made the search for good generators worth while. Ultimately, the properties of any generator will be judged by the use intended for the numbers to be generated, and by the tests applied. A very important feature of congruential generators, which is perhaps inadequately emphasized, is that the arithmetic involved in operating the formula (3.2) is exact, without any round-off error. Thus naïve programming of the formula (3.2) in, say, BASIC can rapidly result in a sequence of unknown properties, because of the use of floating-point arithmetic; this feature is clearly illustrated in Exercise 3.14. This problem is usually solved in computer implementations by machine-code programs which employ integer arithmetic.

In this case the modulus operation can be performed automatically, without division, if the modulus, $m = 2^r$, and r is the computer word size: after $(ax_{i-1} + b)$ is formed, then only the r lowest-order bits are retained; this is the integer 'overspill or carry-out' feature described by Kennedy and Gentle (1980, p. 19).

3.6 Further properties of congruential generators

One might well expect numbers resulting from the formula (3.2) to have unusual dependencies and that this is so is seen from the following illustration:

Let
$$x_{i+1} = 5x_i \quad (\text{mod } m)$$

Here
$$x_{i+1} = 5x_i - h_i m \tag{3.4}$$

in which h_i takes one of the values, 0, 1, 2, 3, 4. Thus pairs of successive values, (x_i, x_{i+1}) give the Cartesian co-ordinates of points which lie on just one of the five lines given by (3.4), and the larger m is, the longer the sequence of generated numbers will remain on any one of these lines before moving to another line. For example, if $x_0 = 1$, $m = 11$, then

$$x_1 = 5, \ x_2 = 3, \ x_3 = 4, \ x_4 = 9, \ x_5 = 1$$

and the line used changes with each iteration.
However, if $x_0 = 1$, $m = 1000$, then

$$x_1 = 5, \ x_2 = 25, \ x_3 = 125, \ x_4 = 625, \ x_5 = 125$$

and the sequence $x_1 \rightarrow x_4$ is obtained from the line

$$x_{i+1} = 5x_i$$

after which the sequence degenerates into a simple alternation pairs of successive values give points which lie on a limited number of straight lines, triplets of successive values lie on a limited number of planes, and so on (see Exercise 3.25).

The mixed congruential generator

$$x_{n+1} = 781 \, x_n + 387 \quad (\text{mod } 1000) \tag{3.5}$$

has cycle length 1000. Figure 3.2 illustrates a plot of u_{n+1} vs. u_n for a sequence of length 500, where $u_i = x_i/1000$, for $0 \le i \le 999$.

The striking sparseness of the points is because of the small value of m used here, which also allows us to see very clearly the kind of pattern which can arise. Thus many users prefer to modify the output from congruential generators before use. One way to modify the output is to take numbers in groups of size g, say, and then 'shuffle' them, by means of a permutation, before use. The permutation used may be fixed, or chosen at random when required. Andrews *et al.* (1972) used such an approach with $g = 500$, while Egger (1979)

u_{n+1}

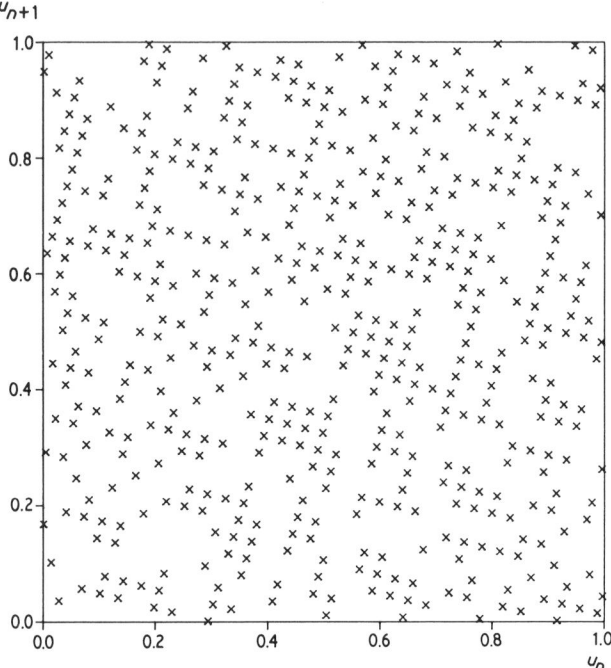

Figure 3.2   A plot of u_{n+1} vs. u_n for half the cycle of the mixed congruential generator of Equation (3.5).

used $g = 100$, a choice also investigated by Atkinson (1980). Page (1967) discusses the construction of random permutations, while tables of these are provided by Moses and Oakford (1963). See also the IMSL routine GGPER described in Appendix 1. An alternative approach, due to MacLaren and Marsaglia (1965) is to have a 'running' store of g numbers from a congruential generator, and to choose which of these numbers to use next by means of a random indicator digit from the range 1 to g, obtained, say, by a separate congruential generator. The gap in the store is then filled by the next number from the original generator, and so on. When this is done for the sequence resulting in Fig. 3.2, we obtain the plot of Fig. 3.3.

For further discussion, see Chambers (1977, p. 173) and Nance and Overstreet (1978). Nance and Overstreet discuss the value of g to be used, and conclude with Knuth (1981, p. 31) that for a good generator, shuffling is often not needed. On the other hand, shuffling can appreciably improve even very poor generators, as demonstrated by Atkinson (1980), a point which is also made in Exercise 3.26. The IMSL routine GGUW employs shuffling with $g = 128$; see Section A1.1 in Appendix 1.

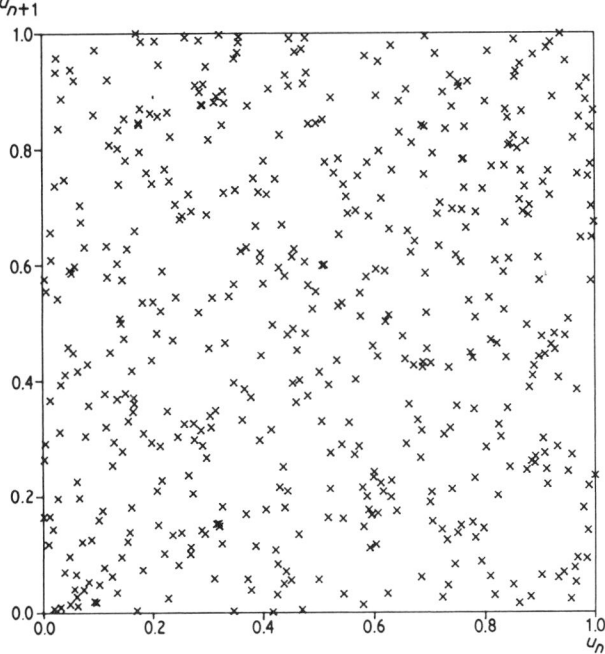

Figure 3.3   The plot resulting from modifying the same sequence that gave rise to Fig. 3.2. The modification entailed choosing the next number 'at random' from a store of length $g = 20$ of numbers from the original sequence, as explained in the text. In this example the random selection was made using Equation (3.1) and a seed of 0.5.

Successive digits in a decimal expansion of a truly random $U(0, 1)$ variable may, as we saw in Section 3.2, be used as uniform random digits. However, this approach is unwise in the case of pseudo-random $U(0, 1)$ variables because of the pattern effects which may arise (see the solution to Exercise 3.21). A disadvantage of congruential generators with $m = 2^k$ is that the low-order bits of generated numbers have short cycles (e.g. Atkinson, 1980). This is not a problem if m is prime (see Exercise 3.31) but then the arithmetic of the method is much more time-consuming on a 'binary' computer than if $m = 2^k$. Ways of reducing computational effort when m is prime are referenced by Law and Kelton (1982, p. 226).

3.7   Alternative methods of pseudo-random number generation

A variety of other methods of pseudo-random number generation exist—see for example, Andrews (1977, p. 170), O'Donovan (1979, p. 33), Law and Kelton (1982, p. 230), Tausworthe (1965) and Craddock and Farmer (1971). Miller and

Prentice (1968), for instance, use the third-order recurrence

$$x_j = x_{j-2} + x_{j-3} \quad (\text{mod } p)$$

in which p is a suitable prime.

As with the congruential methods considered above, it is possible here also to examine the theoretical properties of the resulting sequence (cf. Exercise 3.26).

Different computers have different word-lengths (see Kennedy and Gentle 1980, p. 8), which determine the value of the modulus, m, used in congruential generators. This has resulted in machine-dependent generators, which is undesirable, as it makes it difficult to reproduce results, a positive feature of using pseudo-random numbers. *Portable* generators can result from representing 'large' integers by means of a number of 'short' word-length integers; see also Kral (1972) and Roberts (1982). An alternative approach is given by Wichmann and Hill (1982a, b), who combine three simple multiplicative congruential generators in such a way that the overall cycle-length is the product of the individual cycle lengths (see Exercises 3.17 and 3.18). The result is a portable generator with a cycle length greater than 2.78×10^{13}. As well as providing FORTRAN and Ada listings for their algorithm, they also provide an 82-step program for the Hewlett Packard HP-67 hand-calculator.

3.8 Summary and discussion

The building-blocks of simulation are $U(0, 1)$ random variables, and random digits. We have seen that these may be obtained by the use of physical devices, or arithmetic formulae, and that no method is without its drawbacks. Large-scale simulations take place on computers, for which arithmetic formulae provide the most convenient approach. While any formula may seem to be adequate, and produce reasonable-looking numbers, there is always the danger that the formula could break down at some stage. The advantage of congruential generators is that they can be shown to possess certain desirable features; and to give guaranteed cycle lengths. There is always a chance, however, that because the numbers are pseudo-random, and not truly random, unwanted effects could still arise in any particular application. The answer is clearly to proceed with caution, and to make regular checks for oddities. Certain early generators were blatantly unsuitable, and the possibility remains that these generators are still in use. Well-used computer packages, such as MINITAB (see Ryan, Joiner and Ryan, 1976) do not always specify the generator they employ, which is clearly undesirable. (Indeed, different implementations of the same package may use different generators.) The same is true of certain widely used microcomputers. Possible pitfalls, as may occur here, can be avoided by the use of portable generators, which may be used on any machine, even a hand-calculator. Kennedy and Gentle (1980, p. 165) report that as many as about 30% of papers in the *Journal of the American Statistical Association* in 1978 employed simulation. In such a climate it is extremely important for research papers to specify the algorithm used, and the tests for randomness employed in their investigation. At best, simulation results should be verified using a *different* generator.

In minimal BASIC there are two statements which relate to the work of this chapter. These are:

$$10 \quad \text{RANDOMIZE}$$
$$20 \quad U = \text{RND}$$

The first statement selects a seed in a random fashion, possibly by reference to the current time. If this statement is omitted, the pseudo-random number sequence that is used will always start from the same seed. In the second statement we obtain a realization of a pseudo-random $U(0, 1)$ variable. Both of these statements will occur in programs in later chapters. While the BASIC instructions are as above, the underlying method used will vary from machine to machine, and on many microcomputers a slightly different form from RND is used.

The bibliographies by Sowey (1972, 1978) reveal that random number generation is a wide field of continuing interest. While new generators of proven improved properties may be developed in the future, congruential generators are likely to continue to prove popular and convenient. The need to test random numbers cannot be stressed too strongly, and this is a subject to which we shall return in Chapter 6. We shall now, in Chapters 4 and 5, proceed to see how uniform random numbers may be changed to give random variables of any kind.

*3.9   Proof of the attainment of a full cycle for a particular mixed congruential pseudo-random number generator

In the following, $a, b, c, k, s, t, \alpha, \gamma, \theta, \phi, h_1, h_2$ and h_3 denote positive integers.

THEOREM 3.1
The mixed congruential generator

$$x_{n+1} = ax_n + b \quad (\text{mod } m)$$

with $a = 4c + 1$, b odd and $m = 2^k$, has cycle length m.

PROOF
The basis of the proof is to show that if $x_i = x_j$, for $i \neq j$, then we cannot have $|i - j| < m$. As the cycle length is $\leq m$, then this will prove that the cycle length is m, and the sequence generated within a single cycle is a permutation of the integers from 0 to $(m - 1)$. Without loss of generality, therefore, we shall take $x_0 = 0$, as this simplifies matters.

First of all, note that $x_n = y_n \pmod{m}$,

where $\qquad\qquad y_{n+1} = ay_n + b, \qquad \text{for } n \geq 0$ $\qquad\qquad$ (3.6)

and, by the above, $y_0 = 0$.

From (3.6) we see that

$$y_n = b(1 + a + a^2 + \ldots + a^{n-1}) \qquad \text{for } n \geq 0$$

Now, $x_i = y_i - h_1 2^k$
and if $x_i = x_j$ for some $i > j$, say, then

$$b(a^j + a^{j+1} + \ldots + a^{i-1}) = h_2 2^k$$

i.e., $$b a^j (1 + a + a^2 + \ldots + a^{i-j-1}) = h_2 2^k \qquad (3.7)$$

Let us write $w_n = 1 + a + \ldots + a^{n-1} \qquad$ for $n \geq 1$.

In (3.7), by definition, a and b are odd, and so to prove the theorem we must show that:

$$w_{(i-j)} \neq h_3 2^k \qquad \text{for } (i-j) < 2^k \qquad (3.8)$$

and this we shall now do.

THE CASE $(i-j)$ ODD
If $(i-j)$ is odd, we can write $(i-j) = 2t + 1$, say, for $t \geq 0$.

$$w_{2t} = \left(\frac{1 - a^{2t}}{1 - a} \right) = \{(1 + 4c)^{2t} - 1\}/4c$$

$$= \{(1 + 4c)^t - 1\}\{(1 + 4c)^t + 1)/4c\}$$

$$= \{(1 + 4c)^t + 1\} \left\{ \sum_{i=1}^{t} (4c)^{i-1} \binom{t}{i} \right\},$$

which is even, as $1 + (1 + 4c)^t = 2 + 4c \sum_{i=1}^{t} \binom{t}{i} (4c)^{i-1} \qquad (3.9)$

Thus $w_{2t+1} = w_{2t} + a^{2t+1}$ is odd, as a is odd, and so (3.8) is trivially true.

THE CASE $(i-j)$ EVEN
If $(i-j)$ is even, there exists an s such that $(i-j) = \alpha 2^s$, for some odd, positive integral α, and as $(i-j) < 2^k$, then $s < k$.

$$w_{(i-j)} = w_{\alpha 2^s} = 1 + a + \ldots + a^{\alpha 2^{s-1} - 1} + a^{\alpha 2^{s-1}} + \ldots + a^{\alpha 2^s - 1}$$

$$= w_{\alpha 2^{s-1}} + a^{\alpha 2^{s-1}} w_{\alpha 2^{s-1}}$$

$$= w_{\alpha 2^{s-1}} (1 + a^{\alpha 2^{s-1}}) \qquad (3.10)$$

$$= w_{\alpha 2^{s-2}} (1 + a^{k_1})(1 + a^{k_2})$$

$$\vdots$$

$$= w_\alpha (1 + a^{k_1})(1 + a^{k_2}) \ldots (1 + a^{k_s}),$$

for suitable positive integers, $k_1, k_2, \ldots, k_s$.

Since $a = (1 + 4c)$, we have, from (3.9)

$$w_{\alpha 2^s} = w_\alpha \gamma 2^s,$$

in which w_α and γ are odd positive integers. (We have just proved that w_α is odd if α is odd.) Hence as $s < k$, there does not exist an h_3 such that

$$w_{\alpha 2^s} = h_3 2^k$$

This completes the proof. We note, finally, that it is simple to verify $x_0 = x_m$, since $(x_0 - x_m) = bw_m$, where $m = 2^k$. $w_2 = 1 + a = 2 + 4c$, so for $k = 1$, result (3.7) is true. Let us suppose that

$$w_m = \theta m, \text{ for } m = 2^k \text{ and } k \geq 1 \qquad (3.11)$$

$$w_{2m} = \theta m (1 + a^m), \qquad \text{from (3.10)}$$

and $(1 + a^m)$ is even, from (3.9).

Hence $w_{2m} = \phi(2m)$, and if (3.11) is true for $k \geq 1$, then it is also true for $(k + 1)$. We have seen that it is true for $k = 1$, and so by induction it is true for all $k \geq 1$.

3.10   Exercises and complements

(a) Uses of random numbers

The randomized response technique (RRT) has been much studied and extended. A good introduction is provided by Campbell and Joiner (1973), who motivate the first four questions.

†**3.1**   Investigate the workings of the RRT when the two alternative questions are:

   (i) I belong to group X;
   (ii) I do not belong to group X.

3.2   Describe how you would proceed if the proportion of positive responses to the RRT 'innocent' question is unknown. Can you suggest an innocent question for which it should be possible to obtain the proportion of correct responses without difficulty?

3.3   Investigate the RRT when the randomizing device is a bag of balls, each being one of *three* different colours, say red, white and blue, and the instructions to the (female) respondents are:

   If the red ball is drawn answer the question: 'have you had an abortion'.
   If the white ball is drawn, respond 'Yes'.
   If the blue ball is drawn, respond 'No'.

*3.4 In RRT, consider the implications for the respondent of responding 'Yes', even though it is not known which question has been answered. Consider how the technique might be extended to deal with frequency of activity. Consider how to construct a confidence interval (see, e.g., *ABC*, p. 266) for the estimated proportion.

3.5 100 numbered pebbles formed the population in a sampling experiment devised by J. M. Bremer. Students estimate the population mean weight ($\mu = 37.63$ g) by selecting 10 pebbles at random, using tables of random numbers, and additionally by choosing a sample of 10 pebbles, using their judgement only. The results obtained from a class of 32 biology undergraduates are given below:

Judgement sample means	Random sample means
62.63	31.45
35.85	32.12
55.36	51.93
66.43	24.74
34.96	43.32
37.23	29.41
34.45	42.67
60.53	47.94
49.61	28.76
56.07	56.43
59.02	31.21
50.65	32.73
33.34	55.37
58.62	36.65
47.02	22.44
48.34	40.04
28.56	44.65
26.65	41.43
46.34	39.39
27.86	26.39
39.62	23.88
25.45	35.15
48.82	35.88
66.56	28.03
37.25	31.71
45.98	43.98
32.46	61.49
54.03	31.52
51.89	33.99
62.81	33.78
59.74	49.69
14.05	22.97

Discuss, with reference to these data, the importance of taking random samples.

(b) On uniform random digits

3.6   A possible way of using two unbiased dice for simulating uniform random digits from 0 to 9 is as follows: throw the two dice and record the sum. Interpret 10 as 0, 11 as 1, and ignore 12. Discuss this procedure. (Based on part of an A-level examination question: Oxford, 1978.)

3.7   In Example 3.2 we used a fair coin to simulate events with probability different from 0.5. Here we consider the converse problem (the other side of the coin). Suppose you want to simulate an event with probability $\frac{1}{2}$; you have a coin but you suspect it is biased. How should you proceed? One approach is this: toss the coin twice. If the results of the two tosses are the same, repeat the experiment, and carry on like this until you obtain two tosses that are different. Record the outcome of the second toss. Explain why this procedure produces equi-probable outcomes. Discussion and extensions to this simple idea are given in Dwass (1972) and Hoeffding and Simons (1970).

3.8   In a series of 10 tosses of two distinguishable fair dice, A and B, the following faces were uppermost (A is given first in each case): (1, 4), (2, 6), (1, 5), (4, 3), (2, 2), (6, 3), (4, 5), (5, 1), (3, 4), (1, 2).
  Explain how you would use the dice to generate uniformly distributed random numbers in the range 0000–9999. (Based on part of an A-level examination question: Oxford, 1980.)

†3.9   British car registration numbers are of the form: SHX 792R. Special rôles are played by the letters, but that is not, in general, true of the numbers. Collect 1000 digits from observing car numbers, and examine these digits for randomness (explain how you deal with numbers of the form: HCY 7F).

3.10   Below we give the decimal expansion of π to 2500 places, kindly supplied by T. Hopkins. Draw a bar-chart to represent the relative frequencies of some (if not all!) of these digits, and comment on the use of these digits as uniform random 0–9 digits. Note that Fisher and Yates (1948) adopted a not dissimilar approach, constructing random numbers from tables of logarithms; further discussion of their numbers is given in Exercise 6.8(ii).

3.1415926535 8979323846 2643383279 5028841971 6939937510
5820974944 5923078164 0628620899 8628034825 3421170679
8214808651 3282306647 0938446095 5058223172 5359408128
4811174502 8410270193 8521105559 6446229489 5493038196
4428810975 6659334461 2847564823 3786783165 2712019091
4564856692 3460348610 4543266482 1339360726 0249141273
7245870066 0631558817 4881520920 9628292540 9171536436
7892590360 0113305305 4882046652 1384146951 9415116094
3305727036 5759591953 0921861173 8193261179 3105118548
0744623799 6274956735 1885752724 8912279381 8301194912

9833673362 4406566430 8602139494 6395224737 1907021798
6094370277 0539217176 2931767523 8467481846 7669405132
0005681271 4526356082 7785771342 7577896091 7363717872
1468440901 2249534301 4654958537 1050792279 6892589235
4201995611 2129021960 8640344181 5981362977 4771309960
5187072113 4999999837 2978049951 0597317328 1609631859
5024459455 3469083026 4252230825 3344685035 2619311881
7101000313 7838752886 5875332083 8142061717 7669147303
5982534904 2875546873 1159562863 8823537875 9375195778
1857780532 1712268066 1300192787 6611195909 2164201989

3809525720 1065485863 2788659361 5338182796 8230301952
0353018529 6899577362 2599413891 2497217752 8347913151
5574857242 4541506959 5082953311 6861727855 8890750983
8175463746 4939319255 0604009277 0167113900 9848824012
8583616035 6370766010 4710181942 9555961989 4676783744
9448255379 7747268471 0404753464 6208046684 2590694912
9331367702 8989152104 7521620569 6602405803 8150193511
2533824300 3558764024 7496473263 9141992726 0426992279
6782354781 6360093417 2164121992 4586315030 2861829745
5570674983 8505494588 5869269956 9092721079 7509302955

3211653449 8720275596 0236480665 4991198818 3479775356
6369807426 5425278625 5181841757 4672890977 7727938000
8164706001 6145249192 1732172147 7235014144 1973568548
1613611573 5255213347 5741849468 4385233239 0739414333
4547762416 8625189835 6948556209 9219222184 2725502542
5688767179 0494601653 4668049886 2723279178 6085784383
8279679766 8145410095 3883786360 9506800642 2512520511
7392984896 0841284886 2694560424 1965285022 2106611863
0674427862 2039194945 0471237137 8696095636 4371917287
4677646575 7396241389 0865832645 9958133904 7802759009

9465764078 9512694683 9835259570 9825822620 5224894077
2671947826 8482601476 9909026401 3639443745 5305068203
4962524517 4939965143 1429809190 6592509372 2169646151
5709858387 4105978859 5977297549 8930161753 9284681382
6868386894 2774155991 8559252459 5395943104 9972524680
8459872736 4469584865 3836736222 6260991246 0805124388
4390451244 1365497627 8079771569 1435997700 1296160894
4169486855 5848406353 4220722258 2848864815 8456028506
0168427394 5226746767 8895252138 5225499546 6672782398
6456596116 3548862305 7745649803 5593634568 1743241125

(c) On pseudo-random numbers

3.11 The first pseudo-random number generator was the 'mid-square' proposed by von Neumann (1951). The method is as follows: select a large integer, e.g. 7777. Square it and use the middle four digits as the next integer, square that, and so on. Here we get:

$$7777 \to 60\,\underline{481\,729} \to 4817 \to 23\,\underline{203\,489} \to 2034 \to 4\,\underline{137\,1}56$$
$$\to 1371 \to 1\,\underline{879\,6}41 \to \text{etc.}$$

The above sequence illustrates how we proceed when the squared number does not fill the entire possible field-length of 8. Investigate and comment upon this procedure. Further discussion is provided by Tocher (1975, p. 72) and Knuth (1981, p. 3), who explain the problems that can arise with this method. Craddock and Farmer (1971) provide a modification which avoids the obvious degeneration when the process results in zero.

†3.12 Investigate sequences produced by:

$$u_{n+1} = \text{fractional part of } (\pi + u_n)^5.$$

3.13 Show that for a full-period mixed congruential generator the mean and variance of the values produced by dividing each integer element in the full-period sequence by the modulus m are, respectively, $\frac{1}{2}(1 - 1/m)$, and $(1 + 1/m)/12$.

†3.14 The following BASIC program simulates the mixed congruential generator of Equation (3.4), with $a = 781, b = 387, m = 1000$. Run this

```
10   REM MIXED CONGRUENTIAL GENERATOR
20   INPUT U0
30   LET A = 781
40   LET B = 387
50   FOR I = 1 TO 1000
60     LET U1 = (A*U0+B)/1000
70     LET U1 = (U1-INT(U1))*1000
80     LET U0 = U1
90     PRINT U1
100  NEXT I
110  END
```

program with and without the following change (from Cooke, Craven and Clarke, 1982, p. 70):

75   U1 = INT(U1 + 0.5)

Comment on the results and the reason for using this additional line.

3.15 In pseudo-random number generation using congruential methods we obtain a sequence of integers $\{x_i\}$ over the range $(0, m)$. Approximations to $U(0, 1)$ random variables are then obtained by setting $u_i = x_i/m$. Show that

$$u_{i+1} = (au_i + b/m) \pmod{1}$$

and program this in BASIC (note the lesson of Exercise 3.14 with regard to round-off error). Note also the comments of Knuth (1981, p. 525).

3.16 Show that for a mixed congruential generator, with $a > 1$,

$$x_{n+k} = [a^k x_n + (a^k - 1)b/(a-1)] \pmod{m} \qquad \text{for } k \geq 0, n \geq 0$$

This property is useful in distributed array processing (DAP) programming of congruential generators (Sylwestrowicz, 1982).

*3.17 Show that if U_1 and U_2 are independent $U(0, 1)$ random variables, then the fractional part of $(U_1 + U_2)$ is also $U(0, 1)$. Show further that this result still holds if U_1 is $U(0, 1)$, but U_2 has any continuous distribution.

3.18 (*continuation*) Show that if U_1, U_2 and U_3 are formed independently from congruential generators with respective cycle lengths c_1, c_2 and c_3, then we may take the fractional part of $(U_1 + U_2 + U_3)$ as a realization of a pseudo-random $U(0, 1)$ random variable, and the resulting sequence of $(0, 1)$ variables will have cycle length $c_1 c_2 c_3$ if c_1, c_2 and c_3 are relatively prime. For further discussion, see Neave (1972, p. 6) and Wichmann and Hill (1982a).

3.19 In a mixed congruential generator, show that if $m = 10^k$ for some positive integer $k > 1$, then for the cycle length to equal m, we need to set $a = 20d + 1$, where d is a positive integer.

3.20 Show that the sequence $\{x_i\}$ of Section 3.5, for which $x_0 = 89$, $x_1 = 16$, etc., alternates between even and odd numbers.

†3.21 (a) (Peach, 1961) The mixed congruential generator,

$$x_{n+1} = 9x_n + 13 \pmod{32}$$

has full (32) cycle length. Write down the resulting sequence of numbers and investigate it for patterns. For example, compare the numbers in the first half with those in the second half, write the numbers in binary form, etc.

(b) Experiment with congruential generators of your own.

3.22 (a) Write BASIC programs to perform random shuffling and random replacement of pseudo-random numbers. When might these two procedures be equivalent?

(b) (Bays and Durham, 1976) We may use the next number from a congruential generator to determine the random replacement. Investigate this procedure for the generator

$$x_{n+1} = 5x_n + 3 \pmod{16}; \ x_0 = 1.$$

3.23   A distinctly non-random feature of congruential pseudo-random numbers is that no number appears twice within a cycle. Suggest a simple procedure for overcoming this defect.

3.24   Construct a pseudo-random number generator of your own, and evaluate its performance.

†**3.25**   The much-used IBM generator RANDU is multiplicative congruential, with multiplier 65 539, and modulus 2^{31}, so that the generated sequence is:

$$x_{i+1} = 65\,539 x_i \quad (\text{mod } 2^{31})$$

Use the identity $65\,539 = 2^{16} + 3$ to show that

$$x_{i+1} = (6x_i - 9x_{i-1}) \quad (\text{mod } 2^{31}),$$

and comment on the behaviour of successive triplets (x_{i-1}, x_i, x_{i+1}). See also Chambers (1977, p. 191), Miller (1980a, b) and Kennedy and Gentle (1980, p. 149) for further discussion of this generator. Examples of plots of triplets are to be found in Knuth (1981, p. 90) and Oakenfull (1979).

3.26   (a) The Fibonacci series may be used for a pseudo-random number generator:

$$x_{n+1} = (x_n + x_{n-1}) \quad (\text{mod } m)$$

Investigate the behaviour of numbers resulting from such a series. See Wall (1960) for an investigation of cycle-length when $m = 2^k$.

(b) (Knuth, 1981) In a random sequence of numbers, $0 \le x_i < m$, how often would you expect to obtain $x_{n-1} < x_{n+1} < x_n$? How often does this sequence occur with the generator of (a)? The Fibonacci series above is generally held to be a poor generator of pseudo-random numbers, but its performance can be much improved by shuffling (see Gebhardt, 1967). Oakenfull (1979) has obtained good results from the series

$$x_{n+1} = (x_n + x_{n-97}) \quad (\text{mod } 2^{35})$$

Note that repeated numbers can occur with Fibonacci-type generators (cf. Exercise 3.23).

3.27   (a) For a multiplicative congruential generator, show that if a is an odd power of $8n \pm 3$, for any suitable integral n, and x_0 is odd, then all subsequent members of the congruential series are odd.

(b) As we shall see in Chapter 5, it is sometimes necessary to form $\log_e U$, where U is $U(0, 1)$. Use the result of (a) to explain the advantage of such a multiplicative generator over a mixed congruential generator in such a case.

*3.28 Consider how you would write a FORTRAN program for a congruential generator.

*3.29 (Taussky and Todd, 1956)   Consider the recurrence

$$y_{n+1} = y_n + y_{n-1} \qquad \text{for } n \geq 1,$$

with $y_0 = 0$, $y_1 = 1$.
Show that

$$y_n = \left\{ \left(\frac{\sqrt{5}+1}{2} \right)^n - \left(\frac{1-\sqrt{5}}{2} \right)^n \right\} \bigg/ \sqrt{5}$$

and deduce that for large n,

$$y_n \approx \left(\frac{\sqrt{5}+1}{2} \right)^n \bigg/ \sqrt{5}$$

Hence compare the Fibonacci series generator of Exercise 3.26 with a multiplicative congruential generator. Difference equations, such as that above, occur regularly in the theory of random walks (see Cox and Miller, 1965, Section 2.2).

3.30 The literature abounds with congruential generators. Discuss the choice of a, b, m, in the following. For further considerations, see Kennedy and Gentle (1980, p. 141) and Knuth (1981, p. 170).

(i) a b m
7^5 0 $2^{31}-1$

Called GGL, this is IBM's replacement for RANDU (see Learmonth and Lewis, 1973). Egger (1979) used this generator in combination with shuffling from a $g = 100$ store, and it is the basis of routines GGUBFS and GGUBS of the IMSL library; see Section A1.1 in Appendix 1.

(ii) a b m
16333   25887 2^{15} (from Oakenfull, 1979)

(iii) a b m
3432   6789   9973   (see also Oakenfull, 1979)

(iv) a b m
23   0 $10^8 + 1$ the Lehmer generator

This generator is of interest as it was the first proposed congruential generator, with $x_0 = 47\,594\,118$, by Lehmer (1951).

(v) The NAG generator: GO5CAF:

$$\begin{array}{ccc} a & b & m \\ 13^{13} & 0 & 2^{59} \end{array}$$

See Section A1.1

(vi) $\quad\begin{array}{ccc} a & b & m \\ 171 & 0 & 30269 \end{array}$

This is one of the three component generators used by Wichmann and Hill (1982a, b).

(vii) $\quad\begin{array}{ccc} a & b & m \\ 131 & 0 & 2^{35} \end{array}$ used by Neave (1973).

(viii) $\quad\begin{array}{ccc} a & b & m \\ 2^7+1 & 1 & 2^{35} \end{array}$

This generator is of interest as it is one of the original mixed congruential generators, proposed by Rotenberg (1960).

(ix) $\quad\begin{array}{ccc} a & b & m \\ 397\,204\,094 & 0 & 2^{31}-1 \end{array}$

This is the routine GGUBT of the IMSL library—see Section A1.1

3.31 Show that the cycle length in a multiplicative congruential generator is given by the smallest positive integer n satisfying $a^n = 1 \bmod(m)$. (See Exercise 3.16.)

We stated in Section 3.5, that if $m = 2^k$ in a multiplicative congruential generator, only one-quarter of the integers $0\text{–}m$ are obtained in the generator cycle. However, if m is a prime number then a cycle of length $(m-1)$ can be obtained with multiplicative congruential generators. Let $\phi(m)$ be the number of integers less than and prime to m, and suppose m is a prime number, p. Clearly, $\phi(p) = (p-1)$.

It has been shown (Tocher, 1975, p. 76) that the n above must divide $\phi(m)$. If $n = \phi(p)$ then a is called a 'primitive root' mod (p), and the cycle length $(p-1)$ is attained. Ways of identifying primitive roots of prime moduli are given by Downham and Roberts (1967); for example, 2 is a primitive root of p if $(p-1)/2$ is prime and $p = 3 \bmod (8)$. Given a primitive root r, then further primitive roots r may be generated from: $r = r^k \bmod(p)$, where k and $(p-1)$ are co-prime.

Use these results to verify that the following 5 prime modulus multiplicative congruential generators, considered by Downham and Roberts (1967), have cycle length $(p-1)$.

	$m = p$	a
(i)	67 101 323	8 192
(ii)	67 099 547	8 192
(iii)	16 775 723	32 768
(iv)	67 100 963	8
(v)	7 999 787	32

Extensions and further discussion are given by Knuth (1981, pp. 19–22) and Fuller (1976), while relevant tables are provided by Hauptman *et al.* (1970) and Western and Miller (1968).

*3.32 If a fair coin is tossed until there are two consecutive heads, show that the probability that n tosses are required is

$$p_n = y_{n-1}/2^n \qquad \text{for } n \geq 2$$

where the y_n are given by the Fibonacci numbers of Exercise 3.29 (cf. Exercise 3.26). We see from Exercise 3.29 that as $n \to \infty$, the ratio $y_n/y_{n-1} \to$ the golden ratio, $\phi = (1 + \sqrt{5})/2$, so that the tail of the distribution is approximately geometric, with parameter $\phi/2$. Mead and Stern (1973) suggest uses of this problem in the empirical teaching of statistics. Verify that the distribution above has mean 6.

4

PARTICULAR METHODS
FOR NON-UNIFORM
RANDOM VARIABLES

Some of the results of Chapter 2 may be used to convert uniform random variables into variables with other distributions. It is the aim of this chapter to provide some examples of such particular methods for simulating non-uniform random variables. Because of the important rôle played by the normal distribution in statistics, we shall start with normally distributed random variables.

4.1 Using a central limit theorem

It is because of central limit theorems that the normal distribution is encountered so frequently, and forms the basis of much statistical theory. It makes sense, therefore, to use a central limit theorem in order to simulate normal random variables. For instance, we may simulate n independent $U(0, 1)$ random variables, $U_1, U_2, \ldots, U_n$, say, and then set $N = \Sigma_{i=1}^n U_i$. As $n \to \infty$ the distribution of N tends to that of a normal variable. But in practice, of course, we settle on some finite value for n, so that the resulting N will only be approximately normal. So how large should we take n? The case $n = 2$ is unsuitable, as N then has a triangular distribution (see Exercise 4.8), but for $n = 3$, the distribution of N is already nicely 'bell-shaped', as will be shown later. The answer to this question really depends on the use to which the resulting numbers are to be put, and how close an approximation is desired.

A convenient number to take is $n = 12$, since, as is easily verified, $\mathscr{E}[U_i] = \frac{1}{2}$ and $\text{Var}[U_i] = 1/12$, so that then

$$N = \sum_{i=1}^{12} U_i - 6$$

is an approximately normal random variable with mean zero and unit variance. Values of $|N| > 6$ do not occur, which could, conceivably, be a problem for large-scale simulations. The obvious advantage of this approach, however, is its simplicity; it is simple to understand, and simple to program, as we can see from Fig. 4.1.

```
10   RANDOMIZE                           10   RANDOMIZE
20   INPUT M                             20   INPUT M
30   REM PROGRAM TO SIMULATE 2*M         30   REM PROGRAM TO SIMULATE 2*M
40   REM APPROXIMATELY STANDARD          40   REM STANDARD NORMAL RANDOM VARIABLES
50   REM NORMAL RANDOM VARIABLES         50   REM USING THE BOX-MULLER METHOD
60   REM USING A CENTRAL LIMIT           60   LET P2 = 2*3.14159265
70   REM THEOREM APPROACH                70   FOR I = 1 TO M
80   FOR I = 1 TO 2*M                     80   LET R = SQR(-2*LOG(RND))
90     LET N = 0                          90   LET U = RND
100    FOR J = 1 TO 12                    100    PRINT R*SIN(P2*U),R*COS(P2*U)
110      LET N = N+RND                    110  NEXT I
120    NEXT J                             120  END
130    PRINT N-6
140  NEXT I
150  END
```

Figure 4.1   BASIC programs for simulating 2*M* standard normal random variables.

4.2   The Box–Müller and Polar Marsaglia methods

4.2.1   The Box–Müller method

The last method used a convolution to provide approximately normal random variables. The next method we consider obtains exact normal random variables by means of a one-to-one transformation of two $U(0, 1)$ random variables. If U_1 and U_2 are two independent $U(0, 1)$ random variables then Box and Müller (1958) showed that

$$N_1 = (-2 \log_e U_1)^{1/2} \cos(2\pi U_2)$$

and
$$N_2 = (-2 \log_e U_1)^{1/2} \sin(2\pi U_2)$$

$$(4.1)$$

are independent $N(0, 1)$ random variables.

At first sight this result seems quite remarkable, as well as most convenient. It is, however, a direct consequence of the result of Example 2.4, as we shall now see.

If we *start* with independent $N(0, 1)$ random variables, N_1 and N_2, defining a point (N_1, N_2) in two dimensions by Cartesian co-ordinates, and we change to polar co-ordinates (R, Θ), then

$$N_1 = R \cos \Theta$$
$$N_2 = R \sin \Theta$$

$$(4.2)$$

and in Example 2.4 we have already proved that R and Θ are then independent random variables, Θ with a $U(0, 2\pi)$ distribution, and $R^2 = N_1^2 + N_2^2$ with a χ_2^2 distribution, i.e. an exponential distribution of mean 2. Furthermore, to simulate Θ we need simply take $2\pi U_2$, where U_2 is $U(0, 1)$, and to simulate R we can take $(-2 \log_e U_1)$, where U_1 is $U(0, 1)$, as explained in Exercises 2.1 and 2.2.

We therefore see that Box and Müller have simply inverted the relationship of (4.2), which goes from (N_1, N_2) of a particular kind to (R, Θ) of a particular kind, and instead move from (R, Θ) to (N_1, N_2), simulating Θ by means of $2\pi U_2$, and an independent R from $(-2 \log_e U_1)^{1/2}$, where U_1 is independent of U_2. You are asked to provide a formal proof of (4.1) in Exercise 4.6. As we can see from Fig. 4.1, this method is also very easy to program, and each method considered so far would be easily operated on a hand-calculator.

If the Box–Müller method were to be used regularly on a computer then it would be worth incorporating the following interesting modification, which avoids the use of time-consuming sine and cosine functions.

*4.2.2   The Polar Marsaglia Method

The way to avoid using trignometric functions is to construct the sines and cosines of uniformly distributed angles directly *without* first of all simulating the angles. This can be done by means of a *rejection* method as follows:

If U is $U(0, 1)$, then $2U$ is $U(0, 2)$, and $V = 2U - 1$ is $U(-1, 1)$.

If we select two independent $U(-1, 1)$ random variables, V_1 and V_2, then these specify a point at random in the square of Fig. 4.2, with polar co-ordinates $(\tilde{R}, \Theta)$ given by:

$$\tilde{R}^2 = V_1^2 + V_2^2$$

and $$\tan \Theta = V_2 / V_1$$

Repeated selection of such points provides a random scatter of points in the square, and rejection of points outside the inscribed circle shown leaves us with a uniform random scatter of points within the circle.

For any one of these points it is intuitively clear (see also Exercise 4.11) that the polar co-ordinates $\tilde{R}$ and Θ are independent random variables, and further that Θ is a $U(0, 2\pi)$ random variable. In addition (see Exercise 4.11) $\tilde{R}^2$ is $U(0, 1)$ and so the pair $(\tilde{R}, \Theta)$ are what are required by the Box–Müller method, and we can here simply write

$$\sin \Theta = \frac{V_2}{\tilde{R}} = V_2 (V_1^2 + V_2^2)^{-1/2}$$

$$\cos \Theta \doteq V_1 (V_1^2 + V_2^2)^{-1/2}$$

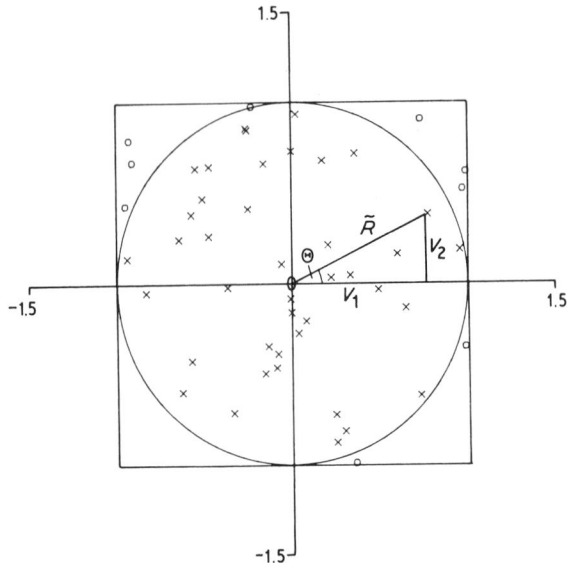

Figure 4.2   An illustration of points (denoted by ○ and ×) uniformly distributed over the square shown. The points denoted by ○ are rejected when one simply requires points uniformly distributed over the unit disc, as in the Polar Marsaglia method.

so that a pair of independent $N(0, 1)$ variables, N_1 and N_2, are given by:

$$N_1 = (-2 \log(\tilde{R}^2))^{1/2} V_2 (V_1^2 + V_2^2)^{-1/2}$$

$$N_2 = (-2 \log(\tilde{R}^2))^{1/2} V_1 (V_1^2 + V_2^2)^{-1/2},$$

i.e. $$N_1 = (-2 \log(V_1^2 + V_2^2))^{1/2} V_2 (V_1^2 + V_2^2)^{-1/2}$$

$$N_2 = (-2 \log(V_1^2 + V_2^2))^{1/2} V_1 (V_1^2 + V_2^2)^{-1/2}$$

resulting in

$$N_1 = V_2 \left(\frac{-2 \log W}{W} \right)^{1/2}$$

$$N_2 = V_1 \left(\frac{-2 \log W}{W} \right)^{1/2}$$

where $W = V_1^2 + V_2^2$.

The philosophy of rejection may seem rather strange at first, as it involves discarding variates obtained at a certain cost and effort, but pairs of variates

(V_1, V_2) are rejected just a proportion $1 - \pi/4$ of the time. The advantage of the rejection method here is that it provides a very simple way of obtaining a uniform scatter of points inside the circle of Fig. 4.2. Another rejection method was described in Example 3.2, and we shall encounter more general rejection methods in the next chapter, which have the same aim and use as here.

A BASIC program for this method is given in Fig. 4.3. Now known as the 'Polar Marsaglia' method, this approach is due originally to Marsaglia and Bray (1964), and is used in the IMSL routine: GGNPM – see Section A1.1.

```
10    RANDOMIZE
20    INPUT M
30    REM PROGRAM TO SIMULATE 2*M STANDARD NORMAL
40    REM RANDOM VARIABLES USING THE POLAR
50    REM MARSAGLIA METHOD
60    FOR I = 1 TO M
70      LET V1 = 2*RND-1
80      LET V2 = 2*RND-1
90      LET R2 = V1*V1+V2*V2
100     IF R2 > 1 THEN 70
110     LET Y = SQR((-2*LOG(R2))/R2)
120     PRINT V1*Y,V2*Y
130   NEXT I
140   END
```

Figure 4.3   BASIC program for simulating $2M$ standard normal random variables, using the Polar Marsaglia method.

4.3   Exponential, gamma and chi-square variates

Random variables with exponential and gamma distributions are frequently used to model waiting times in queues of various kinds, and this is a natural consequence of the predictions of the Poisson process. The simplest way of obtaining random variables with an exponential p.d.f. of e^{-x} for $x \geq 0$ is to set $X = -\log_e U$, where U is $U(0, 1)$, as has already been done in the previous section (see Exercise 2.1).

We have also seen that $Y = X/\lambda$ has the exponential p.d.f. $\lambda e^{-\lambda x}$ for $x \geq 0$ (Exercise 2.2). An alternative approach for simulating exponential random variables will be given later in Exercise 5.35.

We have seen in Section 2.10 that if we have independent random variables, $Y_1, \ldots, Y_n$ with density function $\lambda e^{-\lambda x}$ for $x \geq 0$, then

$$G = \sum_{i=1}^{n} Y_i$$

has a gamma, $\Gamma(n, \lambda)$, distribution. Thus to simulate a $\Gamma(n, \lambda)$ random variable for integral n we can simply set

$$G = -\frac{1}{\lambda} \sum_{i=1}^{n} \log_e U_i$$

where $U_1, \ldots, U_n$ are independent $U(0, 1)$ random variables, i.e.

$$G = -\frac{1}{\lambda} \log_e \left(\prod_{i=1}^{n} U_i \right)$$

Now a random variable with a χ_m^2 distribution is simply a $\Gamma(m/2, \frac{1}{2})$ random variable (see Section 2.10), and so if m is even we can readily obtain a random variable with a χ_m^2 distribution, by the above approach. If m is odd, we can obtain a random variable with a χ_{m-1}^2 distribution by first obtaining a $\Gamma((m-1)/2, \frac{1}{2})$ random variable as above, and then adding to it N^2, where N is an independent $N(0, 1)$ random variable (see Exercise 2.5). Here we are using the defining property of χ^2 random variables on integral degrees of freedom, and use of this property alone provides us with a χ_m^2 random variable from simply setting

$$Z = \sum_{i=1}^{m} N_i^2 \tag{4.3}$$

where the N_i are independent, $N(0, 1)$ random variables. However, because of the time taken to simulate $N(0, 1)$ random variables, this last approach is not likely to be very efficient. Both NAG and IMSL computer packages use convolutions of exponential random variables in their routines for the generation of gamma and chi-square random variables (see Section A1.1). The simulation of gamma random variables with non-integral shape parameter, n, is discussed in Section 4.6.

4.4   Binomial and Poisson variates

4.4.1   Binomial variates

A binomial $B(n, p)$ random variable, X, can be written as $X = \Sigma_{i=1}^{n} B_i$, where the B_i are independent *Bernoulli* random variables, each taking the values, $B_i = 1$, with probability p, or $B_i = 0$, with probability $(1 - p)$. Thus to simulate such an X, we need just simulate n independent $U(0, 1)$ random variables, $U_1, \ldots, U_n$, and set $B_i = 1$ if $U_i \leq p$, and $B_i = 0$ if $U_i > p$. The same end result can, however, be obtained from judicious re-use of a single $U(0, 1)$ random variable U. Such re-use of uniform variates is employed by the IMSL routine, GGBN when $n < 35$ (see Section A1.1). If $n \geq 35$, a method due to Relles (1972) is employed by this routine: in simulating a $B(n, p)$ variate we simply count how many of the U_i are less than p. If n is large then time can be saved by ordering the $\{U_i\}$ and then observing the location of p within the ordered sample. Thus if we denote the ordered sample by $\{U_{(i)}\}$, for the case $n = 7$, and $p = 0.5$ we might have:

In this example we would obtain $X = 3$ as a realization of a $B(7, \frac{1}{2})$ random variable.

Rather than explicitly order, one can check to see whether the sample median is greater than or less than p, and then concentrate on the number of sample values between p and the median. In the above illustration the sample median is $U_{(4)} > p$, and so we do not need to check whether $U_{(i)} > p$ for $i > 4$. However, we do not know, without checking, whether $U_{(i)} > p$ for any $i < 4$. This approach can clearly now be iterated by seeking the sample median of the sample: $(U_{(1)}, U_{(2)}, U_{(3)}, U_{(4)})$ in the above example, checking whether it is greater or less than p, etc. (see Exercise 4.18). A further short-cut results if one makes use of the fact that sample medians from $U(0, 1)$ samples can be simulated directly using a beta distribution (see Exercise 4.17). Full details are given by Relles (1972), who also provides a FORTRAN algorithm. An alternative approach, for particular values of p only, is as follows.

If we write U in binary form to n places, and if indeed U is $U(0, 1)$, then independently of all other places the ith place is 0 or 1 with probability $\frac{1}{2}$ (cf. Section 3.2). Thus, for example, if $U_1 = 0.10101011100101100$, we obtain 9 as a realization of a $B(17, \frac{1}{2})$ random variable, if we simply sum the number of ones. Here the binary places correspond to the trials of the binomial distribution.

If we want a $B(17, \frac{1}{4})$ random variable, we select further an independent $U(0, 1)$ random variable, U_2, say. If $U_2 = 0.10101101100110101$, then place-by-place multiplication of the digits in U_1 and U_2 gives: 0.10101001100100100, in which 1 occurs with probability $\frac{1}{4}$ at any place after the point. In this illustration we therefore obtain 7 as a realization of a $B(17, \frac{1}{4})$ random variable.

This approach can be used to provide $B(n, p)$ random variables, when we can find m and r so that $p = m2^{-r}$ (see Exercise 4.3). Most people are not very adept at binary arithmetic, but quite efficient algorithms could result from exploiting these ideas if machine-code programming could be used to utilize the binary nature of the arithmetic of most computers. However, as we have seen in Chapter 3, pseudo-random $U(0, 1)$ variables could exhibit undesirable patterns when expressed in binary form.

4.4.2   Poisson variates

Random variables with a Poisson distribution of parameter λ can be generated as a consequence of the following result.

Suppose $\{E_i, i \geq 1\}$ is a sequence of independent random variables, each with an exponential distribution, of density $\lambda e^{-\lambda x}$, for $x \geq 0$. Let $S_0 = 0$ and $S_k = \Sigma_{i=1}^{k} E_i$ for $k \geq 1$, so that, from Section 4.3, the S_k are $\Gamma(k, \lambda)$ random variables. Then the random variable K, defined implicitly by the inequalities $S_K \leq 1 < S_{K+1}$ has a Poisson distribution with parameter λ. In other words, we set $S_1 = E_1$, and if $1 < S_1$, then we set $K = 0$. If $S_1 \leq 1$, then we set $S_2 = E_1 + E_2$, and then if $S_2 > 1$, we set $K = 1$. If $S_2 \leq 1$, then we continue, setting $S_3 = E_1 + E_2 + E_3$, and so on, so that we set $K = i$ when, and only when, $S_i \le 1 < S_{i+1}$ for $i \ge 0$.

The BASIC program in Fig. 4.4 shows how easily this algorithm may be programmed, and may also help in demonstrating how it works. Note that we simulate

$$S_k = \sum_{i=1}^{k} E_i \quad \text{by} \quad S_k = -\frac{1}{\lambda} \log\left(\prod_{i=1}^{k} U_i \right)$$

```
10    RANDOMIZE
20    INPUT M,L
30    REM PROGRAM TO SIMULATE M RANDOM
40    REM VARIABLES FROM A POISSON
50    REM DISTRIBUTION OF PARAMETER L
60    LET E1 = EXP(-L)
70    FOR I = 1 TO M
80      LET K = 0
90      LET U = RND
100     IF U < E1 THEN 140
110     LET U = U*RND
120     LET K = K+1
130     GOTO 100
140     PRINT K
150   NEXT I
160   END
```

Figure 4.4   BASIC program for simulating M Poisson random variables.

where as usual the U_i are independent $U(0, 1)$ random variables, as explained in Section 4.3. The comparison $S_k > 1$, then becomes

$$-\frac{1}{\lambda} \log\left(\prod_{i=1}^{k} U_i \right) > 1$$

i.e. $\qquad \log\left(\prod_{i=1}^{k} U_i \right) < -\lambda \qquad$ i.e. $\qquad \prod_{i=1}^{k} U_i < e^{-\lambda}$

and it is this inequality which is being tested in line number 100 of the program.

On first acquaintance, this algorithm has the same 'rabbit-out-of-a-hat' nature as the Box–Müller method. We can certainly show analytically that K thus defined has the required Poisson distribution (see Exercise 4.9), but a consideration of the Poisson process, mentioned in Section 2.7, shows readily the origin of this algorithm, as we shall now see.

In a Poisson process in time (say) of rate λ we have the two important results (see, e.g., *ABC*, chapter 19):

(a) times between events are independent random variables from the exponential p.d.f., $\lambda e^{-\lambda x}$, for $x \ge 0$;
(b) the number of events in any fixed time interval of length t has a Poisson distribution of parameter (λt).

Result (b) tells us that to simulate a random variable with a Poisson distribution of parameter λ, all we have to do is construct a realization of a Poisson process of parameter λ, and then count the number of events occurring in a time interval of unit length. Result (a) tells us how we can simulate the desired Poisson process, by simply placing end-to-end independent realizations of exponential random variables from the $\lambda e^{-\lambda x}$ density. We keep a record of the time taken since the start of the process, and stop the simulation once that time exceeds unity. Figure 4.5 provides an illustration, resulting in $K = 3$, as there have been just three events in the Poisson process in the $(0, 1)$ time interval, occurring at times E_1, $E_1 + E_2$ and $E_1 + E_2 + E_3$ respectively, with the fourth event occurring at time $E_1 + E_2 + E_3 + E_4 > 1$.

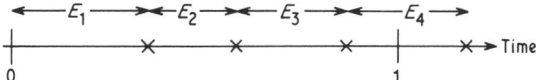

Figure 4.5   Illustration of the simulation of a Poisson process in time, starting at time 0. Four events occur, at times denoted by ×. Inter-event times, E_i, are independent random variables from the $\lambda e^{-\lambda x}$ exponential p.d.f., and the value $K = 3$, the number of events in the $(0, 1)$ time interval, is a realization of a random variable with a Poisson distribution of parameter λ.

*4.5   Multivariate random variables

Particular rules may also be exploited to simulate multivariate random variables. Two examples, one discrete and one continuous, will be considered here.

4.5.1   The bivariate Poisson distribution

This distribution was mentioned briefly in Section 2.16. If three independent random variables, X_1, X_2 and X_3, have Poisson distributions with parameters λ_1, λ_2 and λ_3 respectively, then the derived variables $Y_1 = X_1 + X_3$, $Y_2 = X_2 + X_3$ have a bivariate Poisson distribution. This is readily verified by simply writing down the bivariate moment generating function for Y_1 and Y_2, and observing it is of the form given in Equation (2.6). If we simulate X_1, X_2 and X_3 by the method of Section 4.4.2 then this result readily allows us to simulate random variables Y_1 and Y_2 with a bivariate Poisson distribution. An alternative approach is suggested in Exercise 4.7.

4.5.2   The multivariate normal distribution

This distribution was discussed in Section 2.15. We saw there that if the p-variate random variable $\mathbf{X}$ has the multivariate normal, $N(\mathbf{0}, \mathbf{I})$ distribution, then $\mathbf{Z} = \mathbf{AX} + \boldsymbol{\mu}$ has the multivariate normal $N(\boldsymbol{\mu}, \mathbf{AA'})$ distribution. Hence, if we want to simulate random variables from an $N(\boldsymbol{\mu}, \boldsymbol{\Sigma})$ multivariate normal distribution, then we need only find a matrix $\mathbf{A}$ for which $\boldsymbol{\Sigma} = \mathbf{AA'}$. Ways of doing this are discussed in Exercise 4.14 and in the solution to that exercise. $\mathbf{X}$ is readily simulated, as its elements are independent and $N(0, 1)$. We then set $\mathbf{Z} = \mathbf{AX} + \boldsymbol{\mu}$.

4.6   Discussion and further reading

We have seen in this chapter a utilization of formal relationships between random variables, which enables us to simulate a variety of random variables, using only $U(0, 1)$ variates; other illustrations can be found in the exercises. All of these examples are no more than useful tricks for particular cases. Often very simple algorithms result, as we have seen from some of the BASIC programs presented, and these algorithms could be readily implemented for small-scale simulations, using hand-calculators or microcomputers, for example. In a number of cases, however, the algorithms are less efficient than others which may be devised (see Kinderman and Ramage, 1976, for example), some of which will be considered in the next chapter.

The method of Section 4.4.2 for Poisson variates may become very inefficient if λ is large. In this case we would expect large numbers of events in the Poisson process during the $(0, 1)$ interval, resulting in prohibitively many checks. Atkinson (1979a) compares the algorithm of Section 4.4.2 for simulating Poisson random variables with alternative approaches which will be mentioned in the next chapter, while Kemp and Loukas (1978a, b) make similar comparisons for the bivariate Poisson case. More recent work for the univariate Poisson case is to be found in Atkinson (1979c), Kemp (1982), Ahrens and Dieter (1980, 1982) and Devroye (1981).

Atkinson and Pearce (1976), Atkinson (1977) and Cheng (1977) discuss the simulation of gamma $\Gamma(n, \lambda)$ random variables with *non*-integral shape parameter n, and we consider Cheng's method in Exercise 5.22. More recent work is provided by Cheng and Feast (1979) and Kinderman and Monahan (1980).

Neave (1973) showed that when the standard Box–Müller method is operated using pseudo-random numbers from a particular multiplicative congruential generator, the resulting numbers exhibit some strikingly non-normal properties. This finding was taken up by Chay, Fardo and Mazumdar (1975) and Golder and Settle (1976), and we shall return to this point in Section 6.7.

We have in this chapter only scratched the surface of the relations between random variables of different kinds. The books by Johnson and Kotz (1969, 1970a, 1970b, 1972) provide many more such relationships, and the book by Mardia (1970) provides more information on bivariate distributions.

4.7 Exercises and complements

4.1 Show that a random variable with a $U(0, 1)$ distribution has mean $1/2$ and variance $1/12$.

†**4.2** Consider how you might simulate a binomial $B(3, p)$ random variable using just one $U(0, 1)$ variate, and write a BASIC program to do this.

4.3 Explain how the approach of using a binary representation of $U(0, 1)$ random variables may be used to simulate random variables with a binomial $B(n, p)$ distribution in which $p = m2^{-r}$, for integral $r > 0$, and integral $0 \leq m \leq 2^r$.

4.4 The following result is similar to one in Exercise 2.14: If the independent random variables X_1 and X_2 are, respectively, $\Gamma(p, 1)$ and $\Gamma(r, 1)$, then $Y = X_1/(X_1 + X_2)$ has the beta density,

$$f(y) = \frac{\Gamma(p+r)}{\Gamma(p)\Gamma(r)} y^{p-1} (1-y)^{r-1} \quad \text{for } 0 \leq y \leq 1$$

Use this result to write a BASIC program to simulate such a Y random variable, for integral $p > 1$ and $r > 1$.

4.5 If X_1, X_2, X_3, X_4 are independent $N(0, 1)$ random variables, we may use them to simulate other variates, using the results of Exercise 2.15 as follows:
(a) $Y = |X_1 X_2 + X_3 X_4|$ has an exponential distribution of parameter 1.
(b) $C = X_1/X_2$ has a Cauchy distribution, with density function $1/(\pi(1 + x^2))$.
Use these results to write BASIC programs to simulate such Y and C random variables.

†**4.6** Prove that N_1, N_2, given by Equation (4.1) are independent $N(0, 1)$ random variables.

4.7 When (X, Y) have a bivariate Poisson distribution, the probability generating function has the form

$$G(u, v) = \exp \{\lambda_1 (u-1) + \lambda_2 (v-1) + \lambda_3 (uv-1)\}$$

The marginal distribution of X is Poisson, of parameter $(\lambda_1 + \lambda_3)$, while the conditional distribution of $Y|X = x$ has probability generating function

$$\left(\frac{\lambda_1 + \lambda_3 v}{\lambda_1 + \lambda_3}\right)^x \exp[\lambda_2 (v-1)]$$

Use these results to simulate the bivariate Poisson random variable (X, Y).

*4.8   If $X = \sum_{i=1}^{n} U_i$, where U_i are independent $U(0, 1)$ random variables, show, by induction or otherwise, that X has the probability density function

$$f(x) = \sum_{j=0}^{[x]} (-)^j \binom{n}{j} (x - j)^{n-1}/(n-1)! \qquad \text{for } 0 \leq x \leq n,$$

$$= 0 \qquad \text{otherwise}$$

where $[x]$ denotes the integral part of x.

*4.9   Prove, without reference to the Poisson process, that the random variable K, defined at the start of Section 4.4.2, has a Poisson distribution of parameter λ.

*4.10   Consider how the Box–Müller method may be extended to more than two dimensions.

*4.11   In the notation of the Polar Marsaglia method, show that Θ and $\tilde{R}$, defined by

$$\tan \Theta = V_1/V_2 \qquad \text{and} \qquad \tilde{R}^2 = V_1^2 + V_2^2$$

both conditional on $V_1^2 + V_2^2 \leq 1$, are independent random variables. Show also that R^2 is a $U(0, 1)$ random variable, and Θ is a $U(0, 2\pi)$ random variable.

*4.12   When a pair of variates (V_1, V_2) is rejected in the Polar Marsaglia method, it is tempting to try to improve on efficiency, and only reject one of the variates, so that the next pair for consideration would then be (V_2, V_3), say. Show why this approach is unacceptable.

*4.13   Provide an example of a continuous distribution, with density function $f(x)$, with zero mean, for which the following result is true: X_1 and X_2 are independent random variables with probability density function $f(x)$. When the point (X_1, X_2), specified in terms of Cartesian co-ordinates, is expressed in polar co-ordinates (R, Θ), then R, Θ are not independent.

*4.14   If **S** is a square, symmetric matrix, show that it is possible to write **S** = **VDV'**, where **D** is a diagonal matrix, the ith diagonal element of which is the ith eigenvalue of **S**, and **V** is an orthogonal matrix with ith column an eigenvector corresponding to the ith eigenvalue of **S**. Hence provide a means of obtaining the factorization, $\Sigma = $ **AA'** required for the simulation of multivariate normal random variables in Section 4.5.2. More usually, a Choleski factorization is used for Σ, in which **A** is a lower-triangular matrix. Details are provided in the solution. This is the approach adopted in the IMSL routine GGNSM—see Section A1.1.

*4.15 If X_1, X_2, ..., X_n are independent (column) random variables from a p-variate multivariate normal, $N(0, \Sigma)$ distribution, then $Z = \Sigma_{i=1}^{n} X_i X_i'$ has the *Wishart* distribution, $W(Z; \Sigma, n)$, described, for example, by Press (1972, p. 100). Use this result, which generalizes to p dimensions the result of Equation (4.3), to provide a BASIC program to simulate such a Z. For related discussion, see Newman and Odell (1971, chapter 5).

*4.16 If X_1, X_2 are independent $N(0, 1)$ random variables, show that the random variables X_1 and

$$Y_1 = \rho X_1 + (1 - \rho^2)^{1/2} X_2 \qquad \text{where} \qquad -1 \le \rho \le +1$$

have a bivariate normal distribution, with zero means, unit variances, and correlation coefficient ρ.

†4.17 Let $U_1, U_2, ..., U_{2n-1}$ be a random sample from the $U(0, 1)$ density. If M denotes the sample median, show that M has the $B_e(n, n)$ distribution.

*4.18 (*continuation*) Consider how the result of Exercise 4.17 may be used to simulate a $B(n, p)$ random variable (Relles, 1972).

†4.19 We find, from Section A1.1, that the IMSL computer library has routine GGPON for simulating Poisson variables when the Poisson parameter λ may vary from call to call. Otherwise one might use the IMSL routine GGPOS. Kemp (1982) was also concerned with Poisson variable simulation when λ may vary, and one might wonder why one should want to simulate such Poisson variates. An answer is provided by the following exercise.

The random variable X has the conditional Poisson distribution:

$$\Pr(X = k | \lambda) = \frac{e^{-\lambda} \lambda^k}{k!} \qquad \text{for } k \ge 0$$

If λ has a $\Gamma(n, \theta)$ distribution, show that the unconditional distribution of X is

$$\Pr(X = k) = \binom{n+k-1}{k} \left(\frac{1}{\theta+1}\right)^k \left(\frac{\theta}{\theta+1}\right)^n \qquad \text{for } k \ge 0$$

i.e. $X = Y - n$, where Y has a negative-binomial distribution as defined in Section 2.6. As $n \to 0$ the distribution tends to the logarithmic series distribution much used in ecology, and generalized by Kempton (1975). Kemp (1981) considers simulation from this distribution (see Exercise 4.22).

4.20 (*continuation*) Use the result of the last question to provide a BASIC program to simulate negative-binomial random variables, for integral $n > 1$. This result is sometimes used to explain the frequent use of the negative-binomial distribution for describing discrete data when the Poisson distribution is unsatisfactory. The negative-binomial is a 'contagious' distribution, and much more relevant material is provided by Douglas (1980). An algorithm using the waiting-time definition of the distribution is given in the IMSL routine GGBNR (see Appendix 1).

4.21 Use the transformation theory of Section 2.12 to show that the random variable $Y = e^X$, where X has a $N(\mu, \sigma^2)$ distribution, has the density function

$$f_Y(y) = \frac{1}{y\sigma \sqrt{(2\pi)}} \exp\left(-\frac{1}{2}\left(\frac{\log_e(y) - \mu}{\sigma}\right)^2\right) \qquad \text{for } y \geq 0$$

Y is said to have a *log-normal* distribution. The p.d.f. has the same qualitative shape as that of the $\Gamma(2, 1)$ p.d.f. of Fig. 2.6, and the log-normal distribution is often used to describe incubation periods for diseases (see also Morgan and Watts, 1980) and sojourn times in more general states. We shall, in fact, encounter such a use for this distribution in Example 8.3. Section A1.1 gives IMSL and NAG routines for simulating from this distribution. For full details, see Aitchison and Brown (1966).

4.22 (Kemp, 1981)   The general logarithmic distribution is

$$p_k = -\alpha^k / \{k \log_e(1 - \alpha)\} \qquad k \geq 1, 0 < \alpha < 1$$

Show that its moment generating function is:

$$\log(1 - \alpha e^\theta)/\log(1 - \alpha)$$

and that successive probabilities can be generated from:

$$p_k = \alpha(1 - 1/k)p_{k-1} \qquad \text{for } k \geq 2$$

Show that if X has the conditional geometric distribution

$$\Pr(X = x | Y = y) = (1 - y)y^{x-1} \qquad \text{for } x \geq 1$$

and if $\qquad \Pr(Y \leq y) = \dfrac{\log(1 - y)}{\log(1 - \alpha)} \qquad \text{for } 0 \leq y \leq \alpha$

then X has the logarithmic distribution. Explain how you can make use of this result to simulate from the logarithmic distribution.

5

GENERAL METHODS
FOR NON-UNIFORM
RANDOM VARIABLES

For many uses, simple algorithms, such as those which may arise from particular methods of the kind described in the last chapter, will suffice. It is of interest, however, to consider also *general* methods, which may be used for any distribution, and that we shall now do. In many cases general methods can result in algorithms which, while they are more complicated than those considered so far, are appreciably more efficient.

5.1 The 'table-look-up' method for discrete random variables

For ease of notation, let us suppose that we have a random variable X that takes the values 0, 1, 2, 3, etc., and with $p_i = \Pr(X = i)$ for $i \geq 0$. Thus X could be binomial, or Poisson, for example.

A general algorithm for simulating X is as follows:
Select a $U(0, 1)$ random variable, U.

Set $X = 0$ if $0 \leq U < p_0$, and

$$\text{set } X = j \quad \text{if } \sum_{i=0}^{j-1} p_i \leq U < \sum_{i=0}^{j} p_i \quad \text{for } j \geq 1.$$

We can think of the probabilities $\{p_i, i \geq 0\}$ being put end-to-end and, as $\sum_{i=0}^{\infty} p_i = 1$, filling out the interval $[0, 1]$ as illustrated in Fig. 5.1. We can now see that the above algorithm works by selecting a value U and observing in which probability interval U lands. In the illustration of Fig. 5.1 we have

$$\sum_{i=0}^{1} p_i \leq U < \sum_{i=0}^{2} p_i$$

and so we set $X = 2$.

This algorithm is simply a generalization to more than two intervals of the rule used to simulate Bernoulli random variables in Section 4.4.1. The

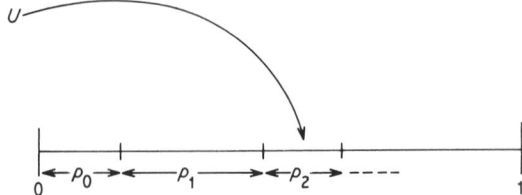

Figure 5.1   An illustration of how the table-look-up method works for simulating a discrete random variable with probability distribution $\{P_i, i \geq 0\}$.

reason the algorithm works is readily explained. We want to simulate the random variable X so that $\Pr(X = i) = p_i$ for $i \geq 0$. As U is $U(0, 1)$, then for any $0 \leq a \leq b$, $\Pr(a \leq U < b) = (b - a)$ (see Section 2.8) and so $\Pr(0 \leq U < p_0) = p_0 = \Pr(X = 0)$, and

$$\Pr\left(\sum_{i=0}^{j-1} p_i \leq U < \sum_{i=0}^{j} p_i\right) = p_j = \Pr(X = j) \qquad \text{for } j \geq 1$$

Thus, for $j \geq 0$, the algorithm returns the value $X = j$ with probability p_j, as required.

The above algorithm is readily modified to cope with discrete random variables with different ranges from that considered above, (i.e. $X \geq 0$) and one such example now follows.

EXAMPLE 5.1
We want to simulate the geometric random variable X, with distribution

$$p_i = \Pr(X = i) = (1 - p)^{i-1} p \qquad \text{for } i \geq 1, 0 < p < 1$$

In order to operate the above algorithm we need successive cumulative sums of the $\{p_i\}$, and in this case, because p_i is of a simple geometric form, then these cumulative sums are also of a simple form. Here,

$$\sum_{i=1}^{j} p_i = \frac{p(1 - (1 - p)^j)}{1 - (1 - p)} = 1 - (1 - p)^j \qquad \text{for } j \geq 1$$

Thus the algorithm becomes:

Set $X = j$ if $1 - (1 - p)^{j-1} \leq U < 1 - (1 - p)^j$ for $j \geq 1$

which is equivalent to: $-(1 - p)^{j-1} \leq U - 1 < -(1 - p)^j$

i.e. $(1 - p)^{j-1} \geq (1 - U) > (1 - p)^j$ (5.1)

Before proceeding, we can here observe that the algorithm entails selecting a $U(0, 1)$ random variable, and then checking the range of $(1 - U)$. Now it is intuitively clear that if U is a $U(0, 1)$ random variable, *then so is* $(1 - U)$, and this result can be readily verified by the change-of-variable theory of Section 2.12 (see Exercise 5.1).

Hence we can reduce the labour of arithmetic slightly if we replace (5.1) by the equivalent test:

$$(1-p)^{j-1} \geq U > (1-p)^j \tag{5.2}$$

Of course, for any particular realization of U, (5.1) and (5.2) will usually give different results; however, the random variable X resulting from using (5.2) will have the same geometric distribution as the random variable resulting from using (5.1). Continuing from (5.2), we set $X = j \geq 1$ if, and only if,

$$(j-1) \log_e (1-p) \geq \log_e U > j \log_e (1-p)$$

so that, recalling that $\log_e (1-p) < 0$, we have $X = j$ if

$$(j-1) \leq \frac{\log_e U}{\log_e (1-p)} < j \qquad \text{for } j \geq 1 \tag{5.3}$$

Finally, we note that we can express (5.3) very simply by setting

$$X = 1 + \left[\frac{\log_e U}{\log_e (1-p)} \right] \tag{5.4}$$

where $[y]$ is used to denote the integral part of y. For further discussion of this result, see Exercise 5.12.

This example therefore uses the general table-look-up algorithm to produce the simple expression of (5.4). This is in contrast to a particular approach which may be used, based upon the definition of the geometric distribution given in Section 2.6. Thus an alternative method would be to test sequentially independent $U(0, 1)$ random variables until one was found to be less than p, and an algorithm using this approach is provided by the IMSL routine GGEOT (see Section A1.1).

This example is unusual in that the cumulative sums of probabilities have a simple form. The next example is far more typical.

EXAMPLE 5.2
If X has a Poisson distribution of parameter 2, its cumulative distribution function is given below to four places of decimals:

i	0	1	2	3	4	5	6	7	8	9
$Pr(X \leq i)$	0.1353	0.4060	0.6767	0.8571	0.9473	0.9834	0.9955	0.9989	0.9998	1.000

Using this table and the table-look-up algorithm, the following eight $U(0, 1)$

random variables can be seen to give rise to the indicated values of X:

U	X
0.0318	0
0.4167	2
0.4908	2
0.2459	1
0.3643	1
0.8124	3
0.9673	5
0.1254	0

This example illustrates why the table-look-up method is so called. Given a table of the cumulative distribution of any discrete random variable, and a supply of $U(0, 1)$ random variables, we can use this method to simulate that random variable. By their very nature, such tables are finite, and if the random variable in question has an infinite range, then the range would have to be truncated for the method to be used. This was done in the above example, where using accuracy of only four decimal places resulted in the range of X being truncated to $[0, 9]$.

Human beings can operate the table-look-up method quite easily, but its implementation on a computer poses some intriguing problems. First of all we can remark that for a computer implementation it is not necessary to store cumulative sums of probabilities—they can be computed each time, as required. Random variables of infinite range need not then have their range truncated, but this approach is usually far too costly in effort because of the repeated duplication of arithmetic each time a new simulation is run. More usually ranges are truncated if necessary, and the resulting finite tables are stored within the computer. The next problem that arises is how to read such stored tables. Computers need specified algorithms which could, for instance, involve reading the table of the cumulative distribution in Example 5.2 from left to right. In such a case, the computer would return $X = 0$ when $U = 0.0318$, with the greatest of ease, but when $U = 0.9673$ it would laboriously check whether $U < 0.1353$, $U < 0.4060$, and so on until it found $0.9473 < U < 0.9834$. Human beings need not be so rigid and have the advantage over computers of being able to change their strategy in the light of superficial evidence on the size of U. By analogy, when looking up a word such as 'wombat' in the dictionary, not many of us would start at the front, with the letter 'A' and then skim through from A to W; rather, we would start from the middle, or somewhere near the end, possibly even working backwards as well as forwards. A more efficient computer algorithm may result if the range of X

were initially subdivided; for example, if $\Pr(X \le \theta) = p \approx 0.5$, say, for some known θ, then if $U > p$ it would not be necessary to chek U against $\Sigma_{i=0}^{j} p_i$ for $j \le \theta$. Such an approach is utilized in the IMSL routine GGDT and the NAG routine GO5EYF (see Section A1.1) and was encountered earlier in Section 4.4.1.

5.2   The 'table-look-up', or inversion method for continuous random variables

We shall now consider the analogue of the above method for continuous random variables. Suppose we wish to simulate a continuous random variable X with cumulative distribution function $F(x)$, i.e. $F(x) = \Pr(X \le x)$, and suppose also that the inverse function, $F^{-1}(u)$ is well-defined for $0 \le u \le 1$.

If U is a $U(0, 1)$ random variable, then $X = F^{-1}(U)$ has the required distribution. We can see this as follows:

If
$$X = F^{-1}(U)$$

then
$$\Pr(X \le x) = \Pr(F^{-1}(U) \le x)$$

and because $F(x)$ is the cumulative distribution function of a continuous random variable, $F(x)$ is a strictly monotonic increasing continuous function of x. This fact enables us to write

$$\Pr(F^{-1}(U) \le x) = \Pr(U \le F(x))$$

But, as U is a $U(0, 1)$ random variable,

$$\Pr(U \le F(x)) = F(x) \qquad \text{(see Section 2.8)}$$

i.e.
$$\Pr(X \le x) = F(x)$$

and so the X obtained by setting $X = F^{-1}(U)$ has the required distribution.

The above argument, which was given earlier in Equation (2.1), is perhaps best understood by considering a few examples. Figure 5.2 illustrates one cumulative distribution function for a truncated exponential random variable. We operate the rule $X = F^{-1}(U)$ by simply taking values of $U(0, 1)$ variates and projecting down on the x-axis as shown, using the graph of $y = F(x)$.

We shall now consider two further examples in more detail.

EXAMPLE 5.3

In the untruncated form, if X has an exponential density with parameter λ, then

$$f(x) = \lambda e^{-\lambda x} \qquad \text{for } x \ge 0, \lambda > 0,$$

and
$$F(x) = 1 - e^{-\lambda x}$$

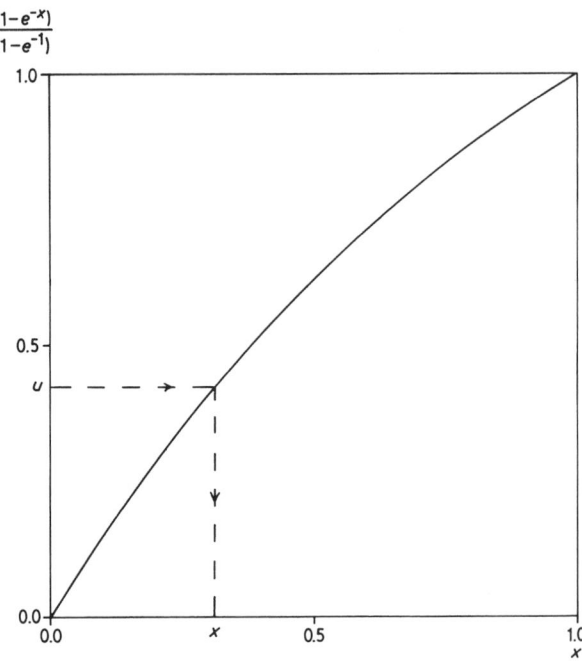

Figure 5.2   Operation of the table-look-up method for a continuous random variable. The curve has the equation $(1 - e^{-x})/(1 - e^{-1})$ and is the cumulative distribution function of a truncated exponential random variable, with probability density function, $e^{1-x}(e-1)^{-1}$ for $0 \le x \le 1$.

To simulate X we set $X = F^{-1}(U)$, i.e., set

$$U = 1 - e^{-\lambda X}, \text{ and solve for } X.$$

This gives

$$X = -\frac{1}{\lambda} \log_e (1 - U)$$

and for the same reasoning as in Example 5.1, we obtain the desired distribution for X from setting

$$X = -\frac{1}{\lambda} \log_e U \tag{5.5}$$

A verification of this result is provided by the solution to Exercise 2.1, and this method is used in the IMSL routine GGEXN and the NAG routine GO5DBF (see Section A1.1).

As with Example 5.1, the result of (5.5) is deceptive in its simplicity, and it is no coincidence that the geometric and exponential distributions play similar rôles, the former in the discrete case and the latter in the continuous case, as discussed in Section 2.10, Exercise 2.23 and Exercise 5.12. It is unfortunately the case that it is often *not* simple to form $X = F^{-1}(U)$. The prime example of this occurs with the normal distribution, which has led to a variety of different approximations to both the normal cumulative distribution function and its inverse. We shall return to the subject of these approximations later in Section 5.7 and Exercise 5.9.

We conclude this section with an example of how this method can be used, in a 'table-look-up' fashion to simulate standard normal random variables.

EXAMPLE 5.4
If we take the same $U(0, 1)$ values as in Example 5.2 then we can use tables of the standard normal cumulative distribution function, $\Phi(x)$ to give the following realizations of an $N(0, 1)$ random variable X:

U	X (to two places of decimals)
0.0318	-1.85
0.4167	-0.21
0.4908	-0.02
0.2459	-0.69
0.3643	-0.35
0.8124	0.89
0.9673	1.84
0.1254	-1.15

Thus, for example, $0.8124 = \Phi(0.89)$, to the accuracy given.

Two points should be made here:

(a) Because of the symmetry of the $N(0, 1)$ distribution, the tables usually only give values of $x \geq 0$ and, correspondingly, values of $\Phi(x) \geq 0.5$, and so when $u < 0.5$ we have to employ the following approach which is easily verified to be correct:
We want x for which $u = \Phi(x)$.
If $u < 0.5$, then by the symmetry of the normal density, $x = -\Phi^{-1}(1 - u)$.
(b) The accuracy of the numbers produced (in this case to two decimal places) depends on the accuracy of the tables, which also determines the degree of the truncation involved.

The 'table-look-up' method for continuous random variables is often called the *inversion* method, and a general algorithm is provided by the IMSL routine GGVCR (see Section A1.1). We shall use these terms interchangeably, though strictly they describe different ways of implementing the same basic method.

5.3   The rejection method for continuous random variables

Suppose we have a method for sprinkling points uniformly at random under any probability density function $f(x)$, and which may give rise to the pattern of points in Fig. 5.3. What is the probability that the abscissa, X say, of any one of these points lies in the range $\alpha \leq X < \beta$, for any $\alpha < \beta$?

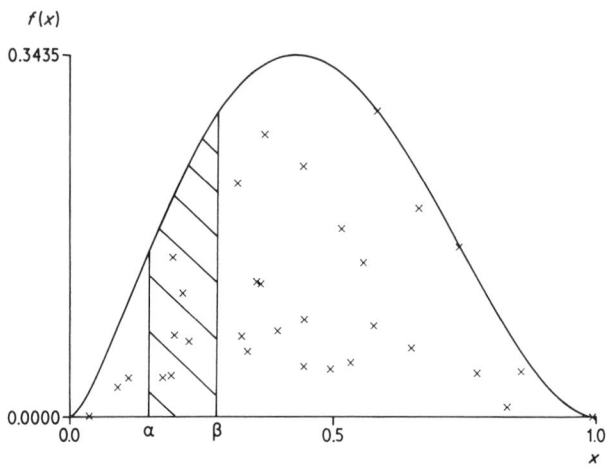

Figure 5.3   An illustration of points × uniformly and randomly distributed underneath the probability density function $f(x)$. For illustration we have used the B_e (2.5, 3) distribution.

The event, $\alpha \leq X < \beta$ is equivalent to the point being in the shaded area shown in Fig. 5.3, and so, because of the assumed uniform distribution of the points, this event has probability

$$\frac{\text{area of shaded area}}{\text{total area under } f(x)}$$

This is

$$\int_{\alpha}^{\beta} f(x)\,\mathrm{d}x \bigg/ \int_{-\infty}^{\infty} f(x)\,\mathrm{d}x$$

i.e.

$$\int_{\alpha}^{\beta} f(x)\,\mathrm{d}x \quad \text{as} \quad \int_{-\infty}^{\infty} f(x)\,\mathrm{d}x = 1,$$

since $f(x)$ is a probability density function.

Thus we are saying that

$$\Pr(\alpha \leq X < \beta) = \int_{\alpha}^{\beta} f(x)\,\mathrm{d}x, \qquad \text{for any } \alpha < \beta$$

where $f(x)$ is a probability density function, i.e. X has probability density function $f(x)$ (see Section 2.3).

Thus, for any probability density function $f(x)$, we can simulate random variables X from this density function as long as we have a method for uniformly and randomly sprinkling points under $f(x)$. Those who have read Section 4.2.2 will have already encountered a similar situation, the solution being in that case to enclose within a square the area to be sprinkled with points. It is a simple matter to distribute points uniformly at random over a square, and in Section 4.2.2, those points not within the area of interest were rejected. The same principle for any density function $f(x)$ results in a general rejection method, attributed to von Neumann (1951). While the rejection method (sometimes also called the 'acceptance–rejection' method) may be used for discrete random variables (see Fishman, 1979, for example), it is usually employed for continuous random variables, the case being investigated here.

If the probability density function $f(x)$ is non-zero over only a finite range, then it is easy to box it in, as shown in Fig. 5.4. Using $U(0, 1)$ random variables it is a simple matter to sprinkle points uniformly and randomly over the rectangle shown, simply by taking points with Cartesian co-ordinates $(\theta + (\alpha - \theta)U_1, \delta U_2)$, where U_1 and U_2 are independent $U(0, 1)$ random variables. Points landing above $f(x)$ are rejected, while for points landing below $f(x)$, we take $\theta + (\alpha - \theta)U_1$ as a realization of X.

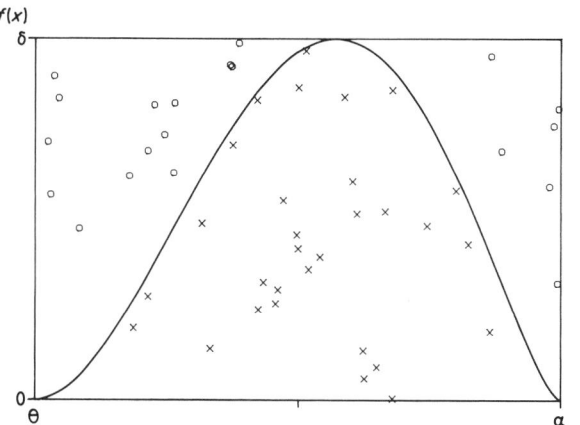

Figure 5.4 Simulating from the probability density function $f(x)$, which is non-zero over the finite range $[\theta, \alpha]$. The method used generates points (denoted by $\circ$ and $\times$) uniformly at random over the rectangle shown. Points denoted by $\circ$ are rejected (cf. Fig. 4.2), while the abscissae of the points $\times$ are accepted as realizations of the random variable with probability density function $f(x)$. For illustration we have used the B_e (3, 2.5) distribution, for which $\theta = 0$, $\alpha = 1$, $\delta = 0.3435$.

As the area of the rectangle in Fig. 5.4 is $\delta(\alpha - \theta)$, and the area under the curve is unity, we see that the probability of accepting a point is $1/(\delta(\alpha - \theta))$, and so the smaller δ is, the larger is the probability of acceptance and, correspondingly, the more efficient the method. This is, of course, why a larger value of δ was not used in Fig. 5.4.

There are two snags with the above approach. A rectangle is, as we have seen, a convenient shape within which to simulate a random uniform spread of points, but it clearly cannot be used if the density $f(x)$ has an infinite range, as we can only simulate uniform random variables over a finite range. Furthermore, the probability of rejection could become quite large: if the density of Fig. 5.4 was replaced by a spiked density, for instance, such as would result from a Laplace distribution, truncated to have a finite range. In such a case the simplicity gained from distributing points uniformly over a rectangular region could be more than offset by the cost of frequent rejection.

Both of these snags can be overcome by using as the enveloping curve a suitable multiple of a *different* probability density function from $f(x)$, as we shall now see. Consider a p.d.f. $h(x)$, with the same range as $f(x)$, *but from which it is relatively easy to simulate.* It is then simple to obtain a uniform scatter of points under $h(x)$, by taking points (X, Y) such that X has density $h(x)$, while the conditional density of Y given $X = x$ is $U(0, h(x))$. For a uniform scatter of points, the conditional p.d.f. of Y clearly must be of this form, while the X co-ordinate must have the property that for any pair (α, β), with $\alpha < \beta$, $\Pr(\alpha \le X < \beta) \sim \int_\alpha^\beta h(x)\,\mathrm{d}x$, i.e. X must have probability density function $h(x)$.

If it were possible to choose $h(x)$ to be of a roughly similar shape to $f(x)$ *and* then to envelop $f(x)$ by $h(x)$, we would obtain the desired scatter of points under $f(x)$ by first obtaining a scatter of points under $h(x)$ and then rejecting just those which were under $h(x)$ but not under $f(x)$. While it is often possible to choose an appropriate $h(x)$ to be of similar shape to $f(x)$, it is clearly not possible to envelop $f(x)$ by $h(x)$, so that, for all x, $f(x) \le h(x)$, since both $f(x)$ and $h(x)$ are density functions, and so $\int_{-\infty}^{\infty} f(x)\,\mathrm{d}x = \int_{-\infty}^{\infty} h(x)\,\mathrm{d}x = 1$. However, the solution to this last obstacle is easily obtained by, effectively, plotting $h(x)$ and the scatter of points obtained under $h(x)$ on stretchable paper, and then uniformly stretching the paper in a direction at right angles to the x-axis until $h(x) \ge f(x)$ for all x. Such stretching clearly does not change the uniformity of the scatter of the points. Mathematically this stretching is done, very simply, by taking as the conditional density of Y given $X = x$, $U(0, kh(x))$, where $k > 1$ is the stretching factor, and where X has probability density function $h(x)$.

Thus for suitable $h(x)$ and k, we have the following algorithm: if we write $g(x) = kh(x)$,

(i) simulate $X = x$ from probability density function $h(x)$;
(ii) simulate Y to be $Ug(x)$, where U is an independent $U(0, 1)$ random variable;

(iii) accept $X = x$ as a realization of a random variable with probability density function $f(x)$ if and only if $Y < f(x)$.

The situation is illustrated in Fig. 5.5.

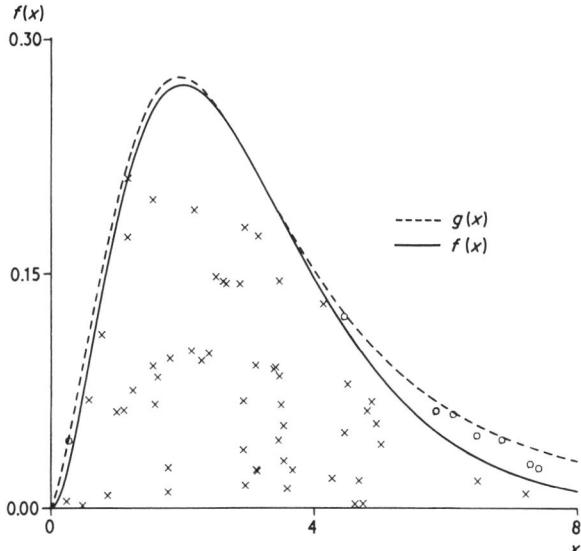

Figure 5.5   The points $\times$ and $\circ$ are uniformly and randomly distributed under the curve $g(x) = kh(x)$, where $k > 1$ is a constant, and $h(x)$ is a density from which it is easy to simulate. The points $\circ$ lie above the density function $f(x)$ and so are rejected. The points $\times$ are accepted and their abscissae are realizations of a random variable with probability density function $f(x)$. See Exercise 5.22 for an explanation of the p.d.f.'s used.

At first sight this algorithm seems unusual and confusing, since the test in (iii) concerns Y, but if the test is satisfied then it is $X = x$ which is accepted. However, in the light of the above discussion, we can now see that (iii) is just a component of testing whether a point constructed randomly and uniformly under $g(x)$ is also under $f(x)$.

The probability of rejection here is

$$\frac{\int_{-\infty}^{\infty} (g(x) - f(x))\,\mathrm{d}x}{\int_{-\infty}^{\infty} g(x)\,\mathrm{d}x} = 1 - \frac{1}{k}$$

reflecting the importance of small k, subject to $k > 1$.

We choose $h(x)$ with shape and convenience in mind. The next two examples provide two approaches for selecting k. The first example uses an exponential envelope to simulate normal random variables. An exponential envelope can only envelop half of the standard normal density, but it can envelop the 'half-normal' density, given by

$$f(x) = \sqrt{\left(\frac{2}{\pi}\right)} e^{-x^2/2} \qquad \text{for } x \geq 0$$

i.e. $f(x) = 2\phi(x)$, for $x \geq 0$. If X has density function $f(x)$, then the random variable

$$\tilde{X} = \begin{cases} X & \text{with probability } \frac{1}{2} \\ -X & \text{with probability } \frac{1}{2} \end{cases}$$

clearly has the standard normal density $\phi(x)$ for $-\infty \leq x < \infty$. We shall therefore simulate from $\phi(x)$ by first simulating from $f(x)$, and then applying the above transformation, from $\tilde{X}$ to X.

EXAMPLE 5.5 *A rejection method for N(0, 1) variables*

Here

$$f(x) = \sqrt{\left(\frac{2}{\pi}\right)} e^{-x^2/2} \quad \text{for } x \geq 0$$

and

$$g(x) = ke^{-x} \qquad \qquad \text{for } x \geq 0$$

One way of choosing k is to consider the condition for equal roots arising from setting

$$ke^{-x} = \sqrt{\left(\frac{2}{\pi}\right)} e^{-x^2/2}$$

as the roots in x of this equation correspond to the intersection of $g(x)$ and $f(x)$. If this equation has no real roots, then k is too large. If the equation has two distinct roots, then k is too small. The case of two equal roots corresponds to the smallest possible value of k, and the two curves touch, as shown in Fig. 5.6.

Setting $k\sqrt{\left(\frac{\pi}{2}\right)} = e^{x-x^2/2}$

results in a quadratic equation in x:

$$x^2 - 2x + 2\log_e\left(k\sqrt{\left(\frac{\pi}{2}\right)}\right) = 0$$

which has equal roots if and only if

$$1 = 2\log_e\left(k\sqrt{\left(\frac{\pi}{2}\right)}\right)$$

i.e.

$$k^2 \frac{\pi}{2} = e$$

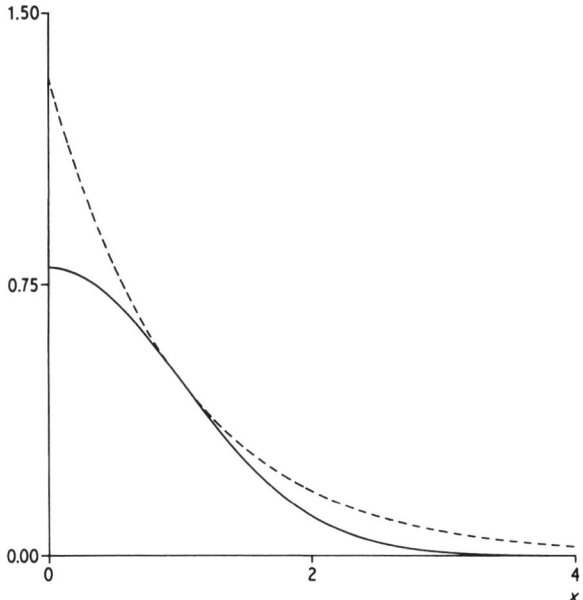

Figure 5.6   An illustration of the optimum choice of k. Here we illustrate the use of the exponential e^{-x} probability density function as the basis of an envelope for the half-normal density (solid line). The enveloping function is given by $\sqrt{\left(\dfrac{2e}{\pi}\right)}e^{-x}$ (dashed line), illustrated for $x \leq 4$, as is $f(x)$.

i.e.
$$k = +\sqrt{\left(\frac{2e}{\pi}\right)} \approx 1.315\,489\,2$$

the equal roots occurring at $x = 1$, which is, in fact, also the point of inflexion for the half-normal density.

The algorithm therefore proceeds as follows:

(a) Simulate X from density function, e^{-x} for $x \geq 0$. We know, from Equation (5.5) above, that we can do this by setting $X = -\log_e U_1$, where U_1 is a $U(0, 1)$ random variable. An alternative approach is given in Exercises 5.33–5.35.

(b) If U_2 is an independent $U(0, 1)$ random variable, set $Y = kU_2 e^{-X}$, i.e. $Y = kU_2 U_1$.

(c) Accept X if and only if,
$$Y < \sqrt{\left(\frac{2}{\pi}\right)}e^{-X^2/2}$$

i.e.
$$kU_1 U_2 < \sqrt{\left(\frac{2}{\pi}\right)}e^{-X^2/2}$$

i.e.
$$U_1 U_2 < \exp(-(1+X^2)/2),$$

since
$$k = \sqrt{\left(\frac{2e}{\pi}\right)}.$$

Thus ultimately the algorithm does not involve k directly.

Finally, of course, we must convert the half-normal random variable X to the standard normal random variable $\tilde{X}$. While this last stage can always be done by selecting a new $U(0,1)$ random variable, and then testing whether it is greater or less than $\frac{1}{2}$, we note that as Y is $U(0, g(X))$, then conditional on

$$Y < \sqrt{\left(\frac{2}{\pi}\right)} e^{-X^2/2},$$

Y has a $U\left(0, \sqrt{\left(\frac{2}{\pi}\right)} e^{-X^2/2}\right)$ distribution,

and the sign of $\tilde{X}$ can be decided by considering whether or not

$$Y < \frac{e^{-X^2/2}}{\sqrt{(2\pi)}}.$$

This is the same idea that was exploited in Exercise 4.2 and Section 4.4.1. A BASIC program for this algorithm is shown in Fig. 5.7.

The reason for using e^{-x} as the p.d.f. for the basis of the envelope here, rather than any other $\lambda e^{-\lambda x}$ p.d.f. can be found from a consideration of the probability of rejection, $1 - 1/k$, and we see in Exercise 5.21 that $\lambda = 1$ minimizes this rejection probability.

```
10   RANDOMIZE
20   INPUT M
30   REM PROGRAM TO SIMULATE M STANDARD NORMAL
40   REM RANDOM VARIABLES, USING A REJECTION METHOD
50   REM WITH A HALF-NORMAL PDF ENVELOPED BY A
60   REM MULTIPLE OF THE EXPONENTIAL PDF WITH
70   REM PARAMETER 1
80   FOR I = 1 TO M
90    LET U1 = RND
100   LET U2 = RND
110   LET X = -LOG(U1)
120   LET B = .5*EXP(-.5-X*X/2)
130   LET C = U1*U2
140   IF C < B THEN 170
150   IF C < 2*B THEN 190
160   GOTO 90
170   PRINT -X
180   GOTO 200
190   PRINT X
200  NEXT I
210  END
```

Figure 5.7   BASIC program for the rejection method illustrated in Fig. 5.6.

The exponential function provides a suitable envelope for the half-normal probability density function, as the rate at which e^{-x} tends to zero as $x \to \infty$ is less than the rate at which $e^{-x^2/2}$ tends to zero as $x \to \infty$.

A general way of finding k is to note that we want k to satisfy $kh(x) \geq f(x)$ for all x, and that we cannot have equality here for all x. k is therefore given by

$$k = \max_{x} \left(\frac{f(x)}{h(x)} \right)$$

if a finite maximum can be found, as then $kh(x) \geq f(x)$ for all x, with equality for at least one x.

A finite maximum will not result if $h(x)$ is unsuitable as a basis for an envelope of $f(x)$. For instance, we could have $h(x) = 0$ when $f(x) > 0$, or we might try setting $f(x) = e^{-x}$ and $h(x) = e^{-x^2/2}$. In this latter case,

$$\log (f(x)/h(x)) = \frac{x^2}{2} - x$$

which increases without bound as $x \to \infty$.

This approach should work, however, if a suitable $h(x)$ has been found. In this example we have

$$\frac{f(x)}{h(x)} = \sqrt{\left(\frac{2}{\pi} \right)} \frac{e^{-x^2/2}}{e^{-x}} = \sqrt{\left(\frac{2}{\pi} \right)} e^{x - x^2/2}$$

$$y = \log (f(x)/h(x)) = \log \left(\sqrt{\left(\frac{2}{\pi} \right)} \right) + x - \frac{x^2}{2}$$

$$\frac{dy}{dx} = 1 - x$$

$$\frac{d^2 y}{dx^2} = -1$$

Thus we maximize $f(x)/h(x)$ by setting $x = 1$, to give, as before, $k = \sqrt{\left(\frac{2e}{\pi} \right)}$.

In Section 4.3 we have already seen one way of simulating $\Gamma(n, \lambda)$ random variables, when n is a positive integer. The next example provides an alternative approach, for the case $n > 1$, using rejection, and an exponential envelope as in the last example. This approach may also be used when $n > 1$ is not integral.

Here we take $\lambda = 1$ for simplicity. If a random variable X results, then the new random variable $Y = X/\lambda$ will have a $\Gamma(n, \lambda)$ distribution, from the theory of Section 2.12 (see Exercise 2.2).

*EXAMPLE 5.6   *A rejection method for $\Gamma(n, 1)$ variables.*

Here
$$f(x) = \frac{x^{n-1} e^{-x}}{\Gamma(n)} \qquad \text{for } x \geq 0, \text{ and } n > 1$$

$$g(x) = k e^{-x/n}/n \qquad \text{for } x \geq 0.$$

As $n > 1$, then as $x \to \infty$, $f(x) \to 0$ faster than $g(x)$, implying that $g(x)$ is a suitable enveloping function for $f(x)$.

Let $y = f(x)/h(x)$. We seek k by maximizing y with respect to x.

$$\log_e y = (n-1)\log_e x - x + \frac{x}{n} + \log_e(n/\Gamma(n))$$

$$\frac{d}{dx}(\log_e y) = \frac{n-1}{x} - 1 + \frac{1}{n}$$

$$\frac{d^2}{dx^2}(\log_e y) = \frac{1-n}{x^2}$$

Thus, as $n > 1$, we maximize y when

$$\frac{n-1}{x} = 1 - \frac{1}{n}$$

i.e. when $x = n$, and so

$$k = n^n e^{1-n}/\Gamma(n)$$

It is now a simple matter to derive the following algorithm. Let U be a $U(0, 1)$ random variable, and let E be an independent exponential random variable with parameter n^{-1}. If

$$t(x) = \left(\frac{x}{n}\right)^{n-1} \exp\left[(1-n)\left(\frac{x}{n} - 1\right)\right] \qquad \text{for } x \geq 0$$

then conditional on $t(E) \geq U$, E has the required gamma p.d.f.

The above method, due originally to G. S. Fishman, is described by Atkinson and Pearce (1976). Of course, any density function $f(x)$ can be enveloped by a wide variety of alternative functions, and an alternative rejection method for simulating gamma random variables is given in Exercise 5.22.

For distributions over a finite range, an alternative approach is to envelop the distribution with a suitable polygon and then use the method of Hsuan (1979). A generalization of the rejection method is given in Exercise 5.29.

5.4   The composition method

Here again we encounter a general method suitable for discrete and continuous random variables. We shall begin our discussion of this method with an illustration from Abramowitz and Stegun (1965, p. 951).

For a binomial $B(5, 0.2)$ distribution we have the following probabilities, given to four places of decimals:

i	p_i
0	0.3277
1	0.4096
2	0.2048
3	0.0512
4	0.0064
5	0.0003

If we take p_0 as an example, we can write

$$p_0 = 0.3277 = 0.9 \times \frac{3}{9} + 0.07 \times \frac{2}{7} + 0.027 \times \frac{7}{27} + 0.003 \times \frac{7}{30}$$

and similarly,

$$p_1 = 0.4096 = 0.9 \times \frac{4}{9} + 0.07 \times \frac{0}{7} + 0.027 \times \frac{9}{27} + 0.003 \times \frac{6}{30}$$

$$p_2 = 0.2048 = 0.9 \times \frac{2}{9} + 0.07 \times \frac{0}{7} + 0.027 \times \frac{4}{27} + 0.003 \times \frac{8}{30}$$

and so on, so that in general,

$$p_i = 0.9\,r_{i1} + 0.07\,r_{i2} + 0.027\,r_{i3} + 0.003\,r_{i4} \qquad \text{for } 0 \le i \le 5 \qquad (5.6)$$

where $\{r_{i1}\}$, $\{r_{i2}\}$, $\{r_{i3}\}$ and $\{r_{i4}\}$ are all probability distributions over the same range, $0 \le i \le 5$.

We can see that in (5.6),

$$0.9 = 10^{-1} \times \text{(sum of digits in first decimal place of the } p_i)$$

$$0.07 = 10^{-2} \times \text{(sum of digits in second decimal place of the } p_i)$$

and so on

while, for example,

$$r_{01} = \frac{0.3}{0.9} = \frac{3}{9}$$

$$r_{11} = \frac{0.4}{0.9} = \frac{4}{9}$$

$$r_{21} = \frac{0.2}{0.9} = \frac{2}{9}$$

$$r_{02} = \frac{0.02}{0.07} = \frac{2}{7}$$

etc.

This explains the derivation of (5.6). We can now use (5.6) to simulate from the $\{p_i\}$ distribution as follows:

(i) Simulate a discrete random variable, R, say, according to the distribution:

j	$\Pr(R = j)$
1	0.9
2	0.07
3	0.027
4	0.003

(ii) If $R = j$, simulate from the $\{r_{ij}\}$ distribution for $1 \leq j \leq 4$. If the resulting random variable is denoted by X,

$$\Pr(X = i) = \sum_{j=1}^{4} \Pr(R = j) r_{ij} \qquad \text{(see for example } ABC, \text{ p. 85)}$$

i.e. $\Pr(X = i) = p_i$

i.e. X has the required binomial distribution.

Of course, a small amount of approximation has taken place here, as we have written the $\{p_i\}$ only to four places of decimals. Nevertheless, this approach may be used for any discrete distribution. While one has to simulate from two distributions $\{\Pr(R = j), 1 \leq j \leq 4\}$ and $\{r_{ij}\}$, most (97%) of the time one is simulating from $\{r_{i1}\}$ and $\{r_{i2}\}$, and these component discrete distributions are of a very simple form. A disadvantage of this method is the need to store the component distributions. In (5.6) we have written the $\{p_i\}$ distribution as a *mixture*, or *composition*, of the $\{r_{ij}\}$ distributions; a further example of this kind is to be found in Exercise 5.42. We shall now consider the analogous procedure for continuous random variables.

It is not unusual to encounter probability density functions which are mixtures of other probability density functions, say

$$f(x) = \alpha f_1(x) + (1 - \alpha) f_2(x) \qquad 0 < \alpha < 1 \tag{5.7}$$

In psychology, for example, bimodal histograms of reaction times are sometimes encountered, which may reflect a tendency for subjects to behave in some standard fashion a proportion α of the time, producing reaction times with probability density function $f_1(x)$, say, but the remainder of the time, possibly due to a loss in concentration, to produce reaction times that tend to be longer than before, with probability density function $f_2(x)$, say. Cox (1966) provides further discussion of this example.

Another example is provided by human height histograms, which could be bimodal due to a mixture of different male and female height histograms. However, samples from such mixtures may not obviously reflect the mixture form of the underlying p.d.f., as is the case in the histogram of Fig. 5.8.

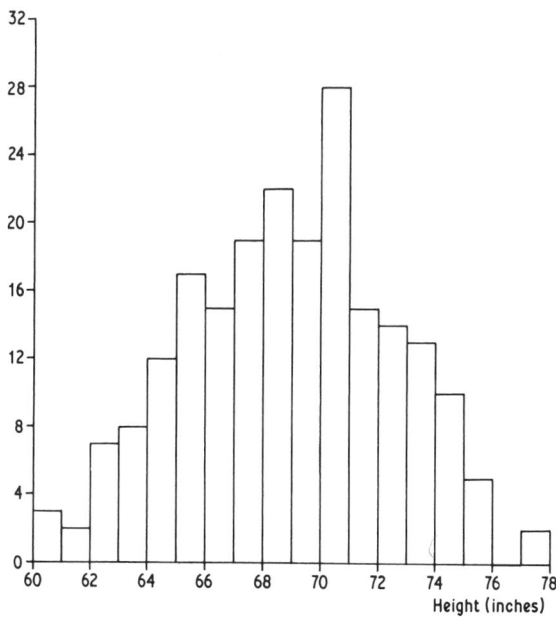

Figure 5.8 Histogram describing the heights of some undergraduates at the University of Kent (taken from Fuller and Lury, 1977, p. 14). For the 142 male undergraduates, heights range from 63 to 77 inches, with a modal height of 70 inches; for the 69 female undergraduates, heights range from 60 to 70 inches, with a modal height of 65 inches.

Indeed, we shall see that in many applications in simulation the mixture form of (5.7) is adopted solely for convenience, even for unimodal distributions such as the normal distribution. The convenience arises if α is fairly large and

$f_1(x)$ is a probability density function which is appreciably easier to simulate from than $f(x)$ itself. If we sample from $f_1(x)$ with probability α, and from $f_2(x)$ with probability $(1-\alpha)$, then because of the relationship of (5.7) we obtain a random variable, X, with probability density function $f(x)$. We see this simply as follows:

$$\Pr(X \le x) = \alpha \Pr(X \le x \mid \text{sample from } f_1(x)) + (1-\alpha)\Pr(X \le x \mid$$
$$\text{sample from } f_2(x))$$

$$= \alpha \int_{-\infty}^{x} f_1(y)\,dy + (1-\alpha) \int_{-\infty}^{x} f_2(y)\,dy$$

$$= \int_{-\infty}^{x} f(y)\,dy \qquad \text{by (5.7)}$$

Let us now consider two examples which illustrate the use of the method.

· EXAMPLE 5.7
Suppose we want to simulate random variables with the p.d.f. of Fig. 5.9.

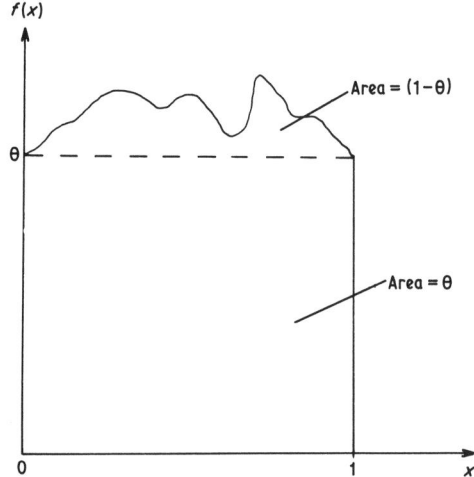

Figure 5.9   An unusual probability density function $f(x)$, from which we shall simulate using a composition, the first density, $f_1(x)$, of which is the $U(0, 1)$ density.

We note here that we can write

$$f(x) = \theta + f(x) - \theta \qquad \text{for any} \quad 0 \le x \le 1$$

i.e., $$f(x) = \theta \times 1 + (1-\theta)\left(\frac{f(x)-\theta}{1-\theta}\right) \qquad (5.8)$$

As $\int_0^1 f(x)dx = 1$, then $\theta < 1$, and (5.8) is of the same form as (5.7). To simulate from $f(x)$, with probability θ we simply select a $U(0, 1)$ random variable, while with probability $(1-\theta)$ we simulate from the p.d.f.,

$$f_2(x) = \left(\frac{f(x) - \theta}{1 - \theta}\right)$$

which has subsumed the features of $f(x)$ which made it a difficult p.d.f. from which to simulate. We can simulate from both $f(x)$ and $f_2(x)$ using a rejection method, but the advantage of the composition of (5.8) is that one only simulates from $f_2(x)$ with probability $(1-\theta)$, and in the illustration of Fig. 5.9, $(1-\theta)$ is appreciably less than 0.5.

EXAMPLE 5.8
A random variable X with the simple beta probability density function

$$f(x) = 6x(1-x) \quad \text{for} \quad 0 \leq x \leq 1$$

can be simulated quite easily by either the table-look-up method, or the rejection method (see Exercise 5.18). Figure 5.10 shows how we may use a

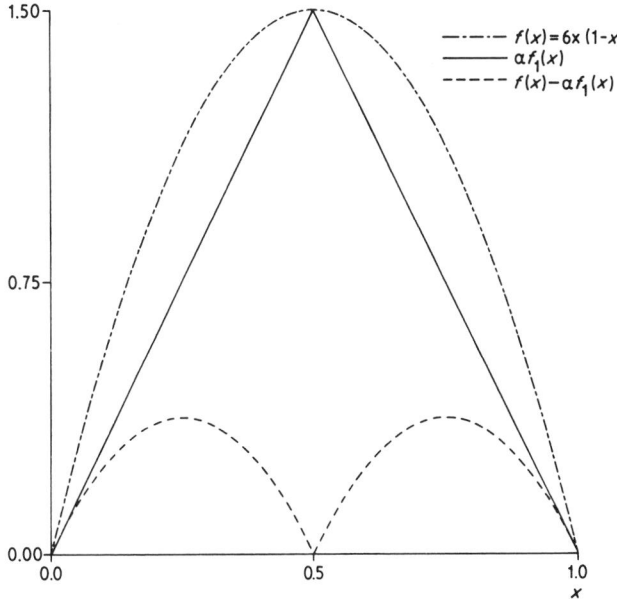

Figure 5.10   The beta density function, $f(x) = 6x(1-x)$ and the basis for a composition.

composition method. Here, $f_1(x)$ is simply the symmetric triangular density over $(0,1)$, i.e.

$$f_1(x) = \begin{cases} 4x & \text{for} \quad 0 \leq x \leq 0.5 \\ 4(1-x) & \text{for} \quad 0.5 \leq x \leq 1 \end{cases}$$

From simple geometrical consideration we see that we must have $\alpha \leq \frac{3}{4}$, and as we want α to be as large as possible, we take $\alpha = \frac{3}{4}$. The second density in the composition is then given by

$$f_2(x) = \frac{f(x) - \alpha f_1(x)}{1-\alpha} = \begin{cases} 12x(1-2x) & \text{for} \quad 0 \leq x \leq 0.5 \\ 12(1-2x)(x-1) & \text{for} \quad 0.5 \leq x \leq 1 \end{cases}$$

We shall leave the reader to consider how we might simulate from $f_2(x)$.

In general, suppose we have a probability density function $f(x)$, and that $f_1(x)$ is a probability density function of roughly similar shape, but that it is appreciably easier to simulate from $f_1(x)$ than from $f(x)$. We shall see shortly why we want $f(x)$ and $f_1(x)$ to be of similar shape.

We can formally write, for any α in the range $0 < \alpha < 1$,

$$f(x) = \alpha f_1(x) + (1-\alpha)\left(\frac{f(x) - \alpha f_1(x)}{1-\alpha}\right)$$

and from the above discussion we see that we can simulate from $f(x)$ by simulating from $f_1(x)$ with probability α, and from $f_2(x) = (f(x) - \alpha f_1(x))/(1-\alpha)$ with probability $(1-\alpha)$. As $f_1(x)$ is chosen to be relatively easy to simulate from, we clearly want α to be as large as possible. The constraint on α is that for all x we must have $f(x) - \alpha f_1(x) \geq 0$, in order to ensure that $f_2(x)$ is also a probability density function (it is easy to see that its integral is unity). Now if

$$\alpha = \min_x \left(\frac{f(x)}{f_1(x)}\right)$$

and a positive, non-zero minimum can be found, then $\alpha \leq f(x)/f_1(x)$, i.e. $f(x) - \alpha f_1(x) \geq 0$, as required, and there is at least one x for which $\alpha f_1(x) = f(x)$, so that a larger value of α cannot be found. This approach to finding α is, of course, analogous to the general approach given in the last section for finding k. Here we can consider α as shrinking $f_1(x)$ so that it just fits completely under $f(x)$, as we have seen in the last two examples. This of course explains why we seek an $f_1(x)$ to be of roughly similar shape to $f(x)$: the more similar in shape $f(x)$ and $f_1(x)$ are, then the larger the shrinking parameter α can be. For rejection, then, we envelop $f(x)$, but for composition it is $f(x)$ itself that plays the enveloping rôle. The method generalizes in a straightforward way, so that we could repeat the procedure for $f_2(x)$, and so on, ending ultimately with the mixture:

$$f(x) = \sum_{i=1}^{n+1} \alpha_i f_i(x)$$

in which
$$\sum_{i=1}^{n+1} \alpha_i = 1$$

and
$$\alpha_i > 0 \quad \text{for} \quad 1 \le i \le n+1,$$

all the $f_i(x)$ are probability density functions, and

$$f_{n+1}(x) = \left(f(x) - \sum_{i=1}^{n} \alpha_i f_i(x) \right) \Bigg/ \left(1 - \sum_{i=1}^{n} \alpha_i \right)$$

In choosing n, one has to counterbalance the difficulty of simulating from the final density function, $f_{n+1}(x)$, with the size of α_{n+1}, and the general desirability of keeping n small.

In the last two examples, all the p.d.f.'s considered had a finite range. As we shall see in the next section, it sometimes happens that $f(x)$ has an infinite range, but $f_1(x)$ has a finite range. In such a case, we have a range of x for which $f_1(x) = 0$, but $f_2(x) > 0$. In fact, as is shown in the next section, we can also have $f_2(x) = 0$ and $f_1(x) > 0$ for certain x.

*5.5   Combining methods for simulating normal random variables

In recent years much ingenuity has been devoted to devising composition methods for the standard normal probability density function. These approaches have also employed rejection, table-look-up and particular methods, and it is fascinating to see all of these different tools put to work on the one problem. As with the rejection method, many different compositions can be formed for any one probability density function, and here we shall just consider one, for the $N(0, 1)$ density. Due to Marsaglia and Bray (1964), the method gives rise to what has been termed their 'convenient' algorithm. Other methods are discussed in Exercises 5.36–5.38.

What many of the different methods proposed for the $N(0, 1)$ p.d.f. have in common, however, is the initial isolation of the *tails* of the normal density function, and the first composition usually taken is:

$$\frac{e^{-x^2/2}}{\sqrt{2\pi}} = \alpha\phi_1(x) + (1-\alpha)\phi_2(x) \tag{5.9}$$

in which

$$\phi_1(x) = \begin{cases} \dfrac{1}{\alpha} \dfrac{e^{-x^2/2}}{\sqrt{(2\pi)}} & \text{for} \quad -3 \le x \le 3 \\[2mm] 0 & \text{for} \quad |x| > 3 \end{cases}$$

and

$$\phi_2(x) = \begin{cases} \dfrac{1}{(1-\alpha)} \dfrac{e^{-x^2/2}}{\sqrt{(2\pi)}} & \text{for} \quad |x| > 3 \\[2ex] 0 & \text{for} \quad -3 \le x \le 3 \end{cases}$$

where

$$1 - \alpha = 2 \int_{-\infty}^{-3} \frac{e^{-x^2/2}}{\sqrt{(2\pi)}} \simeq 0.0027$$

to 4 places of decimals. Here, then, is an instance of the two component p.d.f.'s in (5.7) having different ranges. $|x| > 3$ is used to define the normal p.d.f. tails as 3 is suitably large and, as we shall see, ties in conveniently with the approaches adopted in what follows.

The composition of (5.9) means that most of the time we simulate from the *expanded* normal density, $\phi_1(x)$, over the finite range $|x| \le 3$, while with the very small probability $(1 - \alpha)$ we simulate from the p.d.f. $\phi_2(x)$, formed by expanding the tail areas from the standard normal p.d.f.

Let us consider $\phi_2(x)$ first of all. A random variable X with probability density function $\phi_2(x)$ is simply an $N(0, 1)$ random variable, conditioned to be $|X| \ge 3$. Such random variables result from the Box–Müller or Polar Marsaglia methods of Section 4.2 as follows: in the Box–Müller notation of Equation (4.1), if the exponential variable $-2\log_e U_1 > 9$, then from the geometrical explanation of Section 4.2.1, there is a good chance that at least one of N_1 and N_2 is greater than 3 in modulus, as required. Certainly, if $-2\log_e U_1 < 9$ then neither of N_1 and N_2 will be greater than 3, and so the standard approach of Section 4.2 towards constructing the conditioned normal variables that we require would be very wasteful. However, as is discussed in Exercise 5.24, $Y = 9 - 2\log_e U_1$ is a random variable with the required exponential distribution, but *conditional on being greater than 9*. We can therefore simulate from the p.d.f. $\phi_2(x)$ by replacing $-2\log_e U_1$ in Equation (4.1) by $(9 - 2\log_e U_1)$, but only accepting a resulting N_1 or N_2 value if it is greater than 3 in modulus. Correspondingly, we can modify the Polar Marsaglia method by replacing $(-2\log_e W)$ in Equation (4.2) by $(9 - 2\log_e W)$, and proceeding in the same fashion.

There is more discussion of tail area simulation in the solution to Exercise 5.38. So far we have used the particular approaches of Section 4.2, and the table-look-up method to give exponential random variables of mean 2. Now we shall return to $\phi_1(x)$.

Figure 5.11 illustrates $\phi_1(x)$ and also the probability density function of the random variable

$$Y = 2(U_1 + U_2 + U_3 - 1.5) \qquad -3 \le Y \le 3$$

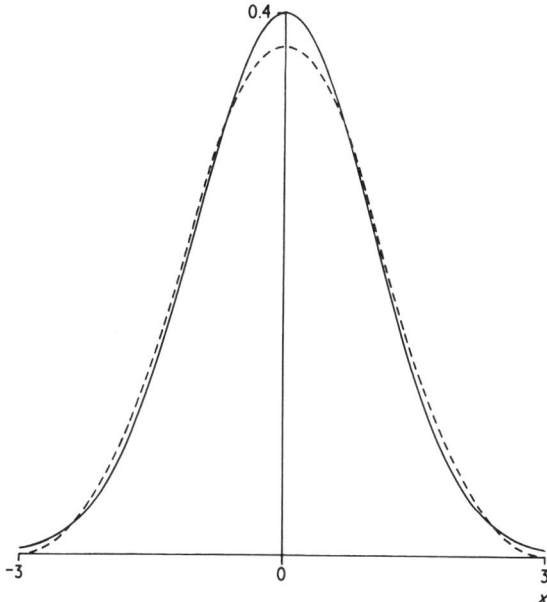

Figure 5.11 The preliminary to a composition method. Here we want to simulate from $\phi_1(x)$ (denoted by a solid line) the standard normal p.d.f. conditioned to the range $|x| \le 3$. It is proposed to use as the first p.d.f. in the composition, $f_1(x)$ (dashed line), the density of $Y = 2(U_1 + U_2 + U_3 - 1.5)$, where U_1, U_2 and U_3 are independent $U(0, 1)$ random variables.

in which U_1, U_2 and U_3 are independent $U(0, 1)$ random variables. (See Exercise 4.8). The two curves are of similar shape, and Y is clearly easy to simulate. We shall now, therefore, seek a composition for $\phi_1(x)$, with the first p.d.f. in the composition being $f_1(x)$, the probability density function of Y, given by (see Exercise 4.8)

$$f_1(x) = \begin{cases} (3 - x^2)/8 & -1 \le x \le 1 \\ (3 - |x|)^2/16 & 1 \le |x| \le 3 \\ 0 & |x| \le 3 \end{cases}$$

Using the approach outlined in the last section, we want to minimize $q(x) = \phi_1(x)/f_1(x)$ with respect to x varying over the range $|x| \le 3$. Because $f_1(x)$ is specified differently over different ranges for x, we shall deal with these different ranges separately.

First let us consider $0 \le x \le 1$. Here,

$$q(x) = \frac{8 \, e^{-x^2/2}}{\alpha(3 - x^2) \sqrt{(2\pi)}}$$

$$l(x) = \log q(x) = \text{constant} - \frac{x^2}{2} - \log(3 - x^2)$$

$$\frac{d}{dx}l(x) = -x + \frac{2x}{(3 - x^2)}$$

$$= 0 \text{ when } x = 0 \text{ and when } 3 - x^2 = 2, \text{ i.e. } x = 1$$

$$\frac{d^2 l(x)}{dx^2} = -1 + \frac{2}{(3 - x^2)} + \frac{4x^2}{(3 - x^2)^2}$$

i.e. $\dfrac{d^2 l(x)}{dx^2}$ is negative when $x = 0$, and positive when $x = 1$.

Next we shall consider the range $1 \le x \le 3$.

Here,
$$q(x) = \frac{16 e^{-x^2/2}}{\alpha(3 - x)^2 \sqrt{(2\pi)}}$$

$$l(x) = \log q(x) = \text{constant} - \frac{x^2}{2} - 2\log(3 - x)$$

$$\frac{d}{dx}l(x) = -x + \frac{2}{(3 - x)}$$

$$= 0 \qquad \text{when} \quad 3x - x^2 = 2$$

i.e. when $x = 1$ and when $x = 2$.

$$\frac{d^2}{dx^2}l(x) = -1 + \frac{2}{(3 - x)^2}$$

i.e. $d^2 l(x)/dx^2$ is negative when $x = 1$, and positive when $x = 2$, revealing a minimum to $q(x)$ when $x = 2$.

Thus for the case of $x \ge 0$, $q(x)$ has a maximum when $x = 0$, a saddle-point when $x = 1$, and a minimum when $x = 2$. We need not consider the case $x \le 0$ separately because of the symmetry present, and so we can conclude that $q(x)$ has minima at $x = \pm 2$ in the range $|x| \le 3$.

Hence if we write

$$\phi_1(x) = \alpha_1 f_1(x) + \alpha_2 f_2(x)$$

$$\alpha_1 = \frac{\phi_1(2)}{f_1(2)} = \frac{16 e^{-2}}{\alpha \sqrt{(2\pi)}}$$

and overall, from considering the compositions for $e^{-x^2}/\sqrt{(2\pi)}$ and $\phi_1(x)$, we simulate from $f_1(x)$ with probability

$$\alpha \alpha_1 = \frac{16 e^{-2}}{\sqrt{(2\pi)}} \approx 0.8638$$

Here we see a dramatic demonstration of the possible power of the composition method: over 86 per cent of the time we can expect to simulate an

$N(0, 1)$ random variable by simply taking a linear function of the sum of three independent $U(0, 1)$ random variables.

In fact there is still more of interest remaining in this example. With probability $\alpha\alpha_2$ we must simulate from the probability density function

$$f_2(x) = \left(\frac{\phi_1(x) - \alpha_1 f_1(x)}{1 - \alpha_1}\right) \qquad -3 \le x \le 3$$

i.e. with probability $\alpha\alpha_2 = \alpha(1 - \alpha_1) = (0.9973 - 0.8638) = 0.1335$.

Figure 5.12 presents a graph of $\phi_1(x) - \alpha_1 f_1(x)$, and the form of the graph suggests proceeding further with the composition for $\phi_1(x)$, by now using a triangular p.d.f. and setting

$$f_2(x) = \beta g(x) + (1 - \beta)h(x)$$

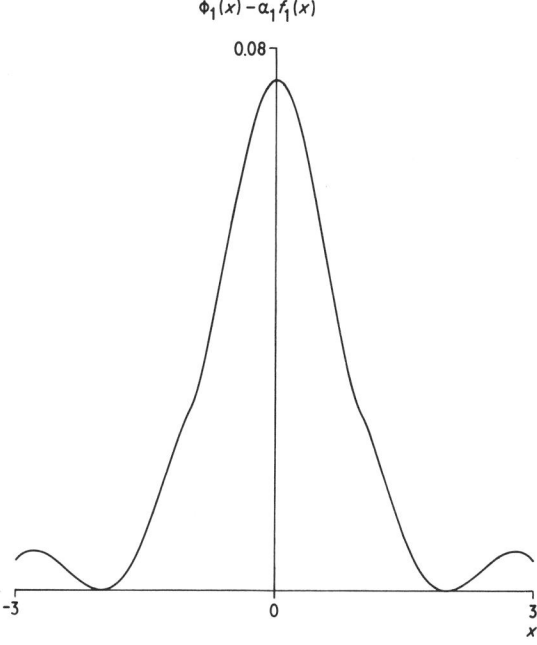

$\phi_1(x) - \alpha_1 f_1(x)$

Figure 5.12 The residual curve $\phi_1(x) - \alpha_1 f_1(x)$ following the composition method envisaged in Fig. 5.11.

where

$$g(x) = (6 - 4|x|)/9 \qquad \text{for } |x| \le 1.5$$

$$= 0 \qquad\qquad \text{for } |x| > 1.5$$

$g(x)$ is simply the probability density function of

$$Y = 1.5(U_1 + U_2 - 1)$$

where U_1 and U_2 are independent $U(0, 1)$ random variables (see Exercise 4.8). In this case, with the aim of determining β, $f_2(x)/g(x)$ cannot be minimized explicitly, but a numerical method such as Newton–Raphson readily provides us with $\beta \approx 0.8292$, the minimum occurring at $x = \pm 0.8739$. We thus simulate from $g(x)$ with probability $\alpha(1 - \alpha_1)\beta = 0.1107$, so that over 97 per cent of the time we use the two simple p.d.f.'s $f_1(x)$ and $g(x)$.

The three compositions that we have dealt with here can be written as one, to give

$$\frac{e^{-x^2/2}}{\sqrt{(2\pi)}} = 0.8638\, f_1(x) + 0.1107\, g(x) + 0.0027\, t(x) + 0.0228\, r(x)$$

$$\text{for} \quad -\infty \le x \le \infty \tag{5.10}$$

where $t(x)$ is the tail-area p.d.f. which we considered earlier, and $r(x)$ is the p.d.f. that remains for $|x| \le 3$.

We simulate from the p.d.f. $r(x)$ only with probability 0.0228, and as $r(x)$ is of a fairly complicated form (shown in Fig. 5.18) we can simulate from it by means of simple rejection, using a rectangular enveloping region over the finite range $|x| \le 3$, with, it can be shown, probability 0.53 of rejection (see Exercises 5.16 and 5.39).

The above derivation of (5.10) should not disguise the fact that (5.10) is a description of one way of dividing up the area under the $N(0, 1)$ probability density function, precisely as was done in a different case in Example 5.7. The end-result is shown in Fig. 5.13.

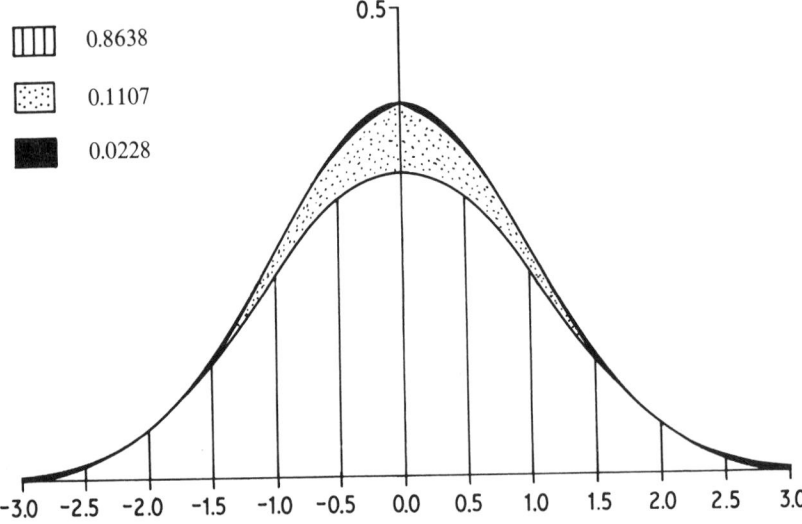

Figure 5.13   A representation of the composition given in Equation (5.10) for the range $|x| \le 3$. The regions shown have the areas indicated above.

5.6 Discussion and further reading

The examples considered in this chapter form only a very small subset of the many interesting and complicated approaches that have been devised in recent years. For instance, having obtained a normal random variable, then one can even use the normal distribution itself as an enveloping distribution; see, for example, Ahrens and Dieter (1974) and Atkinson (1979). More examples are to be found in the exercises.

We have seen that uniform $U(0, 1)$ random variables are the building-blocks for the simulation of any other random variable. Complicated algorithms utilizing composition and rejection methods, and sometimes requiring the storage of a large number of constants, are designed for use on computers that work to high precision. These algorithms are often programmed in machine code and tend to be the most efficient. They are therefore most suitable if one is seeking to provide a computer with an efficient package of programs for simulating a variety of random numbers, which will be used frequently by a large number of individuals. Comparisons of different algorithms, using speed and efficiency, have been undertaken by a number of authors—see, for example, Atkinson and Pearce (1976), Kinderman and Ramage (1976) and Ripley (1983b). Appleton (1976) pointed out that certain methods can be programmed in the programming language APL, to take advantage of APL's vector-handling capabilities, and as an example he found the Box–Müller method to be 30 times faster than the Marsaglia and Bray 'convenient' method, when both were programmed in APL. This is partly due to the fact that APL programs are interpreted, rather than compiled, as are FORTRAN programs. Using FORTRAN, Atkinson and Pearce (1976) found Box–Müller to be roughly twice as slow as the convenient method. Distributed array processing is another factor which could influence the comparison of different algorithms (cf. Exercise 3.16).

Ripley (1983b) provides a list of relatively efficient simple algorithms for a variety of distributions. Most users of main-frame computers will be likely to use library subroutines such as those described in Section A1.1. Because of the time-lag before library subroutines are changed to accommodate new developments, these programs may not always be the most efficient. Each individual clearly has to experiment with the facilities available if it is suspected that long generation times of random variables could render a simulation impractical.

Of course, the simplest way for human beings to simulate random variables is to use tables of realizations of such random variables, such as those by Wold (1954), providing normal random variables, and those by Barnett (1965), which provide exponential random variables. In the absence of such tables, the table-look-up approach is also easily performed by hand if one has suitable tables of cumulative distribution functions, and only a small-scale simulation is envisaged. Some such tables can be found in Harter (1964), Lieberman and Owen (1961), Mardia and Zemroch (1978), Neave (1978), Odeh, *et al.* (1977), Williamson and Bretherton (1963) and Worsdale (1975).

Simulation of discrete random variables by the table-look-up method can be very time consuming. This occurs with the Poisson distribution, for example, if it has large mean, in which case the particular method for this distribution, described in the last chapter, will also be inefficient. There is further discussion of this point in Exercises 5.3 and 5.4. The range-dividing technique, discussed in Section 5.1, can be generalized by dividing the range into $d > 2$ parts, as in Neave (1972), who provides ALGOL programs for several discrete distributions. A faster search procedure is the optimum binary tree search described in Knuth (1968, p. 400). As can be seen from Section A1.1, the NAG library of computer programs simulates all discrete distributions by first of all establishing a reference vector of cumulative sums, and then performing an indexed search by means of the routine G05EYF. The IMSL routine for the table-look-up method for a general discrete distribution is GGDT (see Section A1.1).

The polar Marsaglia method of Section 4.2.2 shows that the ratio V_1 / V_2 of the co-ordinates of a point uniformly distributed over a disc of unit radius and centred on the origin has a Cauchy distribution (see Exercise 5.8). Kinderman and Monahan (1977) have generalized this result to provide a new general method for simulating random variables, viz., the *ratio* method—see Ripley (1983b) for illustrations of its use. A further new general method is the *alias* method for discrete random variables, described in Exercise 5.42.

In this chapter we have only considered univariate random variables, but table-look-up, rejection and composition methods may also be used for multivariate random variables. Kemp and Loukas (1978a, b) consider the table-look-up method for a bivariate Poisson distribution, and the table-look-up method for bivariate Poisson and normal distributions is discussed in Exercises 5.10 and 5.11. Best and Fisher (1979) use a rejection method on the circle, enveloping the von Mises distribution with a wrapped Cauchy distribution.

We shall conclude this chapter with some further discussion of methods for simulating normal random variables.

*5.7 Additional approaches for normal random variables

The table-look-up method for normal random variables is difficult to program for computers because of the intractable form of the standard normal cumulative distribution function, $\Phi(x)$, and its inverse, $\Phi^{-1}(x)$. Various authors have approached this problem by providing approximate methods— see, for example, Zelen and Severo (1966) and Wetherill (1965). Wetherill's approach employs the attractive idea that an efficient algorithm can result from using one approximation to $\Phi^{-1}(x)$ in the middle of the range for x, but another, more complicated algorithm in the tails, which would be used far less frequently. This idea is simply providing a composition method, the components of which are simulated using approximate table-look-up methods. Some other approaches are described below.

Because of the similar shapes of the normal and logistic probability density functions, it is natural to try to approximate the normal cumulative distribution function by the simple logistic cumulative distribution function. In order to obtain a good match over the middle of the range, the logistic cumulative distribution function that may be used is

$$F_1(x) = \left[1 + \exp\left(-2\sqrt{\left(\frac{2}{\pi}\right)}x\right)\right]^{-1} \qquad -\infty \le x \le \infty$$

as this curve has the same slope at $x = 0$ as does $\Phi(x)$. An alternative possibility which might be considered is the logistic cumulative distribution function,

$$F_2(x) = \left[1 + \exp\left(-\frac{\pi x}{\sqrt{3}}\right)\right]^{-1} \qquad -\infty \le x \le \infty$$

corresponding to a random variable with zero mean and unit variance. $F_1(x)$ and $F_2(x)$ are illustrated in Fig. 5.14, for $0 \le x \le 3$.

Table 5.1 is taken from Page (1977), who tries to improve a logistic approximation by adding an extra parameter, resulting in the cumulative distribution function

$$G(x) = \{1 + \exp[-2a_1 x(1 + a_2 x^2)]\}^{-1} \qquad -\infty \le x \le \infty$$

Note that as the coefficient of the new parameter, a_2, is x^3, and not x^2, which may have been considered a more natural choice, then we preserve the property $G(x) + G(-x) = 1$, and the corresponding probability density function is symmetric about $x = 0$.

If $a_1 = \sqrt{(2/\pi)}$, and a_2 is chosen by least squares, then a value of $a_2 = 0.044\,715$ is obtained. A slightly better approximation is obtained by allowing both a_1 and a_2 to be chosen by least squares, but the advantage of keeping $a_1 = \sqrt{(2/\pi)}$ is that if one wanted to approximate $\Phi(x)$ this way on a hand-calculator, only one constant needs to be remembered, most calculators having a 'π' key. To simulate approximate $N(0, 1)$ random variables we need $\tilde{x} = G^{-1}(U)$ (see Exercise 5.9). Some examples are given in Table 5.1.

Hamaker (1978) and Schmeiser (1979) provide further approximations that are suitable for computation on a hand-calculator, and more recent work is described in Bailey (1981) and Lew (1981).

Kinderman and Ramage (1976) use an even simpler p.d.f. for $f_1(x)$, the first p.d.f. in a composition for the standard normal density, than that resulting from the sum of three $U(0, 1)$ random variables. In their case, they used the p.d.f. of the sum of just two $U(0, 1)$ random variables, as illustrated in Fig. 5.15.

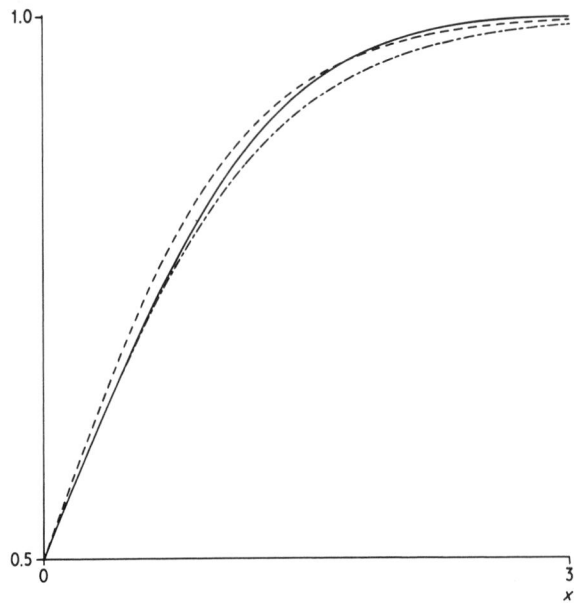

Figure 5.14 An illustration for $0 \le x \le 3$ of two logistic cumulative distribution functions, (—·—): $F_1(x) = \left[1 + \exp\left(-2\sqrt{\left(\frac{2}{\pi}\right)}x\right)\right]^{-1}$ and (———): $F_2(x) = \left[1 + \exp\left(-\frac{\pi x}{\sqrt{3}}\right)\right]^{-1}$, either of which may be used as a rough approximation to the normal cumulative distribution function, $\Phi(x)$, denoted by (———).

Table 5.1 Approximating the standard normal cumulative distribution function $\Phi(x)$. Two possible approximations are $F_1(x)$ and $G(x)$, explained in the text. $\tilde{x}$ results from inverting $G(x)$ (from Page, 1977).

x	$1 - \Phi(x)$	$1 - F_1(x)$	$1 - G(x)$	$\tilde{x}$
0	0.5	0.5	0.5	0
0.1	0.460 172 2	0.460 190 2	0.460 172 5	0.1
0.3	0.382 088 6	0.382 551 9	0.382 096 9	0.3
0.5	0.308 537 5	0.310 478 2	0.308 572 0	0.5001
1.0	0.158 655 3	0.168 573 8	0.158 808 0	1.0006
1.5	0.066 807 2	0.083 657 9	0.066 952 3	1.5011
2.0	0.022 750 1	0.039 485 4	0.022 701 2	1.9991
2.5	0.006 209 7	0.018 174 0	0.006 033 7	2.4901
3.0	0.001 349 9	0.008 266 0	0.001 212 5	2.9693
3.5	0.000 232 6	0.003 739 0	0.000 176 1	3.4332
4.0	0.000 031 7	0.001 687 1	0.000 017 6	3.8800

Figure 5.16 illustrates $\phi(x) - \alpha f_1(x)$ for $-3 \le x \le 3$, which may be simulated by means of rejection, the details of which are discussed in the

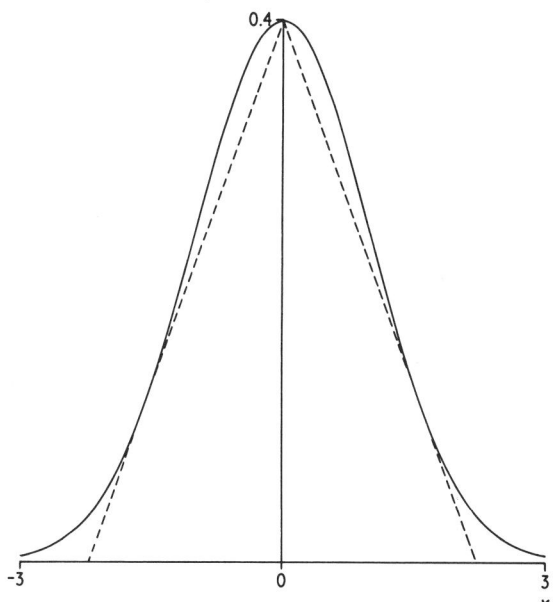

Figure 5.15 The standard normal density function $\phi(x)$ (———) over the range, $(-3, +3)$, and $\alpha f_1(x)$ (------), in the notation of the composition method of the Section 5.4. Here $f_1(x)$ is the probability density function of $\beta(U_1 + U_2 - 1)$, where U_1 and U_2 are independent $U(0, 1)$ random variables. α is chosen so that $\alpha f_1(0) = 1/\sqrt{(2\pi)}$, and β must be chosen to give the triangle illustrated here, i.e. the largest symmetric triangle with height $1/\sqrt{(2\pi)}$ which can be fitted under $\phi(x)$.

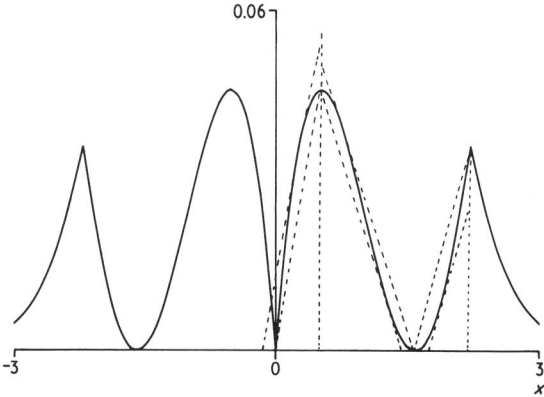

Figure 5.16 A graph of $\phi(x) - \alpha f_1(x)$ for $|x| \leq 3$, resulting from the curves of Fig. 5.15. The dotted lines relate to a particular approach used for the rejection method employed to simulate from the probability density function which is a positive multiple of this curve, as discussed in the solution to Exercise 5.38.

solution to Exercise 5.38. Also presented in Exercise 5.38 is the full algorithm of Kinderman and Ramage (1976).

5.8   Exercises and complements

(a) General

5.1   Use Equation (2.3) to show that when U is a $U(0, 1)$ random variable, then $1 - U$ is also a $U(0, 1)$ random variable.

5.2   When X has the half-normal p.d.f.

$$f_X(x) = \sqrt{\left(\frac{2}{\pi}\right)} e^{-x^2/2} \qquad \text{for} \qquad x \geq 0$$

show that $\tilde{X}$, defined by:

$$\begin{cases} \tilde{X} = X \text{ with probability } \frac{1}{2} \\ \tilde{X} = -X \text{ with probability } \frac{1}{2} \end{cases}$$

has the standard normal p.d.f.

5.3   Simulating Poisson random variables with large mean can be time consuming, whether one uses a particular approach, as in Chapter 4, or a general, table-look-up approach. Discuss one way of tackling this problem, in the context of the distribution of the sum of two independent random variables, each with Poisson distributions. See Exercise 2.8(b).

(b) Table-look-up methods

†**5.4**   Select a Poisson distribution with mode different from zero.
   (a) Simulate from this distribution using the table-look-up method.
   (b) Repeat (a), but employ a θ, as suggested in Section 5.1.
   (c) Repeat (a) but employ two such θ's, thus dividing the range into three parts.
   (d) Repeat (b) after having first ordered the probabilities in increasing order.
   Compare the efficiencies of these four approaches (cf. also Exercise 5.3, and Kemp, 1982).

†**5.5**   (a) Write a computer program to simulate a random variable, X, from

the triangular distribution defined by:

$$f(x) = \begin{cases} 0 \\ x \\ 2-x \\ 0 \end{cases} \qquad F(x) = \begin{cases} 0 & x < 0 \\ x^2/2 & 0 \le x \le 1 \\ 2x - (x^2/2) - 1 & 1 \le x \le 2 \\ 1 & x > 2 \end{cases}$$

using the inversion method. Here $f(x)$ is the probability density function of x, and $F(x)$ is the cumulative distribution function of x. This is the method used in the IMSL routine GGTRA (see Section A1.1).

(b) Compare the efficiency of this program with one which simulates such a random variable by simply summing two independent $U(0, 1)$ random variables.

5.6 Use the table-look-up method to simulate 10 random variables:

(a) from the binomial distribution $B(6, 1/3)$; and
(b) from the normal distribution $N(1, 2)$, using tables of the standard normal cumulative distribution function.

*5.7 Use the table-look-up method to simulate random variables with the simple beta probability density function

$$f(x) = 6x(1 - x) \qquad \text{for } 0 \le x \le 1.$$

*5.8 (a) Explain how to simulate random variables from the Cauchy distribution, with probability density function,

$$f(x) = \frac{1}{\pi(1 + x^2)} \qquad \text{for } -\infty \le x \le \infty$$

using the inversion method. An algorithm using this approach is provided by the IMSL routine, GGCAY (see Section A1.1).

(b) If N_1 and N_2 are independent standard normal random variables then, as we saw in Exercise 2.15(b) and Exercise 4.5(b), their ratio $C = N_1/N_2$ has the Cauchy probability density function of (a) above. Explain how this result may be deduced from (a) and an understanding of the Box–Müller method described in Section 4.2.1.

*5.9 The approximate approach for simulating $N(0, 1)$ random variables described in Section 5.7 involved setting $\tilde{x} = G^{-1}(u)$, where $G(x) = [1 + \exp\{-2a_1 x(1 + a_2 x^2)\}]^{-1}$. Solve for $\tilde{x}$.

*5.10 Discuss how you would use the table-look-up method for simulating from the bivariate Poisson distribution of Exercise 4.7.

*5.11 Discuss how you would use the inversion method for simulating from the bivariate normal distribution.

5.12 If X is a random variable with the exponential, $\lambda e^{-\lambda x}$ p.d.f., for $x \geq 0$, deduce the distribution of the integral part of X, viz., $Y = [X]$. Hence explain why, in (5.4), we obtain a geometric random variable by rounding up an exponential random variable.

†5.13 Use the inversion method to simulate from the following distributions:

(a) logistic: $f(x) = \dfrac{e^{-x}}{(1+e^{-x})^2}$ for $-\infty \leq x \leq \infty$.

Note that this method is implemented in the NAG routine: GO5DCF (see Section A1.1).

(b) Weibull (see Exercise 2.3): $f(\omega) = \dfrac{\beta}{\gamma^\beta} \omega^{\beta-1} \exp\{-(\omega/\gamma)^\beta\}$ for

$0 \leq \omega < \infty, \beta > 0, \gamma > 0$.

Note that this method is implemented in the NAG routine GO5DPF and the IMSL routine GGWIB (see Section A1.1).

(c) Pareto distribution:

$$\Pr(X \leq x) = 1 - \left(\frac{k}{x}\right)^a \qquad \text{for } a > 0, x \geq k > 0.$$

(d) Extreme-value distribution:

$$\Pr(X \leq x) = \exp\{-\exp((\xi-x)/\theta)\} \qquad \text{for } x \geq 0.$$

5.14 Provide a detailed algorithm for simulating from the logarithmic distribution of Exercise 4.22.

*5.15 Barnett (1980) presents the bivariate uniform p.d.f.:

$$f(u, v) = (1-\alpha)[(2uv - u - v)\alpha + 1]\{\Psi(u, v)\}^{-3/2} \qquad (5.11)$$

where $\Psi(u, v) = (\alpha(u+v) - 1)^2 + 4\alpha(1-\alpha)uv$ and $\alpha < 1, 0 \leq u, v \leq 1$. This probability density function is illustrated in Fig. 5.17 for the case $\alpha = -4$. It is constructed from a bivariate distribution of Plackett (1965), which is given implicitly by:

$$\frac{\{F(x, y)\{1 - F_X(x) - F_Y(y) + F(x, y)\}}{\{F_X(x) - F(x, y)\}\{F_Y(y) - F(x, y)\}} = 1 - \alpha$$

From Section 5.2 we can see that if we set $U = F_X(X)$ then U is $U(0, 1)$, and so is $V = F_Y(Y)$. This is an interesting reversal of the aim of Section 5.2, which is to progress from U to X. Verify that this substitution here results in the joint p.d.f. $f(u, v)$ of Equation (5.11). Derive further bivariate uniform distributions in this manner from the following bivariate distributions also presented by Barnett (1980):

$f(u,v)$

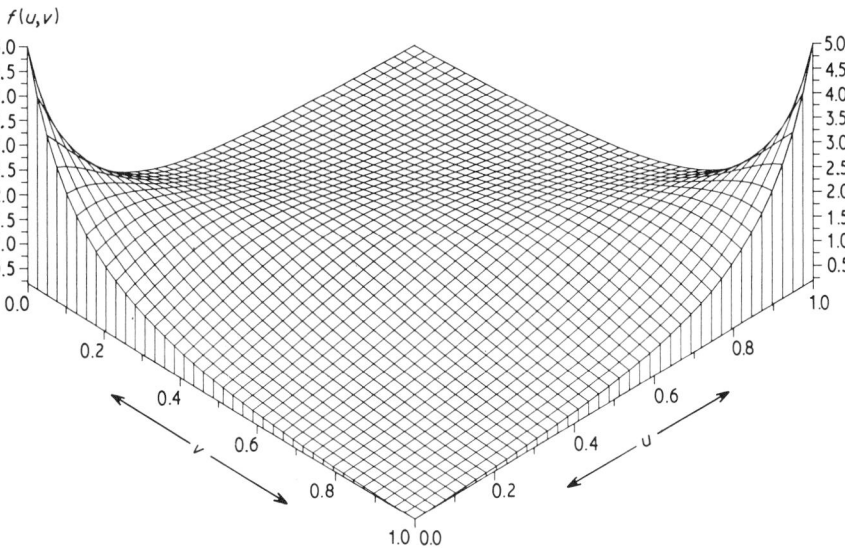

Figure 5.17   Isometric projection of the bivariate uniform density of Equation (5.11), from Morgan (1983).

(a) $F(x, y) = F_X(x)F_Y(y)\{1 - \alpha(1 - F_X(x))(1 - F_Y(y))\}$
for $|\alpha| < 1$.

(b) $f(x, y) = \{(1 + \alpha x)(1 + \alpha y) - \alpha\}\exp(-x - y - \alpha xy)$
for $0 < \alpha < 1$.

(this is a bivariate exponential distribution)

(c) $f(x, y) = \dfrac{1}{2\pi}(1 + x^2 + y^2)^{-3/2}$.

(this is a bivariate Cauchy distribution)

(c) Rejection methods

5.16   Figure 5.18 shows the probability density function $r(x)$ of Equation (5.10), resulting from the composition of (5.10). Explain how you would simulate from $r(x)$ using a rejection method.

5.17   To simulate from the probability density function given by

$$f(x) = \begin{cases} \dfrac{1}{\pi\sqrt{(1 - x^2)}}, & \text{for } -1 \le x \le 1 \\ 0 & \text{otherwise} \end{cases}$$

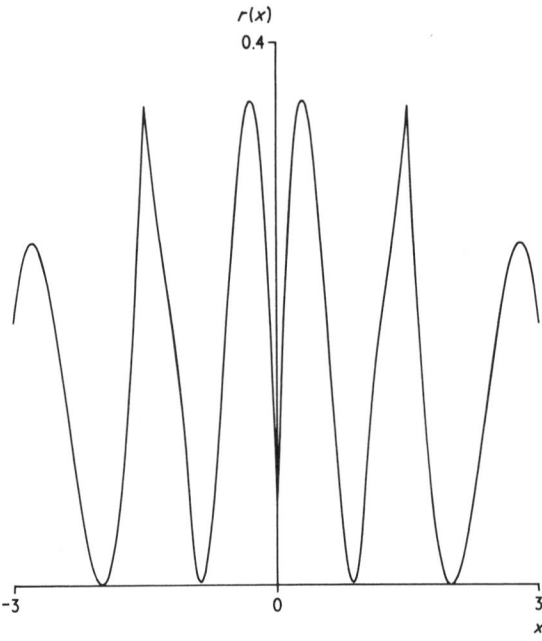

Figure 5.18   The p.d.f. $r(x)$ for $-3 \le x \le 3$, which is sampled with probability 0.0228 in the 'convenient' composition method of Equation (5.10).

we note that this is the probability density function of the random variable $X = \cos(\pi U)$, where U is a $U(0, 1)$ random variable. Use this result to devise a rejection method, based on the first quadrant of a circle and similar to the Polar Marsaglia method, for generating the required random variables.

5.18   Devise a rejection method, with an acceptance probability of not less than 8/9, for simulating random variables from the beta probability density function $f(x) = 6x(1 - x)$ for $0 \le x \le 1$.

†**5.19**   Describe how to simulate a random variable with the logistic probability density function, $e^{-x}(1 + e^{-x})^{-2}$ for $-\infty \le x \le \infty$, using a rejection method based on the exponential envelope, e^{-x} for $x \ge 0$.

5.20   Explain how to simulate normal random variables using a rejection method with an enveloping function based on a logistic p.d.f. Derive the probability of rejection (cf. Exercise 5.31).

†**5.21**   Repeat the approach adopted in Example 5.5 with $k\lambda e^{-\lambda x}$ for $x \ge 0$ as the enveloping function. Show that the probability of rejection is minimized if we take $\lambda = 1$, as in Example 5.5.

*5.22   Cheng (1977) presented a rejection method for simulating from the
$\Gamma(\alpha, 1)$ distribution, where $\alpha > 1$. In his case,

$$h(x) = \lambda \mu x^{\lambda - 1} (\mu + x^{\lambda})^{-2} \qquad \text{for } x \geq 0$$

(a) Consider how you would simulate from the probability density
function, $h(x)$.
(b) If $\mu = \alpha^{\lambda}$ and $\lambda = \sqrt{(2\alpha - 1)}$, determine the probability of rejection,
and the mean number of variable selections until acceptance.
The case $\alpha = 3$ is illustrated in Fig. 5.5, for $x < 8$.

*5.23   Compare the two rejection methods for simulating from a gamma
distribution, given in Example 5.6 and Exercise 5.22.

*5.24   Explain how you would simulate a random variable that has an
exponential distribution of mean 2, conditional on it being greater than
9 (cf. Exercise 5.26).

*5.25   Figure 5.16 presents a p.d.f. to be simulated from by means of a rejection
method. Kinderman and Ramage (1976) used the method of triangles
(see Marsaglia, MacLaren and Bray, 1964), which, in outline, is as
follows.

If a p.d.f. from which one wants to simulate can be sandwiched
between two parallel lines, the X-value for the rejection method is
simulated from an appropriate triangular distribution corresponding to
the upper of the parallel lines. When the corresponding uniformly
distributed Y value is less than the appropriate value on the lower of the
parallel lines, then X is accepted, and it is not necessary to compute the
formula for the curve. If, however, the Y value is greater than the appropriate value on the lower line then it is necessary to compute the
formula for the curve in order to decide on rejection or acceptance.

Discuss the objective of such an approach, and explain its use for the
beta p.d.f. of Exercise 5.18 (cf. comments in the solution to Exercise
5.23).

*5.26   Marsaglia (1964) proposed the following method for simulating
standard normal random variables X, conditional upon $X > a > 0$. Let
U_1, U_2 be two independent $U(0, 1)$ random variables. Set

$$X = (a^2 - 2 \log_e U_1)^{1/2}.$$

Accept X as a realization of the required random variable if $U_2 X < a$.
Otherwise, reject U_1 and U_2, and start again. Verify that X has the
required distribution (cf. Exercise 5.24 and the comments of Section
5.5).

†5.27   If U_1 and U_2 are independent $U(0, 1)$ random variables, show that, conditional upon $(2U_1 - 1)^2 + U_2^2 \le 1$, then $C = (2U_1 - 1)/U_2$ has the Cauchy distribution of Section 2.11. Note that this method is implemented in the NAG routine G05DFF (see Section A1.1).

5.28 If X_1 and X_2 are independent, identically distributed exponential random variables with mean unity, show that, conditional upon $(X_1 - 1)^2 < 2X_2$, then X_1 has a half-normal p.d.f., derived from an $N(0, 1)$ distribution. (This result is due to von Neumann—see Kahn, 1956, p. 39.)

*5.29 Suppose we have a probability density function $f(x)$ which can be written in the form:
$$f(x) = cg(x)r(x)$$
where $g(x)$ is also a p.d.f., $c > 0$ is a constant, and over the range of x, $0 \le r(x) \le m$, for some finite m. Show that we can simulate X from $f(x)$ as follows:

  (i) Simulate X from $g(x)$
  (ii) Accept X if $Um < r(X)$

      where U is an independent $U(0, 1)$ variable. Otherwise reject X and U and start again at (i).

What is the rejection probability? An example is provided by Butcher (1960), in which $f(x)$ is half-normal, and $g(x)$ is exponential. This generalization of the rejection method can give rise to efficient 'switching' algorithms, in which the rôles played by $g(x)$ and $r(x)$ change for different parts of the x-range; see Atkinson and Whittaker (1976), and Atkinson (1979b).

(d) Composition methods

†5.30 Use the composition approach of Section 5.4, as applied in the example of Equation (5.6), to simulate random variables with the Poisson distribution of Example 5.2.

5.31 Explain why it is not possible to simulate normal random variables using a composition, the first element of which, $f_1(x)$, is a logistic density.

†5.32 A continuous random variable X has the 'wedge-shaped' probability density function, $f_1(x) = \alpha - \alpha^2 x/2$, for $0 \le x \le 2/\alpha$ and $\alpha > 0$.

  (a) Explain how you would simulate X.
  (b) It is desired to simulate from the exponential p.d.f. $f(x) = \lambda e^{-\lambda x}$ for $x \ge 0$ and $2\lambda > \alpha$, using a composition, the first p.d.f. of which is to be $f_1(x)$. Derive the shrinking factor for $f_1(x)$, and deduce that, by

suitable choice of α, the probability of simulating from $f_1(x)$ in the composition can be made as large as $2/e$.

5.33 (a) Show that the random variable X, with probability density function

$$f(x) = \frac{e^{m-x}}{(e-1)} \qquad \text{for } (m-1) < x \leq m, \text{ where } m \geq 1$$

is obtained simply by setting $X = m - Y$, where Y has probability density function

$$f(y) = \frac{e^y}{(e-1)} \qquad \text{for } 0 \leq y < 1$$

The cumulative distribution function for X when $m = 1$ is illustrated in Fig. 5.2.

(b) By expanding $f(y)$ as a power series in y, show that we can simulate from $f(y)$ by means of a composition, simulating from probability density function, $(i+1)y^i$ for $0 \leq y < 1$, with probability

$$\frac{1}{(i+1)!(e-1)} \qquad \text{for } i \geq 0.$$

5.34 (*continuation*)   We note that

$$e^{-x} = \frac{(e-1)e^{-m} \times e^{m-x}}{(e-1)} \qquad \text{for any } m \geq 1$$

Explain, with reference to Fig. 5.19, how this result may be used as a basis for a composition method for simulating from the probability density function, e^{-x} for $x \geq 0$.

*5.35 (*continuation*)   Explain the following algorithm, given by Marsaglia (1961), for simulating random variables from the exponential e^{-x} p.d.f.:

(i) Simulate a discrete random variable, I, from the distribution

$$\frac{1}{(i+1)!(e-1)} \qquad \text{for } i \geq 0$$

(ii) Set $W = \max(U_1, U_2, \ldots, U_{I+1})$,
where the $\{U_j\}$ are independent $U(0, 1)$ random variables.

(iii) Simulate a discrete random variable, M, from the distribution

$$(e-1)e^{-m} \qquad \text{for } m \geq 1$$

(iv) Set $X = M - W$.

*5.36 Consider how you would simulate standard normal random variables using a composition method, in which the first p.d.f. in the composition, $f_1(x)$, is of trapezoidal form. See Ahrens and Dieter (1972) for an algorithm based on this approach.

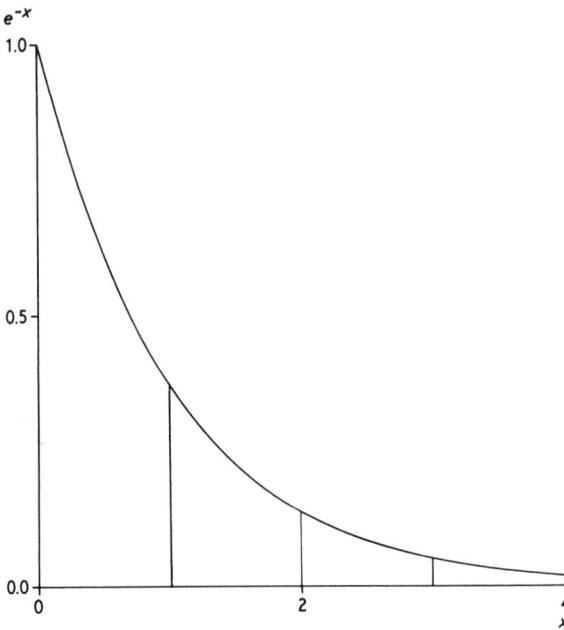

Figure 5.19   A breakdown of the p.d.f. e^{-x} into sections of area, $(e-1)e^{-m}$, for $m \geq 1$, for $x \leq 4$.

*5.37   Figure 5.20 illustrates a portion of the half-normal p.d.f. $2\phi(x)$, and $0.97f_1(x)$, in which $f_1(x)$ is a density composed of 97 rectangles, each of area 1/97. Figure 5.21 illustrates $2\phi(x) - 0.97f_1(x)$. Discuss how these curves may be used to simulate standard normal random variables. This method is due to Lenden–Hitchcock (1980) and is based on a method of Marsaglia, MacLaren and Bray (1964).

*5.38   Kinderman and Ramage (1976) produce the algorithm, given below, for their method discussed in Section 5.7. Explain how the method gives rise to this algorithm. (Note that $\xi = 2.216\,035\,867\,166\,471$ and $f(t) = \phi(t) - 0.180\,025\,191\,068\,563\,(\xi - |t|)$, for $|t| < \xi$. Here we preserve the high accuracy of constants given in the original source.)

Algorithm from Kinderman and Ramage (1976)

    1. Generate u. If $u < 0.884\,070\,402\,298\,758$, generate v and return $x = \xi \times (1.131\,131\,635\,444\,180\,u + v - 1)$.
    2. If $u < 0.973\,310\,954\,173\,898$, go to 4 below.
    3. Generate v, w. Set $t = \xi^2/2 - \log_e w$. If $v^2 t > \xi^2/2$, begin this step

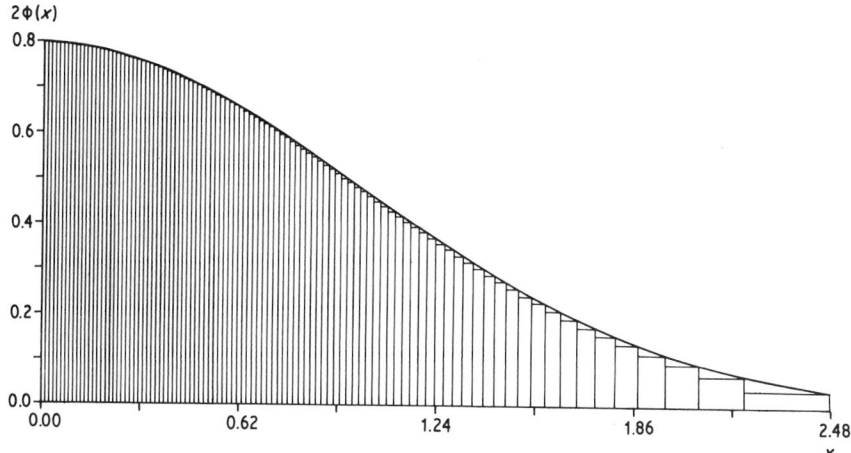

Figure 5.20   Part of the half-normal density,

$$2\phi(x) = \sqrt{\left(\frac{2}{\pi}\right)}e^{-x^2/2},$$

enveloping 97 rectangles, each of area 0.01.

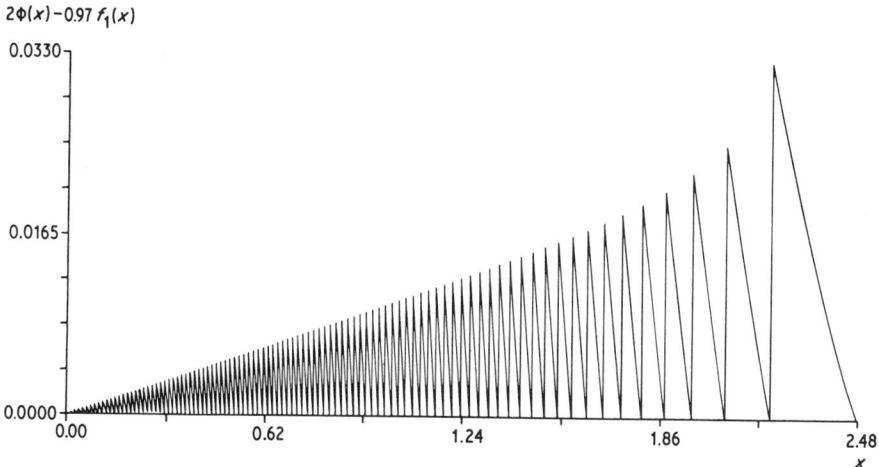

Figure 5.21   A graph of $2\phi(x) - 0.97f_1(x)$, from Fig. 5.20, in which $0.97f_1(x)$ is the envelope of the rectangles shown in Fig. 5.20.

again. Otherwise return $x = (2t)^{1/2}$ if $u < 0.986\,655\,477\,086\,949$ or return $x = -(2t)^{1/2}$ if not.

4. If $u < 0.958\,720\,824\,790\,463$, go to 6 below.

5. Generate v, w. Set $z = v - w$ and $t = \xi - 0.630\,834\,801\,921\,960 \times$

$\min(v, w)$. If $\max(v, w) \leq 0.755\,591\,531\,667\,601$, go to 9. If $0.034\,240\,503\,750\,111|z| \leq f(t)$, go to 9. Otherwise, repeat this step.

6. If $u < 0.911\,312\,780\,388\,703$, go to 8.
7. Generate v, w. Set $z = v - w$ and

$$t = 0.479\,727\,404\,222\,441 + 1.105\,473\,661\,022\,070 \min(v, w).$$

If

$$\max(v, w) \leq 0.872\,834\,976\,671\,790$$

go to 9. If $0.049\,264\,496\,373\,128|z| \leq f(t)$ go to 9. Otherwise, repeat this step.

8. Generate v, w. Set $z = v - w$ and

$$t = 0.479\,727\,404\,222\,441 - 0.59550\,71380\,15940 \min(v, w).$$

If

$$\max(v, w) \leq 0.805\,577\,924\,423\,817$$

go to 9. If $0.053\,377\,549\,506\,886|z| \leq f(t)$, go to 9. Otherwise, repeat this step.

9. If $z < 0$, return $x = t$; otherwise return $x = -t$.

*5.39 For the residual p.d.f. $r(x)$ from the composition of Equation (5.10), show that the probability of rejection is 0.53 when we simulate from $r(x)$ using rejection and an enveloping rectangle.

(e) Additional methods

*5.40 Suppose W is $U(a, b)$, for some $b > a$, and suppose that for $a \leq x \leq b$, $0 \leq g(x) \leq 1$ for some function $g(x)$. Suppose N is the first integer ≥ 1 such that

$$g(W) \geq U_1 \geq U_2 \geq \ldots \geq U_{N-1} < U_N$$

where the $\{U_i\}$ are a sequence of independent, identically distributed $U(0, 1)$ variables. Thus $N = 1$, if and only if $g(W) < U_1$

$$N = 2, \text{ if and only if } g(W) \geq U_1 < U_2$$

etc.

Show that

$$\Pr(N = n | W = w) = \frac{g(w)^{n-1}}{(n-1)!} - \frac{g(w)^n}{n!} \qquad \text{for } n \geq 1$$

and deduce that

$$\Pr(N \text{ is odd} | W = w) = \exp(-g(w))$$

Finally, show that the conditional p.d.f. of W is given by

$$f_W(w \,|\, N \text{ is odd}) = \frac{\exp(-g(w))}{\displaystyle\int_a^b \exp(-g(w))\,dw} \tag{5.12}$$

*5.41   (*continuation*)   Use the results of the last exercise to provide a rejection method to simulate random variables with the p.d.f. of Equation (5.12) over the range (a, b). This is the basis of what is known as *Forsythe's method* (see Forsythe, 1972), which has been used for a variety of distributions (see Atkinson and Pearce, 1976). Of course, the requirement that $g(x) \leq 1$ is, by itself, very restrictive; however, this restriction can be overcome by dividing up the range of x into a number of intervals, and then first of all using a composition to determine the appropriate interval: if $\tilde{g}(x)$ is an increasing function of x, over the range $(0, \infty)$, say, then if the interval (q_i, q_{i+1}) is chosen by the first stage of the composition method, $\{\tilde{g}(q_i + x) - \tilde{g}(q_i)\}$ plays the rôle of $g(x)$ in the last exercise. The $\{q_i\}$ must be chosen so that

$$0 \leq \tilde{g}(q_i + x) - \tilde{g}(q_i) \leq 1 \qquad \text{for } 0 \leq x \leq (q_{i+1} - q_i)$$

One such choice of $\{q_i\}$ gives rise to what is called Brent's GRAND method for $N(0, 1)$ variables (see Brent, 1974). This is the method employed by the NAG routine, G05DDF (see Section A1.1). Further discussion and comparisons with other methods are given by Atkinson and Pearce (1976). One advantage of Forsythe's method is that it avoids time-consuming exponentiation.

*5.42   The random variable X takes the values 1, 2, 3, 4 with the following probabilities:

$$\begin{aligned}
\Pr(X = 1) &= \tfrac{1}{6} = \tfrac{1}{4}(\tfrac{2}{3} + 0 + 0 + 0) \\
\Pr(X = 2) &= \tfrac{1}{12} = \tfrac{1}{4}(0 + \tfrac{1}{3} + 0 + 0) \\
\Pr(X = 3) &= \tfrac{7}{12} = \tfrac{1}{4}(\tfrac{1}{3} + \tfrac{2}{3} + 1 + \tfrac{1}{3}) \\
\Pr(X = 4) &= \tfrac{1}{6} = \tfrac{1}{4}(0 + 0 + 0 + \tfrac{2}{3})
\end{aligned}$$

Thus, by analogy with Equation (5.6), we can write

$$\Pr(X = i) = \frac{1}{4} \sum_{j=1}^{4} r_{ij}$$

where the $\{r_{ij}, 1 \leq i \leq 4\}$ are all probability distributions, for each j, $1 \leq j \leq 4$. The difference as compared with Equation (5.6) is that now random variables with any of the four $\{r_{ij}, 1 \leq i \leq 4\}$ distributions take just one of at most two values, and the distributions in the composition have equal probability of being used. Show that any discrete random variable X over a finite range can be obtained by means of such a composition (Kronmal and Peterson, 1979). This composition results in the *alias* method, so called because if the $\{r_{ij}, 1 \leq i \leq 4\}$ distributions do not select the value $X = j$ then the 'alias' value for X is chosen by the $\{r_{ij}, 1 \leq i \leq 4\}$ distribution. For example, in the above illustration, with probability $\frac{1}{4}$ the component distribution, $\{r_{i2}, 1 \leq i \leq 4\}$ is selected, and then either $X = 2$, with probability $r_{22} = \frac{1}{3}$, or $X = 3$, the alias value, with probability $r_{23} = \frac{2}{3}$. For further discussion, see Peterson and Kronmal (1982). An attractive feature of this method is that it does not require more than two uniform random variables for each value of X. Can you suggest a way in which only one uniform random variable need be used? (See Kronmal and Peterson, 1979.) An algorithm for the alias method is provided by the IMSL routine GGDA (see Section A1.1).

6

TESTING RANDOM NUMBERS

6.1 Introduction

The need for stringent testing of uniform random variables was emphasized in Chapter 3. When tables of random digits were first produced, tests were employed for uniform random digits. More recently, with the development of pseudo-random-number generators, the numbers to be tested are continuously distributed over the range (0, 1). In the latter case, tests for digits are frequently applied to the digit occupying the first decimal place, while in some cases of detailed testing other decimal places are also considered, as in Wichmann and Hill (1982a). An alternative approach, given by Cugini *et al.* (1980), is described in Section 6.3.

We have seen that congruential methods of random number generation are convenient and widely used, but that they can produce sequences of numbers with certain undesirable properties. For any particular application, the need is to determine what may be 'undesirable', so that random numbers should always be tested with an application in mind. This is often easier said than done, but we can see that it could entail testing not only uniform variables, but also variables of other distributions, obtained by methods such as those of the last two chapters. In Chapter 5 in particular, some of the algorithms given are very complicated, and in such cases testing is needed quite simply as a check that there have been no programming errors. In Chapter 4 we saw that particular properties of random variables and processes can be used to generate particular random variables. By the same token, similar properties may be used to test particular random variables, as we shall see in Section 6.6.

A room full of eternally typing monkeys will ultimately produce the plays of Shakespeare, and similarly, a large enough table of uniform random digits will, by the very nature of random digits, contain sections which, by themselves, will certainly fail tests for uniformity. This feature is noted in the table of Kendall and Babbington–Smith (1939a), which contains 100 000, digits. They tested their table as a whole, and also in parts, down to blocks of 1000 digits each. As expected, some of these individual blocks failed certain tests, and a note was added to these blocks, to 'caution the reader from using them by themselves'.

EXAMPLE 6.1
As an illustration of this, let us consider the digits of Table 3.1. For the two halves of the table we obtain the following frequencies for single digits:

Digit	0	1	2	3	4	5	6	7	8	9	Totals
(a)	17	16	13	16	17	16	36	16	20	13	180
(b)	15	19	20	19	14	22	16	20	11	24	180
(a) + (b)	32	35	33	35	31	38	52	36	31	37	360

For the entire table, if the digits were random the expected number for each digit is $360/10 = 36$, and so the departures from 36 observed can be tested by the chi-square test of Section 2.14. Here no parameters have been estimated from the data, and so the number of degrees of freedom is 9. For the entire table we obtain $\chi_9^2 = 9.39$, which is not significant at the 5 % level. However, if we take part (a) of the table above, we find $\chi_9^2 = 22$, just significant at the 1 % level, for a one-tail significance test, or the 2 % level for a two-tail test. As we shall see later, two-tail tests are frequently used for testing random numbers.

In the context of pseudo-random numbers, we have already encountered this same point in Chapter 3, since congruential generators can be devised which have a low first-order serial correlation for their full cycle, but which result in much higher such correlations for fractions of the cycle (see Exercise 6.1). A property of a pseudo-random number generator for its entire cycle provides, effectively, a test of that generator, and a test of a kind that is not possible for physical random number generators. As well as serial correlations, the first- and second-order moments of Exercise 3.13 can be interpreted in this way. Such tests have come to be known as *theoretical* tests, and an elaborate such test is the spectral test of Coveyou and MacPherson (1967). Theoretical tests evaluate the generating mechanisms used, and do not make use of generated numbers. Knuth (1981, p. 89) states that all congruential generators that are thought to be good pass the spectral test, while those that are known to be bad fail it. Oakenfull (1979) and Knuth (1981, p. 102) provide the results of applying this test to a variety of congruential generators. Ultimately, however, we have to test the numbers produced by a generator in the context of their use, and this is done by a variety of *empirical* tests, which are the subject of this chapter. Atkinson (1980) describes when the spectral test is appropriate, and for a number of generators compares the results of theoretical and empirical tests. The same theoretical/empirical comparison is also made by Grafton (1981).

6.2 A variety of tests

When we are dealing with random variables such as Poisson or normal, we want to check that the generated values come from the distributions we think they do. In the case of Poisson variables this could involve checking that the differences between the bar-charts of Fig. 2.3, for example, are not significant, while for normal variables we would be comparing, for instance, the density function of Fig. 2.5(a) with the histogram of Fig. 2.5(b). Methods for making these comparisons will be considered later. In addition, we may well want to consider the serial dependence of the variables, as is done for instance by Barnett (1965) for exponential random variables. Obvious discrepancies can sometimes be spotted by inspection of a convenient graphical display, as can be done for the figures of Chapter 2, but ultimately significance tests must be applied. The scatter plot of Fig. 3.2 is 'obviously' non-random, but what can we say of the scatter of Fig. 3.3? The same question can be asked of the plot of Fig. 6.1, produced by the generator of Equation (3.1).

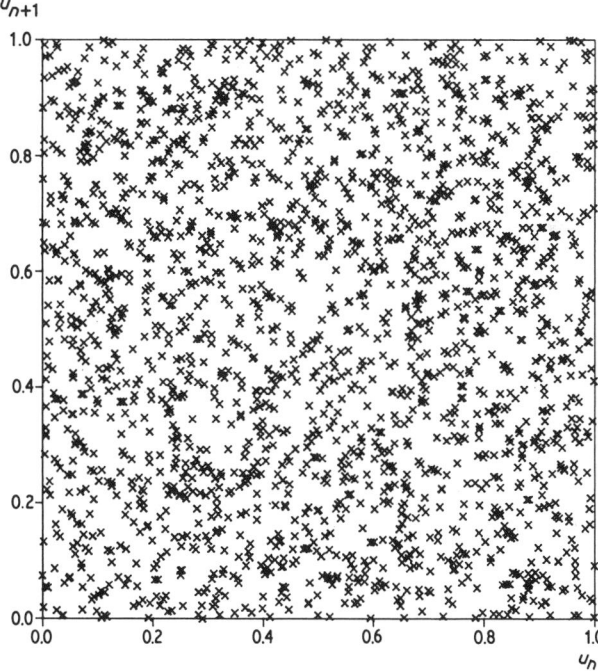

Figure 6.1 A scatter plot of u_{n+1} vs. u_n for a sequence of length 2000 from the generator of Equation (3.1).

Recently a sophisticated approach to judging the randomness of scatter plots has been provided by Ripley (1977, 1981), whose technique itself utilizes repeated simulations, and Atkinson (1977b) applied this technique to numbers resulting from the multiplicative congruential generator

$$x_{i+1} = 5x_i$$

discussed earlier in Equation (3.4).

In inspecting and testing scatter plots such as that of Fig. 6.1, we are implicitly considering how often different one- and two-dimensional intervals are represented. This corresponds to the basic frequency and serial tests which we shall soon describe. Apart from these, however, what other empirical tests should we apply? Thus, hard on the heels of the earlier problems of generator selection and, when appropriate, which method to use for transforming uniform random variables, is the problem of test selection. As mentioned earlier, the glib answer is that tests should be suggested by the use intended for the random variables, and this could result in the application of very specific tests, over and above those already applied to a basic source generator. Different producers of uniform random numbers have answered this question in different ways, and batteries of tests are to be found, for example, in Kendall and Babbington–Smith (1939), Tausky and Todd (1956), Craddock and Farmer (1971), Miller and Prentice (1968) and Wichmann and Hill (1982a).

It is important to realize that there is nothing magical or God-given about a particular set of tests. Clearly an infinity of tests is possible, and as we shall see, numbers which pass one test may fail another. Kendall and Babbington–Smith used just four tests, designed to check frequencies and various forms of sequential dependence, and this basic approach is that adopted by subsequent authors, though conventions have changed with time. We shall now describe certain standard tests for uniform random digits, and then see the results of applying these tests to sequences resulting from a variety of generators.

6.3   Certain tests for uniform random digits

When presented with tables of digits such as those of Tables 3.1, 3.2 and Exercise 3.10, the first reaction of most of us would be to count up the frequencies of occurrence of each digit and compare the observed frequencies with what we would expect for random digits. The statistical yardstick that is usually used in making this comparison is the chi-square test of Section 2.14, and we have already seen such examples of a *frequency* test in Example 6.1, and Exercise 2.24.

Deterministic sequences such as: . . . 89012345678901 . . . satisfy the frequency test (if a one-tail test is used – see later), but blatantly fail tests which consider the ordering of the elements in the sequence. The simplest such test, the *serial* test, takes a sequence of digits: . . . $d_i, d_{i+1}, d_{i+2}, \ldots$, and from considering non-overlapping pairs of digits compares the observed matrix $\{O_{kl}\}$ with the expected $\{e_{kl}\}, 0 \le k, l \le 9$, in which digit l is observed to follow digit k, O_{kl} times, and e_{kl} is the corresponding number to be expected if we have a truly random sequence. If we have n non-overlapping pairs of digits, then $e_{kl} = n/100$, for $0 \le k, l \le 9$. The yardstick here is again a chi-square test, but this time on 99 degrees of freedom. An illustration of a serial test is given in Example 6.2.

Non-overlapping pairs of digits are taken so that the independence requirement of the chi-square test is preserved (see Section 2.14). If overlapping pairs are used, then a modified test, due to Good (1953), may be used. The IMSL routine GTPST, of Section A1.2 performs this test. It is an interesting footnote that Kendall and Babbington–Smith (1939a) used overlapping pairs and then incorrectly applied the standard chi-square test. Deterministic sequences such as that illustrated above do in fact produce *too good* an agreement with expectation in the frequency test, and this is indicated by significantly *small* values of the chi-square goodness-of-fit statistic. Consequently, chi-square tests of randomness are often two-tail tests, unlike customary chi-square tests in which only the upper tail is used as the critical region. An example of a sequence of digits that are too regular is provided by the first 2000 decimal digits of $e = 2.71828 \ldots$. Here the frequency test gives $\chi_9^2 = 1.06$, a value which is significant at the 0.2% level, using a two-tail test. If the first 10 000 decimal digits of e are taken, then we obtain the satisfactory result: $\chi_9^2 = 8.61$; a more detailed breakdown can be found in Stoneham (1965), some of whose results are illustrated in Example 6.7 and Exercise 6.15.

Of course, as stated in Section 2.14, the chi-square test is an asymptotic test, and so is not appropriate if expected cell values are 'small'. The serial test generalizes to the consideration of triples, quadruples and so on, of digits, and the number of cells correspondingly increases geometrically. Thus, especially if one is considering non-overlapping n-tuples, care must be taken in tests of high-dimensional randomness to ensure that expected cell values are large enough for the chi-square test to be valid. An alternative test of randomness in high-dimensional space is the *collision* test described by Knuth (1981, pp. 68–70), and for which a FORTRAN program is given by Hopkins (1983b).

We can test random digits in a less routine way, by looking for patterns. One rudimentary way of doing this is provided by the *gap* test, which is as follows: select any digit, e.g. 7. We can now consider any sequence as consisting of 7's and 'not 7's', i.e., a binary sequence in which $\Pr(7) = 1/10$, and $\Pr(\text{not } 7) = 9/10$, if the sequence is random, and if successive digits are independent then the distribution of the number of digits between 7's is geometric (see

Section 2.6). Thus empirical and observed distributions of numbers of digits between 7's may be compared. For an illustration, see Example 6.2. Like the gap test, the 'coupon-collector' test is also based on a waiting-time, as it considers the number of digits until at least one of each of the digits 0–9 has appeared. This test treats all digits equally, and was first proposed by Greenwood (1955), who found that the test was satisfied by the first 2486 digits in the decimal expansion of $e = 2.71828 \ldots$ and by the first 2035 digits in the decimal expansion of $\pi = 3.14159 \ldots$; details of his test results can be found in Exercise 6.7.

A more obvious way of looking for patterns is provided by the *poker* test, which considers digits in sequences of length 5, and classifies the patterns according to the conventions of the game of poker: all different, two pairs, etc. Further discussion of the coupon-collector and poker tests is provided in Exercises 6.6, 6.7 and 6.15, and Example 6.7. The poker test may be performed by means of the IMSL routine – GTPOK (see Section A1.2).

Example 6.2 gives the results of applying the serial and gap tests to sequences produced by the random number generator of the Commodore PET microcomputer. This generator is not of a standard form, and will not be described here.

EXAMPLE 6.2 *The result of applying the serial and gap tests to the Commodore PET microcomputer random number generator*

(a) SERIAL TEST:

| | | \multicolumn{11}{c}{Following value} | |
		1	2	3	4	5	6	7	8	9	10	11	Totals
	1	15	17	18	27	20	16	21	18	21	20	14	207
	2	30	24	20	18	25	13	18	24	27	25	17	241
	3	25	18	19	23	28	15	14	16	16	16	22	212
Preceding	4	14	24	23	14	22	16	17	16	18	19	19	202
value	5	24	16	16	15	15	23	17	21	24	23	18	212
	6	22	24	22	27	18	8	17	19	31	24	25	237
	7	26	22	21	15	19	24	13	20	19	19	17	215
	8	24	13	18	26	21	16	19	21	19	14	20	211
	9	14	17	24	22	18	18	17	15	18	18	21	202
	10	22	24	26	27	23	25	18	23	25	16	23	252
	11	22	13	24	25	26	18	21	18	25	14	23	229
Totals		238	212	231	239	235	192	192	211	243	208	219	2420

Here we obtain $\chi^2_{120} = 108.8$, which is clearly not significant, and so on the basis of this test we would not reject the hypothesis that the digits were uniform and random.

(b) GAP TEST

Gap size	Actual count	Expected count
0	36	25.90
1	36	23.31
2	23	20.98
3	20	18.88
4	17	16.99
5	6	15.29
6	15	13.76
7	10	12.39
8	11	11.15
9	10	10.03
≥ 10	75	90.31
Totals	259	258.99

Here $\chi_{10}^2 = 19.924$, which is close to significance at the 5 % level (two-tail test), and one would want to repeat this test to see if other samples produced similar results.

Note that these and other test results presented later in this chapter were obtained using the suite of BASIC test programs of Cugini *et al.* (1980). Rather than work with digits, they divided the (0, 1) interval into 11 sections for the serial test, while for the gap test, gaps were recorded between numbers lying in the (0.03, 0.13) interval. Thus for the gap test,

$$25.90 = 259/10,$$
$$23.31 = 25.9 \times 0.9, \text{ etc.}$$

*6.4 Runs tests

A striking feature of a table of digits can be the occurrence of *runs* of the same digit. If such runs occur with greater frequency than one would expect for random digits then one might, for example, expect this feature to result in a significant departure from the geometric distribution of the gap test. One can, however, look at distributions of other types of runs, and this was done by Downham and Roberts (1967).

Runs tests are frequently applied to a sequence of $U(0, 1)$ variates. Here we shall just consider 'runs up'. To illustrate what is meant by a 'run up', consider the following sequence of numbers, given here to 3 decimal places:

(0.134 0.279 0.886) (0.197) (0.011 0.923 0.990) (0.876)

The 'runs up' are indicated in parentheses, so that here we have four such runs, of lengths 3, 1, 3, 1, respectively. We see that a 'run up' ends when the next item in the sequence is less than the preceding item, the next item then starting the next 'run-up'. Levene and Wolfowitz (1944) showed that in a random sequence of n $U(0, 1)$ variates, the expected number of 'runs up' of length $k \geq 1$, R_k, say, is given by:

$$\mathscr{E}[R_k] = \frac{(k^2 + k - 1)(n - k - 1)}{(k + 2)!} \qquad \text{for } 1 \leq k \leq n$$

(See also Knuth, 1981, pp. 65–68, for a derivation of this result.)

Typically, n is taken to be large, so that

$$\mathscr{E}[R_k] \approx \frac{(k^2 + k - 1)}{(k + 2)!} n \qquad \text{for } k \ll n$$

Clearly, for fixed n, $E[R_k]$ decreases as $k \to n$, and it is usual to consider the joint distribution of $(R_1, R_2, \ldots, R_j, S_j)$, for some $j > 1$, where $S_j = \Sigma_{k=j+1}^{n} R_k$; $j = 5$ is frequently adopted. Successive run lengths are not independent, and so a standard chi-square test for comparing observed and expected numbers of runs is inappropriate. The test-statistic used (see Levene and Wolfowitz, 1944) is

$$U = \frac{1}{n} \sum_{i=1}^{6} \sum_{j=1}^{6} (X_i - \mathscr{E}[X_i])(X_j - \mathscr{E}[X_j]) a_{ij} \qquad (6.1)$$

in which $X_k = R_k$ for $1 \leq k \leq 5$, and $X_6 = S_6$,

the $\{a_{ij}\}$ form the inverse of the variance–covariance matrix of the $\{X_k\}$, and for large n are given by:

$$\mathbf{A} \approx \begin{cases} 4529.4 & 9\,044.9 & 13\,568 & 18\,091 & 22\,615 & 27\,892 \\ & 18\,097 & 27\,139 & 36\,187 & 45\,234 & 55\,789 \\ & & 40\,721 & 54\,281 & 67\,852 & 83\,685 \\ & & & 72\,414 & 90\,470 & 111\,580 \\ & & & & 113\,262 & 139\,476 \\ & & & & & 172\,860 \end{cases}$$

the lower half of this matrix being obtained from symmetry. The exact expression is given by Knuth (1981, p. 68). U is referred to chi-square tables on 6 (not 5) degrees of freedom. As with the usual chi-square test, an asymptotic approximation is being made when this test is used, and Knuth recommends taking $n \geq 4000$. An illustration of the outcome of applying this test is given in the following example.

EXAMPLE 6.3   *The result of applying the 'runs up' test to a sequence of length* $n = 5000$ *from the generator* $(131, 0; 2^{35})^{\dagger}$

---

† Note that for convenience we shall henceforth use the notation: $(a, b; m)$ for the congruential generator of Equation (3.2).

Run length (k)	R_k	$\mathscr{E}[R_k]$
1	824	833.34
2	1074	1041.66
3	440	458.33
4	113	131.94
5	42	28.77
≥ 6	7	5.95

$\chi_6^2 = 18.10$, significant at the 2% level, using a two-tail test. Thus here the test rejects the hypothesis that the variables are random and uniform.

Before the work of Levene and Wolfowitz (1944), runs tests were incorrectly used, incorporating the standard chi-square approach. Unfortunately the algorithm by Downham (1970) omitted the $\{a_{ij}\}$ terms of Equation (6.1). That this omission could possibly result in erroneous conclusions is demonstrated by Grafton (1981), who provides a brief comparison between the correct runs test and the spectral test. Grafton (1981) provides a FORTRAN algorithm which tests 'runs down' as well as 'runs up', though the two tests are not independent. See also Section A1.2 for the IMSL routines GTRN and GTRTN. Accounts of the power of runs tests vary, and are clouded by incorrect uses of the tests. Kennedy and Gentle (1980, pp. 171–173) provide the theory for the case of runs up and down.

6.5 Repeating empirical tests

One might expect a poor generator to fail empirical tests, but a failure of an empirical test need not necessarily indicate a poor generator. Conversely, a poor generator can pass empirical tests, and both of these instances are illustrated in the following two examples.

EXAMPLE 6.4
The frequency test was applied to the (781, 387; 1000) generator, starting the sequence with 1. The full cycle was divided into 20 consecutive groups of 50 numbers each. For any group the frequency test was satisfied, but the 20 chi-square statistics took just one of the three values, 10.0, 8.8, 7.2.

EXAMPLE 6.5
The PET generator produced the borderline 5% significance result of Example 6.2(b) under the gap test. Nine subsequent gap tests produced the insignificant statistics of:

9.49, 14.88, 6.50, 13.73, 7.80, 7.80, 4.36, 8.12, 7.80

A similar 'unlucky start' is found with the frequency test applied to the decimal digits of e (Stoneham, 1965).

These difficulties can sometimes be resolved by repeating an empirical test, producing in effect a more stringent test. Chi-square values from repeating tests can be interpreted in a number of ways: a simple graphical representation can be obtained by probability (also called Q–Q) plots (see for example, Chernoff and Lieberman, 1956; Gerson, 1975; and Kimball, 1960), in which a sample of size n from some distribution (chi-square in our case) is ordered and plotted against the expected values of the order statistics. The expected order statistics for chi-square distributions are provided by Wilk *et al.* (1962), and two illustrations are provided by the following two examples.

EXAMPLE 6.6
The RANDU generator, $(65\,539, 0; 2^{31})$, resulted in the probability plot shown in Fig. 6.2 for 30 applications of the 'runs up' test, each applied to a sequence of 5000 numbers.

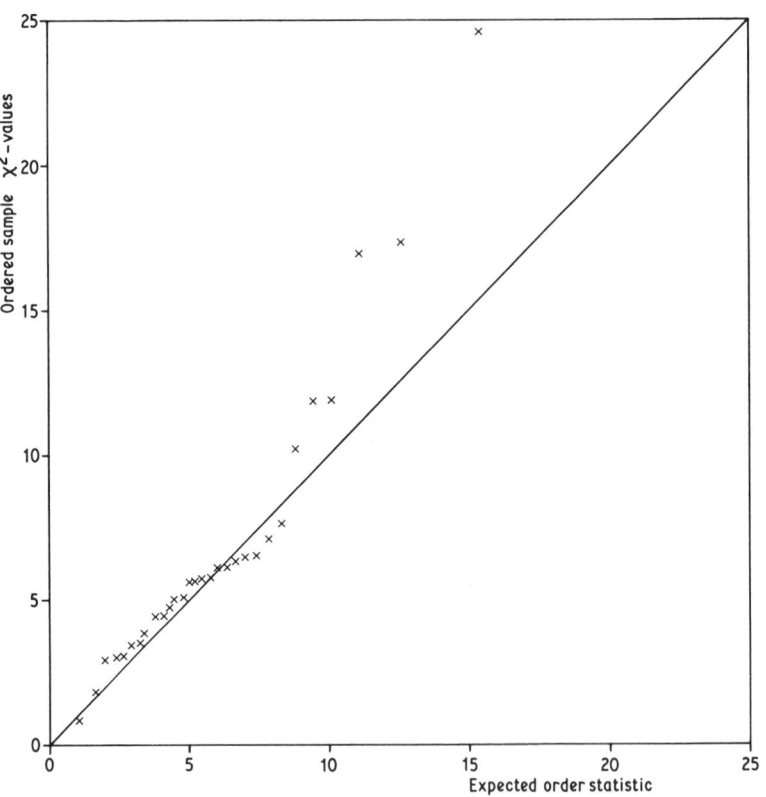

Figure 6.2   A probability plot of 30 test-statistics resulting from the 'runs up' test applied to the RANDU generator. The ordered sample is plotted against the expected order statistics for a sample of size 30 from a χ^2_6 distribution.

EXAMPLE 6.7
Stoneham (1965) made a study of the first 60 000 decimal digits of *e*. The results
of 12 applications of the poker test are illustrated in Fig. 6.3, each test being
applied to a block of 5000 consecutive digits. Some of the detail is presented in
Exercise 6.15.

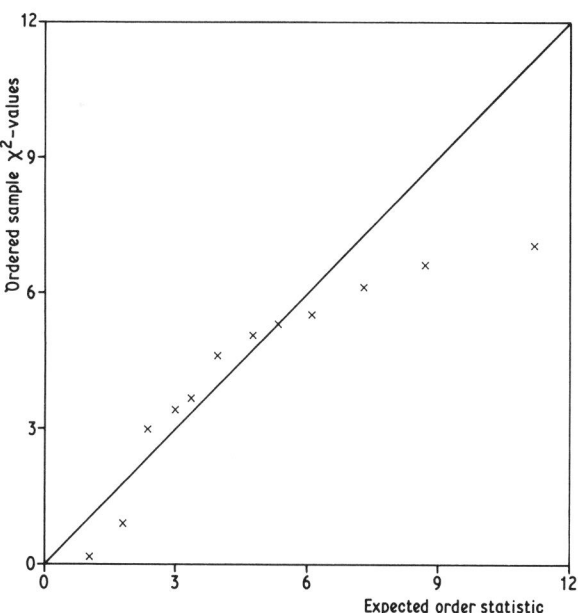

Figure 6.3   A probability plot of 12 test-statistics resulting from the poker test applied
to decimal digits of *e*. The ordered sample is plotted against the expected order statistics
for a sample of size 12 from a χ^2_5 distribution.

Note, however, that Wilk *et al.* (1962) remark that it is difficult to interpret
such plots for fewer than 20 points, reaching their conclusion after applying
the plots to simulated data.

Whether or not these plots indicate significant departures from the
appropriate chi-square distribution can also be judged by means of a further
chi-square test, if the sample size permits, as demonstrated in the next example.

EXAMPLE 6.8
The Cugini *et al.* (1980) program for the frequency test applies the test to 1050
numbers in the (0, 1) range, categorizing them according to 21 equal-length
intervals. The test is then repeated 60 times, and the resulting chi-square
statistics are themselves categorized according to the percentile range into

which the values fall. Applying this test to the PET generator produced the following result:

% range	Actual count	Expected count
0–1	4	0.6
1–5	0	2.4
5–10	2	3
10–25	10	9
25–50	18	15
50–75	15	15
75–90	7	9
90–95	2	3
95–99	2	2.4
99–100	0	0.6
	60	60

Combining the first three and the last three rows allows us to perform a chi-square test, now on 5 degrees of freedom, to this table. We obtain the value of $\chi_5^2 = 1.82$, which is not significant at the 10 % level, and so these numbers pass this empirical test.

An alternative approach is to use the Kolmogorov–Smirnov test, which is described by Hoel (1954, p. 345) and Knuth, 1981 (pp. 45–52) who provides some interesting comparisons of power between the chi-square and Kolmogorov–Smirnov tests. Categorization of the individual chi-square values is unnecessary for the Kolmogorov–Smirnov test, and when applied to the sample of size 12 illustrated in Fig. 6.3, the test does not reveal a significant departure from the expected χ_5^2 distribution at the 5 % level. While the same is true for the sample of size 30 illustrated in Fig. 6.2, in that case the result is significant at the 6 % level.

6.6 Tests of non-uniform random variables

We have already seen, in the last example, an illustration of the chi-square test being used to test a non-uniform (in this case also chi-square) random variable. The same approach may be used, with suitable combining of categories when necessary, for any distribution; see Exercise 6.19 for an illustration. The Kolmogorov–Smirnov test may also be used for any distribution, and, as above, does not require categorization. We shall now briefly consider some particular tests for non-uniform random variables.

6.6.1   Normal variables

Wold (1954) obtained standard normal random variables, to 2 decimal places, by transforming the digits of Kendall and Babbington–Smith (1939); the digits were initially grouped to correspond to $U(0, 1)$ variates and then the table-look-up method of Section 5.2 was used (see Example 5.4). Except for normal variables in the tails of the distribution, for the two-place accuracy needed, just four-decimal place accuracy was necessary for the $U(0, 1)$ variates. The resulting table had 25 000 items, which were tested as a whole, as well as in groups of 500 and 5000. Four tests were employed:

(a) The numbers in a group were summed, and the result referred to the appropriate normal distribution (see Exercise 2.8(a)).
(b) The squares of the numbers in a group were summed, and the result referred to the appropriate chi-square distribution (see Exercise 2.5). As the group sizes are ≥ 500, we can use the approximation that if X has a χ_v^2 distribution, $(\sqrt{(2X)} - \sqrt{(2v - 1)})$ is approximately $N(0, 1)$ (see Exercise 6.24).
(c) From the solution to Exercise 6.4 we see that if R is the range from a random sample of size n from an $N(0, 1)$ distribution, then

$$\Pr(R \leq r) = n \int_{-\infty}^{\infty} (\Phi(x + r) - \Phi(x))^{n-1} \phi(x) \, dx \qquad (6.2)$$

and thus the ranges of such samples of size n can be obtained and compared with what one would expect, using a chi-square test. The distribution of (6.2) is tabulated in Pearson and Hartley (1970, 178–183).
(d) A runs test was applied to the runs of signs only of the sequence of numbers.

As with the Kendall and Babbington–Smith (1939a) tables, a note was appended to each set of numbers that failed any test.

Other tests for normality are discussed by Pearson *et al.* (1977) and Wetherill *et al.* (1984, chapter 8). One of these tests, by Shapiro and Wilk (1965), tests for departures from linearity in the appropriate probability plot.

*6.6.2   Multivariate normal variables

If (X_1, X_2) has the bivariate normal density function of Section 2.15, then the derived univariate statistic,

$$D^2 = \frac{1}{(1 - \rho^2)} \left\{ \left(\frac{X_1 - \mu_1}{\sigma_1}\right)^2 - 2\rho \frac{(X_1 - \mu_1)(X_2 - \mu_2)}{\sigma_1 \sigma_2} + \left(\frac{X_2 - \mu_2}{\sigma_2}\right)^2 \right\} \qquad (6.3)$$

has a χ_2^2 distribution (i.e. exponential of mean 2)—see Exercise 6.5. Healy (1968a) proposed using sample values of D^2 and comparing them with the chi-square distribution they would have if (X_1, X_2) is indeed bivariate normal. Once again a graphical examination can be made with the aid of a probability plot, the expected order statistics in a sample of size n from a χ_2^2 distribution being:

$$\frac{2}{n}, \left\{\frac{2}{n} + \frac{2}{(n-1)}\right\}, \left\{\frac{2}{n} + \frac{2}{(n-1)} + \frac{2}{(n-2)}\right\}, \ldots, \left\{\frac{2}{n} + \frac{2}{(n-1)} + \cdots + \frac{2}{1}\right\}$$

(see Cox and Lewis, 1966, p. 27). In practice the parameters of (6.3) have to be estimated from the data, and there is discussion of this in Barnett and Lewis (1978, pp. 212–215 and 226). This approach can also be extended for general multivariate normal distributions. For additional tests see Mardia (1980), Gnanadesikan (1977, p. 161) and Royston (1983).

6.6.3   Exponential and Poisson variables

A random sample from any exponential density may be illustrated by using the order-statistics of the last section after a preliminary scaling (see Exercise 2.2). Tests for exponential random variables were used by Barnett (1965), who, in contrast to Wold, transformed pseudo-random variables, using a multiplicative congruential generator with $m = 2^{27}$, and the transformation of Equation (5.5). Five tests were then applied to the resulting numbers, including an extension of the test of Cox (1955) for detecting the presence of first-order serial correlation in a sequence of exponential variables. Barnett (1965) also generated and tested χ_1^2 variates by squaring $N(0, 1)$ variables derived by the Box–Müller method of Section 4.2.1. In connection with some of his tests, Barnett was confident that only the right-hand tail of the chi-square distribution need be used for the test critical region.

The mean and variance of Poisson random variables are equal, and the *index of dispersion* test makes use of this result to provide a particular test for the Poisson distribution. If $(x_1, \ldots, x_n)$ is a random sample from a Poisson distribution of parameter λ, then

$$\frac{\sum_{i=1}^{n} (x_i - \bar{x})^2}{\bar{x}}$$

is, approximately, a realization of a χ_{n-1}^2 random variable, where

$$\bar{x} = \frac{\sum_{i=1}^{n} x_i}{n}$$

See for example *ABC*, p. 314.

6.7 Discussion

The very first tabulation, by Tippett (1927), of random digits did not include an account of any systematic testing. By contrast, the testing of random variables has now become a standard procedure, and a description of a variety of computerized algorithms which may be used is given in Section A1.2. A suite of test programs such as that of Cugini *et al.* (1980) indicates the kind of compromise that may be reached in the choice of a suitable subset of empirical tests.

The need to match tests of numbers to the intended application for those numbers is graphically illustrated by the insignificant result of the Kolmogorov–Smirnov test of Example 6.6. The RANDU generator that is tested here has very poor properties when one considers successive triples of numbers, as explained by Exercise 3.25, yet the generator does not fail at the 5% level the repeated runs test of Example 6.6.

The RANDU generator failed the extension of the serial test to three dimensions when this test was applied by Dieter and Ahrens (1974, p. A8): each time the test was applied the resulting chi-square test statistics were roughly 100 standard deviations from the expected chi-square mean for the test. However, only one (a poker test) of the many other empirical tests applied indicated that the generator had poor properties. Caution is clearly the key word. The possible problems with pseudo-random numbers are evident, and true random numbers could be biased in unexpected ways. For instance, Kendall and Babbington–Smith (1938) selected digits from the London telephone directory and found appreciably fewer 5's and 9's than one would expect (see Exercise 6.8). They attributed this to the high acoustic confusion between five and nine (airline pilots use 'fife' and 'niner' respectively), and telephone engineers selecting numbers to try to reduce this effect (for related work, see Morgan *et al.*, 1973, and Exercise 9.10).

Neave (1973) showed that when certain pseudo-random variables were transformed by the Box–Müller transformation of Section 4.2.1, the resulting variables displayed unusual characteristics. For instance, observed frequencies in the intervals $(-\infty, -3.3)$ and $(3.6, \infty)$ were zero, compared with expectations (for 10^6 generated values) of 483 and 159 respectively. However, it has been pointed out subsequently (see, e.g., Golder and Settle, 1976) that this effect is mainly due to the $(131, 0; 2^{35})$ generator used, which was considered earlier in Example 6.3 (see also Exercise 6.25). Atkinson (1980) in fact uses the Box–Müller transformation combined with a test of normal random variables as a test of the underlying generator.

6.8 Exercises and complements

6.1 Compare the bounds on the first-order serial correlation given by Equation (3.3) for a mixed congruential generator with empirical first-

order serial correlations obtained for a sequence of length 1000, for the
following generators:

$$(781, 387; 10^3)$$
$$(6941, 2433; 10^4)$$
$$(5^{17}, 0; 2^{42})$$

Note that an additional test of random numbers is provided by
comparing empirical serial correlations with their expectation for a
random sequence (see, e.g., Cugini *et al.*, 1980).

†6.2 Perform the index-of-dispersion test for the Poisson distribution using
the following sample statistics obtained using the PET generator and
the program of Fig. 4.4

λ	n	$\bar{x}$	s^2
5	500	4.960	4.804
2	500	1.856	1.679
1	500	1.060	1.140
1	500	0.974	1.076
1	500	1.018	1.080
0.5	500	0.438	0.463

Here $s^2 = \sum_{i=1}^{n} (x_i - \bar{x})^2 / (n - 1)$.

†6.3 Two dice were thrown 216 times, and the number of sixes at each throw
were:

No. of sixes	0	1	2	Total
Frequency	130	76	10	216

Test the hypothesis that the probability of a six is $p = 1/6$.
Explain how this test would be modified if the hypothesis to be tested
is that the distribution is binomial with the parameter p unknown.
(Based on an Oxford A-level question, 1978).

6.4 Verify the formula for the distribution function of Equation (6.2), for
the range of a random sample of size n from an $N(0, 1)$ distribution.

*6.5 Use the formula of Equation (2.4) to verify that the random variable of
Equation (6.3) has a χ_2^2 distribution.

*6.6 (*The coupon-collector problem applied to the digits 0–9*) The probability that the full set of digits is obtained for the first time at the jth digit of a sequence is given by:

$$\Pr(j) = 10^{1-j} \sum_{v=1}^{10} (-1)^{v+1} \binom{9}{v-1} (10-v)^{j-1} \qquad \text{for } j \geq 10.$$

Two different ways of proving this result are suggested below.

(i) If the number of digits until the first occurrence of a complete set is denoted by S, then (verify) we can write

$$S = 1 + \sum_{i=2}^{10} X_i$$

where X_i has the geometric distribution of Section 2.6, with $p = ((11-i)/10)$, $2 \leq i \leq 10$.

Show that the probability generating function (see Section 2.16) of S is given by:

$$G(z) = \frac{9! \, z^{10}}{\displaystyle\prod_{i=1}^{9} (10 - iz)}$$

which, further, may be written as:

$$G(z) = \frac{9z^{10}}{10^8} \sum_{i=1}^{9} (-1)^{9-i} \frac{i^8}{(10 - iz)} \binom{8}{i-1}$$

Finally, by expanding this expression as a power series in z, verify the distributional form of S given above.

(ii) An alternative approach uses the theory of occupancy problems, in which r balls are thrown at random into n cells, with $r \geq n$. In our case, each digit corresponds to a ball and each *type* of digit $(0, 1, \ldots, \text{etc.})$ is a cell.

If $u(r, n) = \Pr(\text{no cell is empty when } r \text{ balls are thrown at random into } n \text{ cells})$,

then we see that

$$u(r, 10) = \Pr(S \leq r),$$

and so $\Pr(S = r) = u(r, 10) - u(r-1, 10)$.

Use this approach, coupled with the fact that

$$\Pr(\text{no cell is empty}) = 1 - \Pr(\text{at least one is empty})$$

to obtain $\Pr(S = r)$.

NOTE (Feller, 1957, p. 59) that the median of the distribution of S is 27; $\Pr(S > 50) > 0.05$; $\Pr(S > 75) \approx 0.0037$. Note further that $u(r, n)$ may be used to solve the 'birthday problem': $n = 365$,

r = number of people in a room; if e.g., $r = 1900$, Pr(no day is not represented as a birthday) ≈ 0.135.

†6.7 Greenwood (1955) obtained the following results from the coupon-collector test applied to the first 2486 digits in the decimal expansion of $e = 2.71828 \ldots$ and the first 2035 digits in the decimal expansion of $\pi = 3.14159 \ldots$ (reproduced by permission of the American Mathematical Society)

Number of digits to the full collection	π Observed	π Expected	e Observed	e Expected
10–19	13	11.604	12	14.202
20–23	13	11.720	11	14.344
24–27	9	11.491	14	14.064
28–32	5	11.480	15	14.050
33–39	13	10.195	17	12.477
40+	14	10.510	13	12.863
X^2 values:	6.436		2.826	

Verify the expected values given above and discuss the non-significance of the result for e in relation to the failure of the frequency test by these digits (see Metropolis *et al.*, 1950).

6.8 (i) Kendall and Babbington–Smith (1938) obtained the following distribution of 10 000 digits from the London telephone directory:

Digit	0	1	2	3	4	5	6	7	8	9	Total
Frequency	1026	1107	997	966	1075	933	1107	972	964	853	10 000

Verify that the frequency test results in $\chi_9^2 = 58.582$.

(ii) Fisher and Yates (1948) obtained the following distribution of digits obtained from suitably reading tables of logarithms:

Digit	0	1	2	3	4	5	6	7	8	9	Total
Frequency	1493	1441	1461	1552	1494	1454	1613	1491	1482	1519	15 000

Verify that the frequency test results in $\chi_9^2 = 15.63$, and discuss their decision to remove at random 50 of the 6's, and then replace them with other digits, chosen at random. For additional discussion, see Kendall and Babbington–Smith (1939b).

6.9 Comment on the following test statistics resulting from applying the 'runs up' test to sequences of 5000 numbers from the generators indicated:

	Sequence		
Generator	1	2	3
$(131, 0; 2^{35})$	9.89	2.70	18.10
$(65539, 0; 2^{31})$	13.16	5.59	12.70
$(23, 0; 10^8 + 1)$	6.83	14.62	11.90
$(3025, 0; 67\,108\,864)$	10.87	3.53	3.75
PET	5.16	2.26	8.20
The generator of Equation (3.1)	5.03	4.23	6.90

6.10 Verify that for the sequence of numbers from the $(781, 387; 10^3)$ generator there are no runs up of length greater than 4, and discuss this result.

6.11 Apply tests of this chapter to the digits of Tables 3.1 and 3.2, and of Exercise 3.10. For many years the established decimal expansion for π was that of William Shanks, computed over a 20-year period to 707 decimal places. It was noted that 7 appeared only 51 times. In 1945 it was noticed that Shanks made an error on the 528th decimal, and all subsequent decimals are wrong. In the correct series the frequency of 7's is as one would expect (Gardner, 1966, p. 91).

6.12 Consider how you might construct a sequence of numbers which pass the frequency test, but which fail the serial, gap, poker and coupon-collector tests.

6.13 A test which is sometimes used (see, e.g., Cugini *et al.*, 1980) is the 'maximum-of-t' test. Here numbers are taken in disjoint groups of size t, and the largest number is recorded for each group. The resulting maxima are then compared with the expected distribution.

$$\text{If } M = \max(U_1, U_2, \ldots, U_n)$$

where the $\{U_i\}$ are independent, identically distributed $U(0, 1)$ random variables, show that M has the density function

$$f_M(x) = nx^{n-1} \qquad \text{for } 0 \le x \le 1.$$

What is the distribution of M^n?

†**6.14** A further test is the permutation test, in which, again, numbers are taken in groups of size t. Here the ordering of the numbers is recorded, and the empirical distribution of the orderings compared with expectation, which allots a probability of $1/t!$ for independent uniformly distributed numbers. (The possibility of tied values is not considered.) The following results were obtained for the PET generator and the case $t = 4$.

Permutation	Number of cases	Permutation	Number of cases
1	9	13	11
2	17	14	12
3	16	15	10
4	9	16	7
5	6	17	9
6	6	18	13
7	10	19	9
8	12	20	12
9	10	21	7
10	10	22	8
11	8	23	10
12	11	24	8

Assess the significance of these results using a chi-square test.

***6.15** (Stoneham, 1965)   The detail of six of the poker test results presented in Example 6.7 and Fig. 6.3 is given below:

Block	Hands					
	All different	One pair	Two pairs	One triple	One triple and one pair	4 or 5 of the same kind
1	316	506	98	70	5	5
2	307	499	108	80	6	0
3	317	503	90	72	13	5
4	299	511	114	58	12	6
5	299	498	99	84	18	2
6	307	503	111	67	8	4
Theoretical frequencies for random digits	302.4	504	108	72	9	4.6

Verify the theoretical frequencies and the resulting chi-square values:

Block	1	2	3	4	5	6
χ_5^2	3.42	6.62	5.53	4.61	5.32	7.05

6.16 Invent a test of your own for uniform random numbers.

6.17 Investigate and test the random number generators that are available to you. This can be quite revealing. Miller (1977a, 1977b) and Bremner (1981) have revealed errors in Texas hand-calculator multiplicative congruential generators. Furthermore, Bremner has pointed out that the RND function available in the University of Kent implementation of BASIC is (3025, 0; 67 108 864), and not (3125, 0; 67 108 864), as intended, and for which test results were available! (See Pike and Hill, 1965). Nevertheless, the (3025, 0; 67 108 864) generator passes the empirical tests of Cugini *et al.* (1980). (See also Exercise 6.27.)

6.18 (Cooper, 1976) The Box–Müller method involves computing the functions, log, square-root, sin and cos, for each pair of normal random variables generated. If (as in Barnett, 1965) the aim is to simulate χ_1^2 variables, show how the number of functions computed can be reduced.

6.19 Use a chi-square test to compare the p.d.f. and histogram of Fig. 2.5. The frequencies illustrated by the histogram are:

$$2, 4, 8, 18, 19, 12, 14, 14, 5, 2, 2$$

Repat this approach for other appropriate figures from Chapter 2, reading the frequencies from the histograms/bar-charts.

6.20 Wold defined the *P*-value for each test as the two-tail probability of being as, or more, extreme as the resulting value of the test-statistic. Thus, for example, the sum of the first 500 numbers was $S = 159.97$, $\Phi(159.97/\sqrt{5000}) = 0.9882$, and $P = 2(1 - 0.9882) = 0.0237$. In addition to the tests already described, he wrote:

'For each type of test, the distribution of *P*-values obtained from the 50 page sets has been compared with the expected distribution, which is rectangular over the interval (0, 1). On the whole, the agreement with the expected distribution is good. The deviations have been tested by the χ^2 method, grouping the distribution in 10 equal intervals. The *P*-values obtained for the 4 tests are 49.4, 13.7, 29.0 and 91.1 % respectively. The agreement was also tested by the method of Kolmogoroff, mentioned above, a method not involving grouping, with the results $P = 15.5, 42.6, 26.6$ and 98.9 %'.

Discuss his approach and conclusions (cf. Section 6.5).

6.21 The 30 test statistics illustrated in Fig. 6.2 are given below:

0.84	1.82	2.92	3.01	3.06	3.43	3.51	3.84	4.43	4.45
4.74	5.02	5.09	5.61	5.64	5.73	5.77	6.11	6.12	6.33
6.47	6.52	7.09	7.62	10.20	11.84	11.88	16.93	17.32	24.59

Use a chi-square test to assess whether these values come from a χ_6^2 distribution. The Kolmogorov–Smirnov test of Example 6.6, applied to these data, was made with the aid of the NAG FORTRAN routine GO1BCF, which evaluates the right-hand tail areas, assuming these values form a random sample from χ_6^2; and GO8CAF, using the option: null = 1, which performs a Kolmogorov–Smirnov test of whether these tail areas are uniform (cf. Exercise 6.20). Tail areas for chi-square densities are also given in Pearson and Hartley (1972, p. 160). For computational formulae see Kennedy and Gentle (1980, Section 5.7).

6.22 Use the results of Exercise 2.8 to construct particular tests for exponential and Poisson variables. How might you make use of the result of Exercise 4.17?

*6.23 Use the results of Exercise 4.14 to simulate bivariate normal random variables, and test them using the approach of Section 6.6.?

*6.24 In Section 6.6.1 we used the result:
If X has a χ_v^2 distribution, for large v, then
$N = \sqrt{(2X)} - \sqrt{(2v-1)}$ is approximately $N(0, 1)$.
Why is this? (Note that the results of Section 2.12 are exact, while here we are seeking an approximate relationship.)

*6.25 Neave (1973) combined a multiplicative congruential generator with only the sine form of the Box–Müller transformation (see Section 4.2.1), obtaining

$$N = (-2 \log_e U_1)^{1/2} \sin(2\pi U_2)$$

in which $x_2 = ax_1 \pmod{m}$,

and $\qquad U_1 = x_1/m; \; U_2 = x_2/m.$

Show that we can write N in the form:

$$N = (-2 \log_e U)^{1/2} \sin(2\pi \, aU)$$

Chay *et al.* (1975) suggested using the $\{U_i\}$ from the multiplicative congruential generator in the opposite order to that above, resulting in:

$$N = (-2 \log_e U_2)^{1/2} \sin(2\pi U_1)$$

Show that $x_1 = a^* x_2 \pmod{m}$
where $aa^* = 1 \pmod{m}$

so that the 'Chay interchange' is equivalent to changing the multiplier in the generator, and keeping to the original sequence. Kronmal (1964) applied the Box–Müller transformation to pseudo-random numbers from two mixed congruential generators, one for U_1 and another for U_2, and found that the resulting numbers passed a variety of tests.

6.26   Write computer programs to perform the tests considered in this chapter.

†**6.27**   Conduct empirical tests of the $(25\,173, 13\,849; 2^{16})$ generator. T. Hopkins has pointed out that this generator, proposed by Grogono (1980), performs badly on the spectral test. The choice of multiplier here appears to be particularly unfortunate, as literally hundreds of alternative multipliers give rise to a much better result on the spectral test. Consider, for example, $(13\,453, 13\,849; 2^{16})$.

6.28   What will the result be if the frequency test is applied to the entire cycle of a full-period mixed congruential generator? What are the implications of this result?

***6.29**   Given a sequence of pseudo-random numbers, how would you test whether or not a cycle was present?

7

VARIANCE REDUCTION
AND INTEGRAL ESTIMATION

7.1 Introduction

In the preceding chapters we have seen how to generate uniform random variables, and we have considered ways of transforming these to produce other common random variables. Having tested our random variables, we are well prepared for using them in a simulation exercise. However, before pressing on in a bull-at-a-gate fashion it is worth while first of all considering whether the efficiency of the approach to be adopted could be increased. Andrews (1976) writes:

> 'In a recent Monte Carlo study of a regression problem the computing cost was about £250. The cost of generating the required 160 000 Gaussian [normal] deviates was 50p, a negligible amount relative to the total cost. I have found that variance reduction methods often apply. As these affect sample size they affect the remaining £249.50. Modest gains in efficiency result in large savings; very efficient methods can often be found.'

Thus variance reduction is a way of improving value for money, and it can result in much greater savings than those involved in just changing from one algorithm to another for generating variates. As we shall see, there are many different variance-reduction techniques, and a ready way of illustrating these techniques arises in the context of integral estimation using random numbers. The above quotation used the term 'Monte Carlo'; this is now frequently employed as a more evocative synonym for simulation when random variables are employed. 'Monte Carlo' frequently also has an implied connotation of some variance-reduction method having been used. (See, for example, Cox and Smith, 1961, p. 128; Gross and Harris, 1974, p. 383; and Schruben and Margolin, 1978, for related discussion.) It is an item of folklore that this term was introduced as a code-word for secret simulation work in connection with the atomic bomb during the Second World War (Rubinstein, 1981, p. 11).

The basic idea of variance reduction is contained in the following example.

EXAMPLE 7.1   *Buffon's cross*
The Buffon needle experiment has already been described in Exercise 1.2. If a thin needle of length l is thrown at random on to an infinite horizontal table with parallel lines a distance $d \geq l$ apart, then the probability that the needle will cross a line is given by $2l/\pi d$. This probability may be estimated by the proportion of crossings in an experiment consisting of a number of successive throws of a needle, and knowledge of l and d then enables us to estimate π. From the data of Exercise 1.2, we see that π is not very precisely estimated in this way (cf. the precision of Exercise 3.10), even for as many as 960 throws of the needle. Soldiers recovering from wounds sustained during the American Civil War had the time, and apparently also the interest (Hammersley and Handscomb, 1964, p. 7), for multiple repeats of the needle experiment, but present-day experimenters are unlikely to be so patient.

One way to speed the process up is to throw more than one needle each time, and then the picking up of the needles is facilitated if the needles are joined together. In its simplest form, this is accomplished by fusing two needles of equal length at right-angles at their centres, to form a cross. If Z denotes the total number of lines intersected from a single throw of the cross, we can write $Z = X + Y$, where X and Y separately denote the number of crossings of each of the two needles. The distribution of X, and equivalently Y, is unaffected by the presence of the other needle, and so $\mathscr{E}[X] = \mathscr{E}[Y] = 2l/(\pi d)$, and $\mathscr{E}[Z] = 4l/(\pi d)$. The best approach is to take $l = d$ (see Exercise 1.2), and let us, in this case set $\theta = 2/\pi$. X and Y are simple binomial random variables and so (see Table 2.1), $\text{Var}(X) = \text{Var}(Y) = \theta(1 - \theta)$.

It can be shown that the distribution of Z is given by:

$$\Pr(Z = 0) = 1 - \frac{2\sqrt{2}}{\pi}; \; \Pr(Z = 1) = 4(\sqrt{2} - 1)/\pi;$$

$$\Pr(Z = 2) = 4(1 - 1/\sqrt{2})/\pi$$

This enables us to evaluate $\text{Var}(Z) = \text{Var}(X) + \text{Var}(Y) + 2\,\text{Cov}(X, Y)$, yielding, ultimately,

$$\text{Cov}(X, Y) = 2(\pi(2 - \sqrt{2}) - 2)/\pi^2 \approx -0.0324$$

reflecting the fact that the needles are fixed together, and X and Y are not independent: $\text{Corr}(X, Y) = -0.14$.

In the original Buffon experiment, $\text{Var}(\hat{\theta}) \approx 0.2313/n$, where n denotes the number of throws of the needle. In the case of the cross, from one throw, $\hat{\theta} = \frac{1}{2}(X + Y)$, and so, because of the term $\frac{1}{2}$ we immediately have a reduction in $\text{Var}(\hat{\theta})$, for

$$\text{Var}(\hat{\theta}) = \tfrac{1}{4}(\text{Var}(X) + \text{Var}(Y) + 2\,\text{Cov}(X, Y)) = \tfrac{1}{2}\text{Var}(X) + \tfrac{1}{2}\text{Cov}(X, Y)$$

and $\text{Cov}(X, Y) < 0$

Thus fixing the two needles together has a utility over and above the added ease of collecting the needles. For n throws of the cross, $\text{Var}(\hat{\theta}) \approx 0.0995/n$. So we see that using a cross, rather than a single needle, is a variance-reduction technique; it results in greater precision, i.e. an estimator of smaller variance. Against this gain must be offset the labour of correctly fixing the needles to form a cross (though this is more easily done by etching a cross on a clear perspex disc), and the computation of the new theory. These losses occur once only, and would clearly be worth while if a very large experiment were envisaged. There is no reason why further gains should not be obtained from the use of more than two fused needles, and Kendall and Moran (1963, p. 72) provide the result for the case of a star shape; see Hammersley and Morton (1956) for details. In the case of a star, further additional labour (small for a cross) is involved in counting the number of crossed lines. The converse to changing the needle is changing the grid, and Perlman and Wichura (1975) provide the theory for the case of square and triangular grids. Further discussion and elaboration are to be found in Mosteller (1965, pp. 86–88) and Ramaley (1969), as well as Exercises 7.1–7.4.

The extended example above illustrates the basic features of a variance-reduction method, and we shall encounter these features again in the next section. Of course the above example is artificial in that we already know π, which permits a simple evaluation of the variance reduction achieved.

A fundamental aspect of the above example, and others to follow, is the estimation of a parameter θ by an estimator $\hat{\theta}$, with

$$\mathscr{E}[\hat{\theta}] = \theta \qquad \text{and} \qquad \text{Var}(\hat{\theta}) \propto n^{-1}.$$

In both the needle and the cross cases, $\hat{\theta}$ is proportional to a sum of random variables, and hence for large n (and typically n is large in such experiments), central limit theorems apply, so that $\hat{\theta} \approx N(\theta, \kappa/n)$, for appropriate κ. Thus as well as simply producing the estimate $\hat{\theta}$, we can also obtain approximate confidence intervals for θ; for example, a 95% confidence interval is $(\hat{\theta} \pm 1.96\,\kappa^{1/2}/n^{1/2})$, when the normal approximation is valid. The width of this interval is $\propto n^{-1/2}$, so that, for instance, to halve an interval width one has to quadruple the number of observations. Because of this feature it is clearly desirable to employ a variance-reduction technique that results in as small a value for κ as possible.

7.2   Integral estimation

A definite integral, such as $\Phi(x)$, which cannot be explicitly evaluated, can be obtained by a variety of numerical methods. Some of these are described by Conte and de Boor (1972), and algorithms are available for programmable hand-calculators, as well as within computer subroutine libraries such as

NAG. For numerical evaluation of integrals in a small number of dimensions one would therefore be unlikely to use simulation. However, simulation methods can be viable for high-dimensional integration, say in the dimensional range 6–12 (Davis and Rabinowitz, 1975, p. 314). In this section we refer solely to simple one-dimensional integrals, as they provide a convenient vehicle for illustrating some basic methods of variance reduction. It is in any case interesting to see how random numbers may be used to evaluate deterministic integrals. In the following we shall again consider estimation of π, but now through the representation:

$$\frac{\pi}{4} = \int_0^1 \sqrt{(1-x^2)}\,dx \tag{7.1}$$

each side of (7.1) being the area of a quadrant of a circle of radius unity.

7.2.1   Hit-or-miss Monte Carlo

The integral of (7.1) is the area of a quadrant of the circle, radius 1 and centre 0. If that quadrant is enclosed by a unit square, and points thrown independently at random on to the square, then the proportion R/n of n points thrown that land within the quadrant can be used as an estimate of the probability, $\pi/4$, of any point landing within the quadrant; see Fig. 7.1(a). Thus $4R/n$ can be used as an estimate of π. Now R is a random variable with a $B(n, \pi/4)$ distribution, with $\mathrm{Var}(R) = n\frac{\pi}{4}(1-\frac{\pi}{4})$ so that $\mathrm{Var}(4R/n) = 16\,\mathrm{Var}(R)/n^2 = \pi(4-\pi)/n$ $\approx 2.697/n$.

7.2.2   Crude Monte Carlo

As $\dfrac{\pi}{4} = \displaystyle\int_0^1 \sqrt{(1-x^2)}\,1\,dx$, we can write

$\pi/4 = \mathscr{E}\left[\sqrt{(1-U^2)}\right]$, where U is a $U(0, 1)$ random variable, and so if we take a random sample, $U_1, U_2, \ldots, U_n$, we can estimate $\pi/4$ by:

$$I = \sum_{i=1}^n \sqrt{(1-U_i^2)}/n$$

This approach is termed 'crude' Monte Carlo.

Clearly, $\mathrm{Var}(I) = n^{-1}\,\mathrm{Var}\left(\sqrt{(1-U^2)}\right)$

$$= n^{-1}\left(\int_0^1 (1-x^2)\,dx - \left(\int_0^1 \sqrt{(1-x^2)}\,dx\right)^2\right)$$

$$= n^{-1}\left(\frac{2}{3} - \frac{\pi^2}{16}\right)$$

$$\approx 0.498/n$$

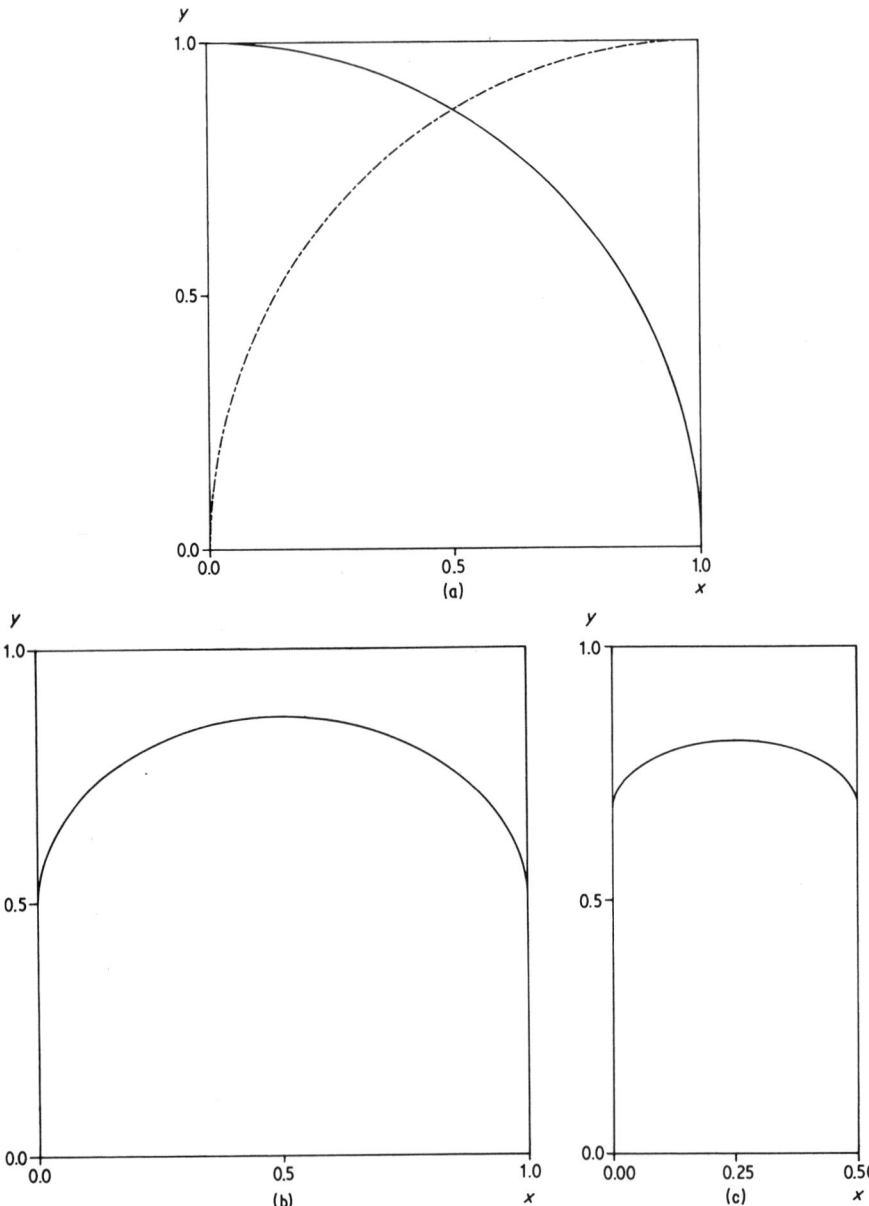

Figure 7.1   A graphical demonstration of variance reduction

(a) ——— $y = \sqrt{(1 - x^2)}$; — · — · $y = \sqrt{(1 - (1 - x)^2)}$

(b) $y = \frac{1}{2}\{\sqrt{(1 - x^2)} + \sqrt{(1 - (1 - x)^2)}\}$

(c) $y = \frac{1}{4}\{\sqrt{(1 - x^2)} + \sqrt{(1 - (1 - x)^2)} + \sqrt{(1 - (\frac{1}{2} - x)^2)} + \sqrt{(1 - (\frac{1}{2} + x)^2)}\}$

The area under the curves of (a) and (b) is $\pi/4$; the area under the curve of (c) is $\pi/8$.

To construct a point at random within a unit square, we only need to take a point with Cartesian co-ordinates (U_1, U_2), where U_1 and U_2 are independent $U(0, 1)$ variables. In order to compare the 'hit-or-miss' and crude Monte Carlo approaches to estimating π we can take $2n$ $U(0, 1)$ variables in each case, resulting in the respective variances of estimators of π:

$$2.697/n \quad \text{and} \quad 0.0498 \times 16/2n = 0.398/n$$

So we see that the variance in the hit-or-miss case is roughly seven times larger than in the crude case, indicating that crude Monte Carlo is much more efficient.

7.2.3 Using an antithetic variate

Let $$H = \tfrac{1}{2}\{\sqrt{(1 - U^2)} + \sqrt{(1 - (1 - U)^2)}\},$$

where U is $U(0, 1)$. $(1 - U)$ is the 'antithetic' variate to U. As both U and $(1 - U)$ are $U(0, 1)$ random variables,

$\mathscr{E}[H] = \pi/4$, but now

$$\begin{aligned}
\text{Var}(H) &= \tfrac{1}{4}\{\text{Var}(\sqrt{(1 - U^2)}) + \text{Var}(\sqrt{(1 - (1 - U)^2)}) \\
&\quad + 2\,\text{Cov}(\sqrt{(1 - U^2)}, \sqrt{(1 - (1 - U)^2)})\} \\
&= \tfrac{1}{2}\{\text{Var}(\sqrt{(1 - U^2)}) + \text{Cov}(\sqrt{(1 - U^2)}, \sqrt{(1 - (1 - U)^2)})\} \\
&= \frac{1}{2}\left(\frac{2}{3} - \frac{\pi^2}{16}\right) + \frac{1}{2}\mathscr{E}\left[\sqrt{((U + 1)U(U - 1)(U - 2))} - \frac{\pi^2}{16}\right]
\end{aligned}$$

It can be shown that

$$\mathscr{E}[\sqrt{((U + 1)U(U - 1)(U - 2))}] = \frac{\pi}{4}\left\{\frac{71}{96} - 6\sum_{k=2}^{\infty} \frac{(2k - 3)!(2k - 1)!(12)^{-2k}}{(k - 2)!(k - 1)!k!(k + 1)!}\right\}$$

$$\approx 0.5806$$

leading to: $$\text{Var}(H) = 0.0052.$$

Thus if $2n$ $U(0, 1)$ variates were used to estimate π using this antithetic approach, the resulting estimator would have variance $0.042/n$. The crude estimator variance is just over nine times larger than this, while the hit-or-miss estimator variance is roughly 64 times larger. In real terms this means that to obtain the same precision using the hit-or-miss and antithetic approaches, we need 64 times as many uniform variates in the hit-or-miss approach. Of course there may be losses in the different types of arithmetic involved between these two different approaches. We commented on this aspect in our comparison of the Buffon needle and cross, and we can now see that the second needle of the cross produced an antithetic variate, resulting in the negative correlation between X and Y in Section 7.1. The idea of using antithetic variates was formally introduced by Hammersley and Morton (1956), who explained the idea through the example of Buffon's needle.

*7.2.4 Reducing variability

The reduction in variability obtained by the use of an antithetic variate as above is simply seen from a comparison of the curves of Fig. 7.1(a) and 7.1(b): the ranges of the y-values are, respectively, 1, $(\sqrt{3}-1)/2$, while the areas under the curves are each $\pi/4$. We can clearly reduce variability even further by using the curve of Fig. 7.1(c), enabling us to estimate π using:

$$H = \{ \sqrt{(1-U^2)} + \sqrt{(1-(1-U)^2)} + \sqrt{(1-(\tfrac{1}{2}-U)^2)} \\ + \sqrt{(1-(\tfrac{1}{2}+U)^2)}\} \tag{7.2}$$

where now U is $U(0, 0.5)$. This process can be continued without end, rather like the testing of random numbers. Again a compromise has to be reached, in this case between variance reduction and increase in computation. For more discussion of this approach, see Morton (1957) and Shreider (1964, p. 53).

The variability in $y = \sqrt{(1-x^2)}$ can be reduced in a number of additional ways. For instance, we can write

$$y = \{1-x^2\} + \{\sqrt{(1-x^2)} - (1-x^2)\} \qquad \text{for } 0 \le x \le 1 \tag{7.3}$$

as suggested in Simulation I (1976, p. 41). In (7.3), of course, both the components of y can be integrated explicitly, but if one knows how to integrate $\{1-x^2\}$ and not $\sqrt{(1-x^2)}$, then the decomposition of (7.3) replaces the variability of $\sqrt{(1-x^2)}$ by the smaller variability of $\{\sqrt{(1-x^2)} - (1-x^2)\}$. A decomposition of y can also be obtained without introducing a new function, simply by splitting up the range of x, and evaluating the integral as the sum of the integrals over the separate parts of the x-range. This is called *stratified sampling*, and is familiar to students of sampling theory (see Barnett, 1974, p. 78). For illustration, suppose the function to be integrated is $y = f(x)$, over the range (0, 1), (see Fig. 7.2).

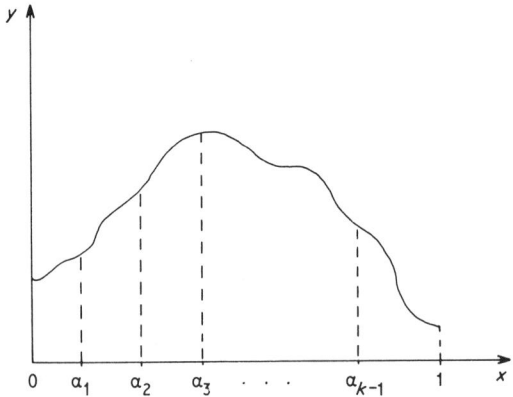

Figure 7.2 Stratified sampling for Monte Carlo integral estimation.

We shall break the range of integration into k pieces, of length $(\alpha_j - \alpha_{j-1})$, for $1 \leq j \leq k$, with $0 = \alpha_0 < \alpha_1 < \ldots < \alpha_k = 1$. Clearly, the variability of y within each piece is less than the variability of y over the full range. Estimation of the sub-integrals may be done in each case by, for example, crude Monte Carlo. If we use n_j $U(0, 1)$ variates for the jth interval, then we can estimate $\theta = \int_0^1 f(x)\, dx$ by:

$$\hat{\theta} = \sum_{j=1}^{k} \sum_{i=1}^{n_j} \frac{(\alpha_j - \alpha_{j-1})}{n_j} f(\alpha_{j-1} + (\alpha_j - \alpha_{j-1})U_{ij})$$

in which the U_{ij} are independent $U(0, 1)$ random variables. Thus

$$Y_{ij} = \alpha_{j-1} + (\alpha_j - \alpha_{j-1})U_{ij} \quad \text{is} \quad U(\alpha_{j-1}, \alpha_j)$$

as required for crude Monte Carlo estimation within the jth interval. The terms $(\alpha_j - \alpha_{j-1})$ are weights which are needed to ensure that $\hat{\theta}$ is unbiased. We see that

$$\mathscr{E}[\hat{\theta}] = \sum_{j=1}^{k} \sum_{i=1}^{n_j} \frac{(\alpha_j - \alpha_{j-1})}{n_j} \int_0^1 f(\alpha_{j-1} + (\alpha_j - \alpha_{j-1})x)\, dx$$

$$= \sum_{j=1}^{k} \left(\frac{\alpha_j - \alpha_{j-1}}{n_j} \right) \int_{\alpha_{j-1}}^{\alpha_j} \frac{f(x)\, dx}{(\alpha_j - \alpha_{j-1})} \left(\sum_{i=1}^{n_j} 1 \right)$$

$$= \sum_{j=1}^{k} \int_{\alpha_{j-1}}^{\alpha_j} f(x)\, dx = \int_0^1 f(x)\, dx = \theta$$

In order to examine the precision of this approach we need the variance of $\hat{\theta}$:

$$\text{Var}(\hat{\theta}) = \sum_{j=1}^{k} \sum_{i=1}^{n_j} \frac{(\alpha_j - \alpha_{j-1})^2}{n_j^2} \text{Var}(f(\alpha_{j-1} + (\alpha_j - \alpha_{j-1})U_{ij}))$$

which leads readily to:

$$\text{Var}(\hat{\theta}) = \sum_{j=1}^{k} \frac{1}{n_j} \left\{ (\alpha_j - \alpha_{j-1}) \int_{\alpha_{j-1}}^{\alpha_j} f^2(x)\, dx - \left(\int_{\alpha_{j-1}}^{\alpha_j} f(x)\, dx \right)^2 \right\} \quad (7.4)$$

In using stratified sampling, one has to choose k, $\{\alpha_j\}$, $\{n_j\}$. Increasing k, the number of pieces, or strata, is likely to increase precision, but results in more arithmetic labour, and as before, a compromise is usually reached. For given k and $\{\alpha_j\}$, one can try to choose the $\{n_j\}$ to minimize the variance of (7.4). We can write

$$\text{Var}(\hat{\theta}) = \sum_{j=1}^{k} \frac{a_j}{n_j}$$

say, and we want to choose the $\{n_j\}$ to minimize $\text{Var}(\hat{\theta})$, subject to a restriction such as $\Sigma_{j=1}^{k} n_j = n$, for fixed n. We can incorporate the constraint by introducing a Lagrange multiplier, λ, and then minimizing the Lagrangian:

$$L = \sum_{j=1}^{k} \frac{a_i}{n_j} - \lambda \left(n - \sum_{j=1}^{k} n_j \right)$$

Stationary values are obtained by setting

$$\frac{\partial L}{\partial n_j} = 0 \qquad \text{for } 1 \leq j \leq k,$$

i.e. $a_j = \lambda n_j^2$, for $1 \leq j \leq k$.

Standard theory (see Exercise 7.6) verifies that this stationary value is indeed a minimum, obtained by selecting the

$$n_j \propto a_j^{1/2} = (\alpha_j - \alpha_{j-1}) \sqrt{\{\text{Var}(f(\alpha_{j-1} + (\alpha_j - \alpha_{j-1})U_{ij})\}} \qquad (7.5)$$

Unfortunately, as we can see from (7.4), a_j involves the very integral we are seeking, and so is unknown. However, the message of (7.5) is clear, suggesting that the larger strata, and strata with more variable function values, should receive relatively more variates, as one would expect. Thus one could, as a rough rule of thumb, choose the $\{\alpha_j\}$ to correspond, as closely as possible, to parts of the curve with a constant range of y-values, and then allot the $\{n_j\}$ in proportion to $\{(\alpha_j - \alpha_{j-1})\}$. Alternatively, one could conduct a preliminary experiment to estimate the unknown variances in (7.5), and then use those estimates in deciding upon the $\{n_j\}$ for a full, subsequent investigation.

*7.2.5 Importance sampling

In stratified sampling, proportions n_j/n of n $U(0, 1)$ variables are transformed to the range (α_{j-1}, α_j). This is not dissimilar from selecting the U_{ij} variates which go to form $\hat{\theta}$ from the composition density function:

$$\Psi(x) = \sum_{j=1}^{k} \left(\frac{n_j}{n} \right) \eta_j(x) \qquad (7.6)$$

in which $\eta_j(x) = \begin{cases} 1/(\alpha_j - \alpha_{j-1}) & \text{for } \alpha_{j-1} \leq x < \alpha_j \\ 0 & \text{otherwise} \end{cases}$ for $1 \leq j \leq k$.

The continuous analogue of (7.6) is found in *importance sampling*, so called because, as with (7.6), $f(x)$ is evaluated at the important parts of the range more frequently than otherwise. Continuing with the above illustration, let us further suppose $f(x) > 0$ for $0 \leq x \leq 1$, and also that $g(x)$ is a probability density function over this range.

$$\theta = \int_0^1 f(x) \, dx = \int_0^1 \frac{f(x)}{g(x)} g(x) \, dx$$
$$= \mathscr{E}[f(X)/g(X)], \text{ when } X \text{ has probability density function } g(x).$$

If $(X_1, \ldots, X_n)$ constitutes a random sample from $g(x)$, then we can estimate θ by:

$$\hat{\theta} = \frac{1}{n} \sum_{i=1}^{n} \frac{f(X_i)}{g(X_i)}$$

which has variance given by

$$\mathrm{Var}(\hat{\theta}) = \frac{1}{n} \left\{ \int_0^1 \frac{f^2(x)}{g(x)} dx - \theta^2 \right\}$$

$$= 0 \text{ if } g(x) = \theta^{-1} f(x) \tag{7.7}$$

which would clearly be a good form to adopt for $g(x)$. Unfortunately, this entails knowledge of the unknown θ before $g(x)$ can be specified. However, if $g(x)$ is of roughly the same shape as $f(x)$ then (7.7) will hold approximately, and so we would expect the $\hat{\theta}$ that results to have small variance.

EXAMPLE 7.2 *An illustration of importance sampling*

Evaluation of $\quad \Phi(x) = \int_{-\infty}^x \frac{e^{-y^2/2}}{\sqrt{(2\pi)}} dy = \int_{-\infty}^x \phi(y) \, dy$

We have seen already in Example 5.4 how this distribution function may be used in simulation, and we know that it is not possible to evaluate the integral explicitly. A density curve of similar shape to $\phi(y)$ is the logistic:

$$f(y) = \frac{\pi \exp(-\pi y/\sqrt{3})}{\sqrt{3}(1 + \exp(-\pi y/\sqrt{3}))^2}$$

with mean 0 and variance 1, already encountered in Section 5.7.

$$\Phi(x) = \int_{-\infty}^x \frac{k\phi(y)}{f(y)} \frac{f(y)}{k} dy$$

where k is chosen so that $f(y)k^{-1}$ is a density function over the range $(-\infty, x)$.

Thus $\quad f(y)k^{-1} = \frac{\pi \exp(-\pi y/\sqrt{3})(1 + \exp(-\pi x/\sqrt{3}))}{\sqrt{3}(1 + \exp(-\pi y/\sqrt{3}))^2} \tag{7.8}$

and if Y is a random variable with the density function of (7.8) then,

$$\Phi(x) = (1 + \exp(-\pi x/\sqrt{3}))^{-1} \mathscr{E}[\phi(Y)/f(Y)]$$

We can therefore estimate $\Phi(x)$ by:

$$\hat{\theta} = \frac{1}{n}(1 + \exp(-\pi x/\sqrt{3}))^{-1} \sum_{i=1}^{n} \phi(Y_i)/f(Y_i)$$

where $\{Y_i, 1 \le i \le n\}$ is a random sample from the density of Equation (7.8), conveniently simulated by means of the inversion method of Section 5.2 as follows. If U is a $U(0, 1)$ random variable,

set $\qquad U = F(Y) = (1 + \exp(-\pi x/\sqrt{3}))/(1 + \exp(-\pi Y/\sqrt{3}))$

then we seek $Y = F^{-1}(U)$, resulting in:

$$Y = -\frac{\sqrt{3}}{\pi} \log_e \{(1 + \exp(-\pi x/\sqrt{3}))U^{-1} - 1\}$$

A BASIC program to evaluate $\Phi(x)$ in this way is given in Fig. 7.3 for a selection of x-values, and results from using this program are shown in Table 7.1.

```
10   RANDOMIZE
20   REM PROGRAM TO CALCULATE PHI(X)
30   REM USING IMPORTANCE SAMPLING
40   INPUT N
50   LET P1 = 1.813799364
60   LET X = -2.5
70   FOR K = 1 TO 4
80     LET X = X+.5
90     LET S = 0
100    FOR I = 1 TO N
110      LET R = (1+EXP(-X*P1))/RND
120      LET Y = -(LOG(R-1))/P1
130      LET P2 = (EXP(-Y*Y/2))/2.506628275
140      LET Q = (P2*(1+EXP(-Y*P1))^2)*EXP(Y*P1)
150      LET S = S+Q
160    NEXT I
170    LET S = S/N
180    LET S = S/(P1*(1+EXP(-X*P1)))
190    PRINT X,S,N
200  NEXT K
210  END
```

Figure 7.3	A BASIC program to evaluate $\Phi(x)$ using importance sampling. Note that $\pi/\sqrt{3} \approx 1.813\,799\,364$, and $\sqrt{(2\pi)} \approx 2.506\,628\,275$.

An application of importance sampling in queueing theory is provided by Evans *et al.* (1965).

Table 7.1

	Estimated $\Phi(x)$			Actual $\Phi(x)$ (from tables) to 4 d.p.
x	$n = 100$	$n = 1000$	$n = 5000$	
-2.0	0.0222	0.0229	0.0229	0.0227
-1.5	0.0652	0.0681	0.0666	0.0668
-1.0	0.1592	0.1599	0.1584	0.1587
-0.5	0.3082	0.3090	0.3081	0.3085

We shall not here consider Var$(\hat{\theta})$ for this example, but see the solution to Exercise 7.7.

7.3 Further variance-reduction ideas

7.3.1 Control variates

With antithetic variates, negative correlation was used to reduce variance. One can also use positive (or negative) correlation with some additional, *control* variate. As with stratified sampling, comparisons can be made here with elements of sampling theory.

Suppose X is being used to estimate a parameter θ, and $\mathscr{E}[X] = \theta$. If Z is a random variable with known expectation μ, then for any positive constant c, we can write

$$Y = X - c(Z - \mu)$$

Thus Y, like X, is an unbiased estimator of θ, as $\mathscr{E}[Y] = \mathscr{E}[X] = \theta$. Whether or not Y is a better estimator of θ than X depends on the relationship between X and Z. Now $\mathrm{Var}(Y) = \mathrm{Var}(X) + c^2\,\mathrm{Var}(Z) - 2c\,\mathrm{Cov}(X, Z)$, and so if $\mathrm{Cov}(X, Z) > c\,\mathrm{Var}(Z)/2$, then $\mathrm{Var}(Y)$ will be less than $\mathrm{Var}(X)$, indicating that Y is the better estimator. The maximum variance reduction is obtained when $c = \mathrm{Cov}(X, Z)/\mathrm{Var}(Z)$, and while $\mathrm{Cov}(X, Z)$ (and possibly also $\mathrm{Var}(Z)$) may not be known, it could be estimated by means of a pilot investigation. However, many investigators have simply taken $c = \pm 1$ as appropriate.

We shall see an example of the use of a control variate in Section 9.4.2. More than one control variate may be used, and a variety of different approaches have been employed to obtain the desired correlation between the variate of interest, X, and the control variate, Z. See, for example, Law and Kelton (1982, Section 11.4). Improvements in the use of control variates are considered by Cheng and Feast (1980). A recent application is provided by Rothery (1982), in the context of estimating the power of a non-parametric test, and an illustration from queueing theory is given in the following example.

EXAMPLE 7.3 (*Barnett, 1965, p. XVII*): *Machine interference*
A mechanic services n machines which break down from time to time. We suppose that machines break down independently of one another, and that for any machine, breakdowns are events in a Poisson process of rate λ. We suppose also that the time taken to repair any machine is a constant, μ. The interference arises if queues of broken-down machines form. This process can be solved analytically, but it was presented by Barnett as an illustration of the use of a control variate. It is clearly simple to simulate the process, and to estimate the 'machine availability', S, over a time period of length t, by,

$$\hat{S} = \frac{\text{total cumulative running time for all machines}}{nt}$$

As the control variate, Barnett used the estimate, $\hat{L}$, of $1/\lambda$, given by

$$\hat{L} = \mu \times \frac{\text{total cumulative running time for all machines}}{\text{total cumulative repair time for all machines}}$$

It was estimated empirically that the correlation between $\hat{S}$ and $\hat{L}$ was $\approx$ $+0.95$ for a variety of values of n, t and the product $\lambda\mu$. Thus S was estimated by

$$\hat{S}_1 = \hat{S} - c\left(\hat{L} - \frac{1}{\lambda}\right)$$

and c was chosen as indicated above for maximum variance reduction, using estimates of second-order moments obtained from a pilot study. It was estimated that $\text{Var}(\hat{S})/\text{Var}(\hat{S}_1) \approx 9.87$. (Further discussion of this example is given in Exercises 7.22 and 7.23.)

The standard approach for estimating $\theta = \mathscr{E}[X]$ is by forming

$$\hat{\theta} = \frac{\sum\limits_{i=1}^{n} X_i}{n}$$

where $\{X_i, 1 \leq i \leq n\}$ forms a random sample from the distribution of X. This is completely analogous to the averaging approaches used in the previous examples in integral estimation, which is to be expected, since here θ is a mean value which, for continuous random variables, can be written as an integral, and vice versa. As was pointed out in Section 7.2, integrals of low dimensionality are probably best evaluated by a numerical method which does not involve simulation. However, while one can certainly think of the estimation of a mean of a random variable in terms of evaluating an integral, in this case the integrand is itself almost certainly going to be a function of the (unknown) mean, and so simulation methods are then appropriate.

When a model is to be simulated under different conditions, and comparisons made between the different simulations, then the variation between the simulations can be reduced by using *common random numbers* in the different simulations. This is a very popular method of variance reduction and, as with many uses of control variates, it relies on an induced positive correlation for its effect. We shall return to the use of common random numbers in Chapter 9. A good example, involving the comparison of alternative queueing mechanisms, is given by Law and Kelton (1982, p. 352).

*7.3.2 Using simple conditioning

The principle here is best illustrated by means of an example.

EXAMPLE 7.4   (*Simon*, 1976)

Suppose we want to estimate the mean value θ of a random variable X, which has the beta $B_e(W, W^2 + 1)$ distribution, where W itself is a random variable, with a Poisson distribution of known mean, η. The obvious approach is to simulate n X-values and simply average them. However, this involves simulation from Poisson and beta distributions, and an alternative approach is as follows.

We know, from Section 2.11, that

$$\mathscr{E}[X|W = w] = w/(w^2 + w + 1)$$

Furthermore, $$\sum_{w=0}^{\infty} \mathscr{E}[X|W = w]\frac{e^{-\eta}\eta^w}{w!} = \theta$$

(here we are using a property of conditional expectation—see Grimmett and Stirzaker, 1982, p. 44).

So we may estimate θ by

$$\hat{\theta} = \frac{1}{n}\sum_{i=1}^{n} w_i/(w_i^2 + w_i + 1)$$

where the $\{w_i, 1 \leq i \leq n\}$ form a random sample from the Poisson distribution, parameter μ; a procedure which does not, in fact, involve simulating X. Thus this approach certainly saves labour. Discussion of how the variance of the above estimator may be further reduced is given in Exercise 7.20. In a different context, Lavenberg and Welch (1979) use conditioning to reduce variance in a particular queueing network, and their example is reproduced by Law and Kelton (1982, p. 364).

7.3.3   The M/M/1 queue

It is interesting to see how variance-reduction techniques that have been clearly expressed for simple procedures, such as the evaluation of one-dimensional integrals, may be employed in more complicated investigations. We shall here consider the M/M/1 queue. This model of a simple queue has already been encountered in Exercise 2.27, which also provided a BASIC program for the simulation of the queue.

We are often interested in the average customer waiting-time in a queue. The waiting time of the nth customer, from arrival at the queue until departure, W_n, may be very simply expressed as:

$$\left.\begin{array}{ll} W_n = (W_{n-1} - I_n + S_n) & \text{if } W_{n-1} \geq I_n \\ W_n = S_n & \text{if } W_{n-1} < I_n \end{array}\right\} \tag{7.9}$$

where S_n is the service-time of customer n, and

I_n is the time between the arrival of the nth and $(n-1)$th customers, for $n \geq 2$.

Note here that we take $W_1 = S_1$, i.e., the first customer arrives to find an empty queue. Figure 7.4 provides a BASIC program for simulating this queue, for which the service and inter-arrival times are both exponential, with respective parameters $\mu = 1$, $\lambda = 0.6$. We see that the average waiting-time is computed for 200 customers. The process is then repeated 100 times so as to provide an estimate of the variance of the average waiting time. The program of Fig. 7.4 provides a much simpler way of estimating average waiting-time than direct use of the program of Exercise 2.27, and we shall return to this point in Chapter 8.

```
10  REM BASIC PROGRAM TO ESTIMATE THE AVERAGE
20  WAITING TIME OF THE FIRST 200 CUSTOMERS
30  REM AT AN M/M/1 QUEUE, STARTING EMPTY, USING (7.9)
40  LET L=.6
50  LET M=1
60  RANDOMIZE
70  LET T1=0
80  LET T2=0
90  FOR J=1 TO 100
100   LET S2=0
110   LET W=0
120   FOR I=1 TO 200
130     LET U=RND
140     LET S=(-LOG(U))/M
150     LET U=RND
160     LET T=(-LOG(U))/L
170     IF W<T THEN 200
180     LET W=W+S-T
190     GOTO 210
200     LET W=S
210     LET S2=S2+W
220   NEXT I
230   LET T1=T1+(S2/200)
240   LET T2=T2+(S2/200)^2
250 NEXT J
260 LET V=(T2-(T1*T1)/100)/99
270 PRINT "VARIANCE OF AVERAGE WAITING TIME = ",V
280 PRINT "MEAN = ",T1/100
290 END
```

Figure 7.4   A BASIC program to estimate the average waiting-time of the first 200 customers at an M/M/1 queue, starting empty. The procedure is repeated 100 times. Note that in lines 140 and 160 the method of Equation (5.5) is used.

This is an example where an antithetic-variate approach could prove useful. Figure 7.5 provides another BASIC program for simulating this queue. In this case we duplicate each block of 200 customers, and in the duplicate block each original U is replaced by $(1 - U)$, with the result that long service times are replaced by short services times, and vice versa, and similarly also for inter-arrival times. Each block average therefore still estimates the same average waiting time, but the two duplicate block averages might now be expected to have a negative correlation. Table 7.2 illustrates the results of the start of a run of the program of Fig. 7.5, and we can see here the anticipated relationship developing between the two sets of W_n values. Proofs that variance reduction will occur when antithetic variates are used in this, and more general, queueing

```
10    REM ILLUSTRATION OF VARIANCE REDUCTION USING
20    REM ANTITHETIC VARIATES IN AN M/M/1 QUEUE
30    DIM R(400)
40    LET L = .6
50    LET M = 1
60    LET I1 = 0
70    RANDOMIZE
80    LET U1 = 0
90    LET U2 = 0
100   LET T1 = 0
110   LET T2 = 0
120   LET N = 50
130   FOR J = 1 TO N
140    FOR I = 1 TO 400
150     LET R(I) = RND
160    NEXT I
170   LET S2 = 0
180   LET W = 0
190   LET K = 0
200   FOR I2 = 1 TO 200
210    LET K = K+1
220    LET U = R(K)
230    LET S = (-LOG(U))/M
240    LET K = K+1
250    LET U = R(K)
260    LET T = (-LOG(U))/L
270    IF W < T THEN 300
280    LET W=W+S-T
290    GOTO 310
300    LET W = S
310    LET S2 = S2+W
320   NEXT I2
330   IF I1 = 1 THEN 360
340   LET S5 = S2/200
350   GOTO 400
360   LET S5 = (S5+S2/200)/2
370   LET T1 = T1+S5
380   LET T2 = T2+S5*S5
390   GOTO 460
400   FOR I = 1 TO 400
410    LET R(I) = 1-R(I)
420   NEXT I
430   REM THIS FORMS THE ANTITHETIC VARIATES
440   LET I1 = 1
450   GOTO 170
460   LET I1 = 0
470   NEXT J
480   LET V = (T2-T1*T1/N)/(N-1)
490   PRINT "VARIANCE OF AVERAGE WAITING TIME=",V
500   PRINT "ESTIMATE OF MEAN WAITING TIME =",T1/N
510   END
```

Figure 7.5   A BASIC program to estimate the average waiting-time of the first 200 customers in an M/M/1 queue, starting empty. The procedure is based upon Equation (7.9) and uses antithetic variates, as explained in the text.

models are provided by Mitchell (1971) and others (see Kleijnen, 1974, p. 190), who also provide empirical investigations, as do Law and Kelton (1982, p. 356).

A variety of results from running the programs of Figs 7.4 and 7.5 are given in Table 7.3. We can see, by considering the results from different runs, that the estimate of efficiency gain can vary appreciably, but in all comparisons there is a gain in efficiency. Use of equations (7.9) does in fact contravene a basic rule for variance reduction, already encountered in (7.3) (see also Exercises 7.8 and

Table 7.2   An illustration of the use of Equation (7.9) to compute waiting times in an M/M/1 queue, and the effect of replacing service (S_n) and inter-arrival times (T_n) by their antithetic counterparts.

	Main block				Antithetic block		
n	S_n	T_n	W_n	n	S_n	T_n	W_n
1	1.35	—	1.35	1	0.30	—	0.30
2	0.20	0.40	1.15	2	1.70	1.34	1.70
3	0.75	1.89	0.75	3	0.64	0.22	2.12
4	0.17	0.36	0.56	4	1.88	1.42	2.58
5	0.43	0.97	0.43	5	1.05	0.60	3.03
6	1.83	0.39	1.87	6	0.18	1.34	1.23

7.20), as we shall now explain. We can write

$$W_n = Q_n + S_n \qquad \text{for } n \geq 1$$

where Q_n is the time spent by the nth customer queueing before being served. Q_n and S_n are independent, and $\text{Var}(Q_n) < \text{Var}(W_n)$. Therefore in order to estimate $\mathscr{E}[W_n]$ it is more efficient to estimate $\mathscr{E}[Q_n]$ and then add on the known $\mathscr{E}[S_n] = 1/\mu$. This can be seen from a comparison of Tables 7.3(a) and (b). This comparison also suggests, however, that the use of (7.9) combined with an antithetic-variate approach can increase efficiency, relative to the use of (7.10) below combined with antithetic variates. Note that

$$Q_n = \max(Q_{n-1} + S_n - I_n, 0) \qquad (7.10)$$

Table 7.3   (a) Sample variance of the estimator of the mean waiting-time of the first 200 customers in an M/M/1 queue, starting empty and with $\mu = 1$. 100 replications were used in each case, with 50 matched pairs when antithetic variates were employed. In this case the waiting-times were simulated including the service-times, i.e. using Equation (7.9)

λ		No variance reduction	Using antithetic variates
	Run		
	1	0.1338	0.0760
0.5	2	0.1426	0.0736
	3	0.2174	0.0561
		0.1646	0.0686
	1	0.3102	0.1338
0.6	2	0.4010	0.1851
	3	0.5233	0.1676
		0.4115	0.1622

(*contd.*)

Table 7.3   (*contd.*)

λ		No variance reduction	Using antithetic variates
	1	2.3254	0.2434
0.7	2	5.4315	0.4734
	3	1.0977	0.3401
		2.9515	0.3523
	1	2.7757	1.2451
0.8	2	6.6679	1.6539
	3	6.7081	2.1598
		5.3839	1.6863

Corresponding average waiting-times of the first 200 customers

λ		No variance reduction	Using antithetic variates	Theoretical value in equilibrium (see Exercise 7.24)
	Run			
	1	1.889	1.998	
0.5	2	2.035	2.020	2.0
	3	2.030	1.984	
		1.985	2.001	
	1	2.513	2.464	
0.6	2	2.467	2.408	2.5
	3	2.533	2.543	
		2.504	2.472	
	1	3.320	3.041	
0.7	2	3.497	3.199	3.33
	3	3.051	3.241	
		3.289	3.160	
	1	4.244	4.392	
0.8	2	4.416	4.404	5.0
	3	5.152	4.451	
		4.604	4.416	

(b) The following results are obtained by simulating the waiting-times without the service-times, i.e. using Equation (7.10). First of all we give the sample variances, as in (a).

λ	Run	No variance reduction	Using antithetic variates
	1	0.1314	0.0927
0.5	2	0.1547	0.0902
	3	0.1225	0.0466
		0.1362	0.0765
	1	0.3228	0.2487
0.6	2	0.6141	0.1521
	3	0.5180	0.1182
		0.4850	0.1730
	1	0.8762	0.3201
0.7	2	1.1808	0.5234
	3	0.9866	0.6251
		1.0145	0.4895
	1	6.4527	1.3440
0.8	2	3.2618	2.0066
	3	5.4324	1.6008
		5.049	1.6703

Corresponding average waiting-times of the first 200 customers in an M/M/1 queue, starting empty and with $\mu = 1$, as above. Values are obtained by computing the average waiting-time, excluding service, and then adding on the known mean service-time.

λ	Run	No variance reduction	Using antithetic variates	Theoretical values in equilibrium (see Exercise 7.24)
	1	2.028	2.019	
0.5	2	2.000	2.034	2.0
	3	1.936	1.989	
		1.988	2.014	
	1	2.354	2.533	
0.6	2	2.550	2.418	2.5
	3	2.559	2.401	
		2.488	2.451	

λ		No variance reduction	Using antithetic variates	Theoretical values in equilibrium (see Exercise 7.24)
0.7	1	3.129	3.300	
	2	3.134	3.165	3.33
	3	3.198	3.282	
		3.154	3.249	
0.8	1	4.594	4.444	
	2	4.447	4.445	5.0
	3	4.576	4.531	
		4.539	4.473	

As one might expect, whether we are using Equations (7.9) or (7.10), the amount of variance reduction achieved depends on the relationship between λ and μ: if λ is appreciably smaller than μ, then the queue will frequently be empty, reducing the negative correlation. The values of λ and μ also affect the rate at which a steady-state system is reached (for the case $\lambda < \mu$ – see Exercise 7.24). Barnett's (1965) tables of exponential random variables provide values of $-\log_e(1 - U)$ as well as $-\log_e U$, with just such antithetic investigation in mind (see Example 7.22).

An alternative approach to antithetic variance reduction in simple queues was applied by Page (1965), who used the following idea. Suppose we are simulating an $M/M/1$ queue, constructing service and inter-arrival times respectively from:

$$S = -\frac{1}{\mu}\log_e(U_1)$$

$$T = -\frac{1}{\lambda}\log_e(U_2)$$

for independent $U(0, 1)$ variables U_1 and U_2.
A duplicate run can be made with

$$\tilde{S} = -\frac{1}{\mu}\log_e(U_2)$$

$$\tilde{T} = -\frac{1}{\lambda}\log_e(U_1)$$

In this case U_1 values giving rise to large service times in the original run will be translated into large waiting times in the duplicate run, and vice versa. Page showed that

$$\text{Corr}\,((S - T),\,(\tilde{S} - \tilde{T})) = -2\rho/(1 + \rho^2) \qquad \text{where } \rho = \lambda/\mu.$$

*7.3.4   A simple random walk example

In random walks we are interested in the distribution of the position of a particle which moves along a line according to probability rules. A simple example results when the particle moves between absorbing barriers at 0 and a, a being a positive integer, according to the rules specified in Fig. 7.6. Exercise 1.4 provided an example of a random walk with a reflecting barrier at 0.

The particle position can be used to describe features of more complicated processes such as the population size of a colony of bacteria; the particular example of Fig. 7.6 is often called the 'gambler's ruin' problem, as the particle position can be taken as the capital of one of two gamblers, with combined capital of a units. In the game played by the gamblers, money changes hands in single units according to the probabilities p and q, and the game ends when one of the gamblers loses all his/her capital, corresponding to the particle reaching one of the barriers. Various features of this walk are of interest, such as the values $\{d_k,\ 1 \le k \le a - 1\}$, where d_k is the average number of steps to termination of this walk, when the walk starts at k.

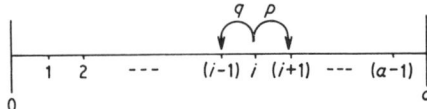

Figure 7.6   Illustration of a simple random walk. When the particle is at i (i an integer in the range $1 \le i \le a - 1$), then, independently of the past, it moves to $(i + 1)$ with probability p, and to $(i - 1)$ with probability $q = (1 - p)$. Once either of the barriers at 0 and a is reached, then the walk terminates.

It can be shown that the $\{d_k\}$ satisfy:

$$\begin{aligned}
d_k &= 1 + p d_{k+1} + q d_{k-1} \qquad \text{for } 1 \le k \le a - 1 \\
d_0 &= d_a = 0
\end{aligned} \tag{7.11}$$

(see, e.g., Bailey, 1964, p. 27, and Exercise 7.30). While these equations have an explicit solution, given in Exercise 7.31, the $\{d_k\}$ can be estimated by simulation (cf. Exercise 3.29). Thus one can select a value of k, and simulate n walks, starting at k, running until absorption. d_k may then be estimated by the sample mean time to absorption. A very simple variance-reduction idea which may be used here is outlined by Barnett (1962a). If a walk starting at k passes through some point j at a later stage, say after r steps, then if it takes n steps for the original walk to end, we also have, from this walk, an example of a walk taking $(n - r)$ steps to absorption, starting from j. Thus a single walk starting from any k can provide information on mean times to absorption for walks starting from points other than k. Clearly by this method the mean-time estimators for different values of k will not be independent. For discussion of this see Barnett (1962a), who used this approach for a two-dimensional random walk without an explicit solution; and Morgan and Robertson (1980), who considered an intractable one-dimensional problem (see Exercise 7.32).

7.4   Discussion

The aim of this chapter has been to underline the importance of variance reduction, and to introduce some of the methods that are used. Further methods and illustrations will be encountered later. Quite apart from its importance, variance reduction is attractive because of the extra information that can often be squeezed out of single random variables, this process frequently requiring a 'flash of insight', in Barnett's (1976) words. Because of the dramatic gain in value that can result, variance-reduction techniques are sometimes also termed 'swindling' (see Simon, 1976, and Schruben and Margolin, 1978, p. 524). Of course, this idea of obtaining as much value as possible from a single random variable has already been encountered, for instance in Exercise 4.2, and Example 5.5. We can also note the analogies between the rejection method of Section 5.3, and importance sampling and hit-or-miss Monte Carlo, as well as between the composition method of Section 5.4, and stratified sampling and the method of extracting an easier integral, in Section 7.2.4.

Much more detail of variance reduction is provided by Law and Kelton (1982, chapter 11), and Kleijnen (1974, chapter 3). Frequently it is necessary to run a pilot study in order to assess the possible value of a variance-reduction technique, for in complicated systems theoretical justification for employing such a technique is usually not possible. Indeed, it is unfortunately the case that in some applications variances have inadvertently been increased from using a 'reduction' approach. Cheng (1982) points out the importance of *high* negative correlation when antithetic variates are used, and suggests a modified procedure which has been applied successfully to a variety of models.

We saw in Section 7.1 that in many applications simulation estimators of parameters are expressed as sums of independent random variables. In these and related cases it is a simple matter to estimate variances of estimators, which are vital for the interpretation of estimates. In Section 7.2.2 the variance of the estimator involved the parameter that was being estimated. In that case, for illustration, the known parameter value was used in calculating the variance, while in practice the variance itself would only be an estimate. Knowledge of the answer (π) has undoubtedly also affected the reporting of experiments involving Buffon's needle, as well as the decision of when to stop. As Mantel (1953) points out, Lazzerini's experiment, conducted in 1901, produced $\hat{\pi} = 3.141\,592\,9$ after 3408 throws, but ending the experiment one throw sooner or later inevitably loses half the decimal place accuracy. When one is estimating waiting-times in queues, one is averaging *dependent* random variables, which complicates variance estimation. A number of possible approaches for such cases are described and compared by Moran (1975), and we shall return to this topic in Chapter 8.

Additional reductions in variance may be obtained by the judicious combination of different methods (see Exercise 7.23 and Schruben and Margolin, 1978), though here again one must proceed with caution, as Kleijnen (1974, Section III.8) has shown. 'Proceed with care' is therefore clearly the watchword for variance reduction, as it was with the use of pseudo-random numbers. However, as with the use of pseudo-random numbers, the benefits from using an appropriate variance-reduction technique can be substantial. Finally, we may note that common random variables and antithetic variates are variance-reduction techniques of *general* applicability, in so far as the same approach is adopted, whatever the problem. In contrast, methods such as importance sampling, stratified sampling, and the use of control variates all have to be individually tailored to particular problems.

7.5   Exercises and complements

(a) On Buffon's needle

7.1   What is the mean number of lines crossed if $l > d$? For discussion of this case, see Mosteller (1965, p. 88) and Mantel (1953).

†7.2   (Gani, 1980)   Suppose the centre of the needle lands at a distance x from a line, and that the needle makes an angle θ with the direction of the lines. Map out the sample space for (x, θ), and by identifying the subset of the sample space corresponding to the needle crossing a line, show that, for $l \le d$, $\Pr(\text{needle crosses a line}) = 2l/\pi d$. When $l > d$, show that this probability must be corrected by the amount

$$\frac{2}{\pi}\cos^{-1}(d/l) - \frac{2l}{\pi d}\sqrt{(1 - d^2/l^2)}.$$

*7.3   (Perlman and Wichura, 1975)   The case of a single needle thrown on to a double grid. Here we have grids, A and B, say, each of parallel lines a distance d apart, the grids being at right angles to each other. This problem was originally studied by Laplace. Let $r = d/l$, and let p_{AB} = $\Pr(\text{needle crosses an A-line and a B-line})$, $p_{A\bar{B}}$ = $\Pr(\text{needle crosses an}$

A-line but not a B-line), etc. Show that

$$p_{AB} = \frac{r^2}{\pi}$$

$$p_{A\bar{B}} = p_{\bar{A}B} = \frac{2r}{\pi} - \frac{r^2}{\pi}$$

$$p_{\bar{A}\bar{B}} = 1 - \frac{4r}{\pi} + \frac{r^2}{\pi}$$

Separate evaluation of these probabilities may be used to estimate π, in different ways, and which one to use is an intriguing question. If in n throws of the needle there are $n_{\bar{A}\bar{B}}$, $n_{\bar{A}B}$, $n_{A\bar{B}}$ and n_{AB} throws in the four possible categories, then (verify)

$$\frac{n_{AB}}{n} \text{ is an estimator of } \frac{r^2}{\pi}$$

$$\frac{n_{A\bar{B}} + n_{\bar{A}B} + 2n_{AB}}{n} \text{ is an estimator of } \frac{4r}{\pi}$$

$$\frac{n - n_{\bar{A}\bar{B}}}{n} \text{ is an estimator of } \frac{4r - r^2}{\pi}$$

If $r = 1$, Perlman and Wichura show that the variances of the resulting estimators are, respectively, $5.63/n$, $1.76/n$, $0.466/n$.

†**7.4**   (Continuation)   The following data were collected from class experiments by E. E. Bassett:

	n	$n_{\bar{A}\bar{B}}$	$n_{\bar{A}B}$	$n_{A\bar{B}}$	n_{AB}
experiment 1	400	16	112	125	147
experiment 2	990	64	315	304	307

Use these data and the estimators of the last exercise to provide a variety of estimates of π.

Further data are provided by Kahan (1961), who describes practical problems such as the blunting of the needle with use, and Gnedenko (1976, pp. 36–39), who also considers the throwing of a convex contour. Historical background is found in Holgate (1981), who conjectures on how Buffon obtained his solution. Mantel (1953) obtains an estimator of π from the estimation of a variance, rather than a mean.

7.5   (Holgate, 1981)   Another problem studied by Buffon was the 'Jeu du franc-carreau': a circular coin of radius b is thrown on to a horizontal square grid of side $2a$. Show that if $b/a = (1 - 2^{-0.5})$ then the coin is as likely as not to land totally within a square.

(b) Integral estimation

*7.6 Show that the stationary value of the Lagrangian

$$L = \sum_{j=1}^{k} \frac{a_j}{n_j} - \lambda \left(n - \sum_{j=1}^{k} n_j \right)$$

given by: $n_j \propto a_j^{1/2}$ is a minimum.

†7.7 When the computer program of Fig. 7.3 is run for values of $x = 0.5, 1.0, 1.5$ and 2.0, the following values result:

	Estimated $\Phi(x)$			$\Phi(x)$
x	$n = 100$	$n = 1000$	$n = 5000$	to 4 d.p.
0.5	0.6999	0.6905	0.6898	0.6915
1.0	0.8477	0.8405	0.8423	0.8413
1.5	0.9253	0.9386	0.9337	0.9332
2.0	0.9867	0.9822	0.9761	0.9773

We obtain better accuracy with the results of Table 7.1. Use the argument of Section 7.2.4 to explain why we might expect this.

7.8 Hammersley and Handscomb (1964, p. 51) define the relative efficiency of two Monte Carlo methods for estimating a parameter θ as follows: The efficiency of method 2 relative to method 1 is:

$$(n_1 \sigma_1^2)/(n_2 \sigma_2^2)$$

where method i takes n_i units of time, and has variance σ_i^2, $i = 1, 2$. Write BASIC programs to estimate the integral:

$$I = \int_0^1 e^{-x^2} dx$$

by hit-or-miss, crude and antithetic variate Monte Carlo methods, and compare the efficiencies of these three methods by using a timing facility. Suggest, and investigate, a simple preliminary variance-reduction procedure. Investigations of variance reduction when $I = \int_0^1 g(x) dx$ are given by Rubinstein (1981, pp. 135–138).

7.9 Write a BASIC program to estimate the integral of Exercise 7.8 using a stratification of four equal pieces, and 40 sample points. How should you distribute the sample points?

7.10   Explain how the use of stratification and antithetic variates may be combined in integral estimation.

*7.11   Verify that $\int_0^1 \sqrt{\{(x+1)x(x-1)(x-2)\}} \, dx \approx 0.5806$.

7.12   Given the two strata, $(0, \sqrt{3}/2)$, $(\sqrt{3}/2, 1)$, for evaluating $\int_0^1 \sqrt{(1-x^2)} dx$, how would you allot the sampling points?

†7.13   In crude Monte Carlo estimation of $\int_0^1 \sqrt{(1-x^2)} dx$, how large must n be in order that a 95 % confidence interval for the resulting estimator of π has width v? Evaluate such an n for $v = 0.01, 0.1, 0.5$.

*7.14   Daley (1974) discusses the computation of integrals of bivariate and trivariate normal density functions. Describe variance-reduction techniques which may be employed in the evaluation of such integrals using simulation. For related discussion, see Simulation I (1976, Section 13.8.3) and Davis and Rabinowitz (1975, Section 5.9).

7.15   Repeat Exercise 7.8, using the pseudo-random number generator of Equation (3.1).

(c)   General variance reduction

7.16   Show that the maximum variance reduction in Section 7.3.1 is obtained when $c = \text{Cov}(X, Z)/\text{Var}(Z)$.

7.17   (Kleijnen, 1974, p. 254)   Suppose one is generating pseudo-random uniform variables from the $(a, 0; m)$ generator, with seed x_0. Show that the corresponding antithetic variates result from using $(m - x_0)$ as the seed.

7.18   Suppose one wants to estimate $\theta = \text{Var}(X)$, and $X = U + V$, when U, V are independent random variables, and $\text{Var}(U)$ is known. Clearly here a simulation should be done to estimate $\text{Var}(V) < \text{Var}(X)$. Use this result to estimate $\text{Var}(M)$, where M is the median of a sample of size n from a $N(0, 1)$ distribution. You may assume (see Simon, 1976) that $\bar{X}$ and $(M - \bar{X})$ are independent, where $\bar{X}$ denotes the sample mean.

7.19   (*Continuation*)   Conduct an experiment to estimate the extent of the variance reduction in Exercise 7.18.

7.20   (Simon, 1976)

(i) Verify that $\left(\dfrac{W}{W^2 + W + 1}\right) = \dfrac{1}{(W+1)} - \dfrac{1}{(W+1)(W^2 + W + 1)}$.

(ii) If W has the Poisson distribution of Example 7.4, show that $E[1/(W+1)] = (1 - e^{-\eta})/\eta$.

(iii) Show that

$$\tilde{\theta} = \left(\frac{1-e^{-\eta}}{\eta}\right) - \frac{1}{n}\sum_{i=1}^{n}\{(W_i+1)(W_i^2+W_i+1)\}^{-1}$$

is an unbiased estimator of θ, and that $\text{Var}(\tilde{\theta}) < \text{Var}(\hat{\theta})$.

(d) Queues

*7.21 Investigate the use of the mean service-time as a control variate for the mean waiting-time in an $M/M/1$ queue.

7.22 For any simulation giving rise to the estimate $\hat{S}$ in the machine-interference study of Example 7.3 we can construct an antithetic run, replacing each $U(0, 1)$ variate U in the first simulation by $(1 - U)$ in the second. If we denote the second estimator of S by $\hat{S}'$, then a further estimator of S is:

$$\hat{S}_2 = \tfrac{1}{2}(\hat{S} + \hat{S}').$$

Barnett (1965) found empirically that the correlation between $\hat{S}$ and $\hat{S}'$ was ≈ -0.64, a high negative value, as one might expect. Estimate the efficiency gained (see Exercise 7.8) from using this antithetic-variate approach (cf. Fig. 7.5).

7.23 (*Continuation*) Barnett (1965) considered the further estimator of S:

$$\hat{S}_3 = \tfrac{1}{2}\{(\hat{S} + \hat{S}') - k(\hat{L} + \hat{L}' - 2/\lambda)\}$$

where $\hat{L}'$ is the estimator of $1/\lambda$ from the antithetic run. Discuss this approach, which combines the uses of control and antithetic variates. Show how k should be chosen to maximize the efficiency gain, and compare the resulting gain in efficiency with that obtained from using control variates and antithetic variates separately.

7.24 In an $M/M/1$ queue, when $\lambda < \mu$, then after a period since the start of the queue, the queue is said to be 'in equilibrium', or to have reached the 'steady state'. The distribution of this period depends on λ, μ and the initial queue size. In equilibrium the queue size Q has the geometric distribution

$$\Pr(Q = k) = \rho(1-\rho)^k \qquad \text{for } k \geq 0$$

where $\rho = \lambda/\mu$, and is called the 'traffic intensity'. Use this result to show that the customer waiting-time in equilibrium (including service time) has the exponential density: $(\mu - \lambda)e^{(\lambda - \mu)x}$, and hence check the theoretical equilibrium mean values of Table 7.3. Further, comment on the disparities between the values obtained by simulation and the theoretical equilibrium mean values. For related discussion, see Rubinstein (1981, p. 213) and Law and Kelton (1982, p. 283).

Unfortunately, it is the cases when ρ is near 1, for $\rho < 1$, that are often of practical importance, but also the most difficult to investigate using simulation.

7.25   Exponential distributions are, as we have seen, often used to model inter-arrival and service-times in queues. Miss A. Kenward obtained the data given below during a third-year undergraduate project at the University of Kent. Illustrate these data by means of a histogram, and use a chi-square test to assess whether, in this case, the assumption of exponential distributions is satisfactory (cf. Exercise 2.26).

The following data were collected from the sub-post office in Ashford, Kent, between 9.00 a.m. and 1.00 p.m. on a Saturday in December, 1981.

Inter-arrivals

Time in seconds	0–10	10–20	20–30	30–40	40–50	50–60	60–70	70–80	80–90	90–100
No. of arrivals	179	108	79	37	32	21	10	13	8	4

Time in Seconds	100–110	110–120	120–130	130–140	140–150	150–160	160–170
No. of arrivals	5	1	3	2	1	0	2

Service times

Time in minutes	0–0.5	0.5–1	1–1.5	1.5–2	2–2.5	2.5–3	3–3.5	3.5–4
No. of customers	63	32	21	10	7	6	0	2

Time in minutes	4–4.5	4.5–5	5–7	7–7.5	7.5–8
No. of customers	0	1	0	1	1

*7.26   (Gaver and Thompson, 1973, p. 594) Sometimes service in a queue takes a variety of forms, performed sequentially. For example, if one has two types of service: payment for goods (taking time T_1), followed by packing of goods (taking time T_2), then the service time $S = T_1 + T_2$. It is an interesting exercise to estimate $\mathscr{E}[S]$ by simulation, using antithetic variates. In an obvious notation, this would result in:

$$\hat{S} = \frac{1}{2n} \sum_{i=1}^{n} (T_{1i} + T_{2i} + T'_{1i} + T'_{2i}),$$

in which T'_{ji} is an antithetic variate to $T_{ji}, j = 1, 2, 1 \le i \le n$. If T_1 and T_2 are independent, exponential variables, with density e^{-x}, then show that the usual approach of taking, for example, $T'_{1i} = \log_e (1 - U)$, where $T_{1i} = \log_e U$, results in:

$$\text{Var}(\hat{S}) = \frac{1}{n}\left\{1 + \int_0^1 \log_e x \log_e(1-x)\,dx\right\} = \frac{1}{n}\left(2 - \frac{\pi}{6}\right)^2 \approx 0.36/n.$$

Compare this value with that which results from a usual averaging procedure. Of course in this simple example the distribution of S is known to be $\Gamma(2, 1)$ (see Section 2.10). However, we have here the simplest example of a *network*, and for a discussion of more complicated networks see Gaver and Thompson (1973, p. 595), Rubinstein (1981, pp. 151–153) and Kelly (1979).

7.27 Investigate further the findings of Table 7.3 by means of a more extensive simulation. Validate your conclusions by also using the generator of Equation (3.1).

7.28 Ashcroft (1950) provides an explicit solution to the machine-interference problem with constant service-time, while Cox and Smith (1961, pp. 91–109) and Feller (1957, pp. 416–420) provide the theory and extensions for the case of service-times with an exponential distribution. Discuss how you would simulate such a model. Bunday and Mack (1973) consider the complication of a mechanic who patrols the machines in a particular order.

7.29 (Page, 1965) In the simulation of Fig. 7.5, let D_n and D'_n be defined by:

$$D_n = S_n - I_n$$
$$D'_n = S'_n - I'_n$$

Show that Corr $(D, D') = -0.645$ (cf. Exercise 7.26).

(e) Gambler's ruin

7.30 From a consideration of the first step taken by the particle in the gambler's ruin problem of Section 7.3.4, verify the relationships of Equation (7.11).

*7.31 Show that the solution of Equation (7.11) is given by:

(i) the case $p \neq q$:

$$d_k = \frac{k}{(q-p)} - \frac{a}{(q-p)}\frac{(1-(q/p)^k)}{(1-(q/p)^a)} \qquad 0 \leq k \leq a.$$

(ii) the case $p = q = \frac{1}{2}$:

$$d_k = k(a-k) \qquad 0 \leq k \leq a.$$

*7.32 Write a BASIC program to simulate the gambler's ruin problem of Section 7.3.4, employing the variance-reduction technique of that section, and compare estimated values of $\{d_k\}$ with the theoretical values given in the last exercise.

8

MODEL CONSTRUCTION
AND ANALYSIS

As discussed in Chapter 1, there are many uses of simulation in statistics, and a variety of these will be considered in Chapter 9. In this chapter we shall concentrate on the problems that arise when we use simulation to investigate directly the behaviour of models.

8.1   Basic elements of model simulation

8.1.1   Flow diagrams and book-keeping

In Table 7.2 we used Equation (7.9) to calculate the waiting-time of customers in an M/M/1 queue. An alternative approach is simply to chart the occurrence of the different events in time, and from this representation deduce the desired waiting-times. This is done for the main block simulation of Table 7.2 in Fig. 8.1.

In order to construct the time sequence of Fig. 8.1 from the service and inter-arrival times we have used the approach that is now standard in many simulations of this kind. This involves using a 'master-clock' which records the current time, and which changes not by constant units, as in Exercise 1.4, for example, but by the time until the next event. The time until the next event is obtained by reference to an 'event-list' which provides the possible events which may occur next, and which also designates when they will occur. Thus in the illustration of Fig. 8.1, when the master-clock time is 0.40 the second customer has just arrived, and the event list has two events: the arrival of customer 3 (after a further time of 1.89) and the departure of customer 1 (after a further time of 0.95). This queue operates on a 'first-in-first-out' policy (FIFO), so that at this stage the event list need not contain the departure of customer 2 as customer 1 is still being served. Customer 1 departs before customer 3 arrives and the master-clock time becomes 1.35, and so on.

Customer	Service time	Time to arrival of next customer
1	1.35	0.40
2	0.20	1.89
3	0.75	0.36
4	0.17	0.97
5	0.43	0.39

The sequence of events in time is then as shown below: Ai (Di) denotes the arrival (departure) of customer i, $i \geq 1$:

Type of event	Time from start of simulation	Queue size	Total waiting times
A1	0	1	
A2	0.40	2	
D1	1.35	1	1.35
D2	1.55	0	1.15
A3	2.29	1	
A4	2.65	2	
D3	3.04	1	0.75
D4	3.21	0	0.56
A5	3.62	1	
A6	4.01	2	
D5	4.05	1	0.43
⋮	⋮	⋮	⋮

Figure 8.1   Details of the main block simulation of an M/M/1 queue, from Table 7.2, the first customer arriving at time 0 to find an empty queue and an idle server.

This procedure is usually described by means of a 'flow-diagram', which elaborates on the basic cycle:

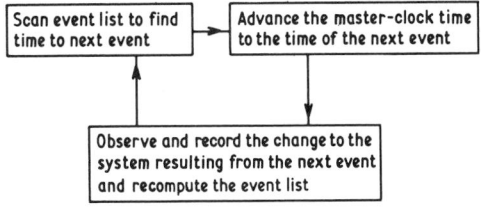

Naturally there have to be starting configurations and stopping conditions, as well as 'book-keeping' procedures for calculating statistics of interest, such as the waiting times in Fig. 8.1. In the simulation of continuous-time Markov processes, an alternative approach may be used which is described in Section 8.3, and which was employed in the program of Exercise 2.27. Another alternative approach is to advance the clock by fixed units of time, each time checking for changes to the system, but this approach is not regarded as generally useful (Law and Kelton, 1982, p. 51).

8.1.2 Validation

In Table 7.3 we were interested in the effect of using antithetic variates on the sample variance of an estimator of mean waiting-time in an $M/M/1$ queue. In addition we computed average waiting-times and compared them with the known theoretical values for an $M/M/1$ queue in equilibrium. Had the comparisons been poor, say out by a factor of 10, then we would have seriously questioned the validity of our model and/or programming, and sought a possible error. This is a simple example of *validation*, which is akin to the work of Chapter 6 on testing random numbers. Thus in validation one is checking that the model is working as intended. Further discussion is given in Naylor and Finger (1967) and Law and Kelton (1982, chapter 10). When a model is simulated via a computer program, then validation is very much like the debugging of any computer program. However, detailed checking of the simulation output of a model against observed data from the real-life system that is being modelled, can provide challenging time-series problems. One approach, adopted by Hsu and Hunter (1977) is to fit standard Box–Jenkins models to each output series, and then test for differences between the parameters of the two fitted models. A more complicated approach examines the similarities between the serial correlations in both series by comparing their spectra (see Naylor, 1971, p. 247).

8.2 Simulation languages

Not only do model simulations have features such as master-clocks and event-lists in common, but any model simulation is likely to require random variables from various distributions, as well as book-keeping routines for calculating and printing summary statistics. Simulation languages have been developed which automatically provide these required facilities for computer simulation. This allows much faster programming than if such facilities have to be programmed laboriously in languages such as BASIC or FORTRAN.

One of the most popular simulation languages is GPSS (which stands for General Purpose Simulation System), described, for example, by O'Donovan (1979). In order to program a simple FIFO queue in this language it suffices to describe the experience of a single customer passing through the queue. Distributions for inter-arrival and service-times can be selected by means of a simple index and the entire simulation can then be effected by a dozen or so instructions. Additionally, such languages can perform certain checks for errors. A defect of GPSS is that it can only simulate discrete random variables, causing one to have to reverse the common procedure of approximating discrete random variables by continuous random variables, if the latter are required for a simulation.

Other simulation languages allow more detail to be provided by the user, as in the FORTRAN-based GASP IV, and the ALGOL-basd SIMULA. Law and Kelton (1982) present a FORTRAN-based language, SIMLIB, that is similar to but simpler than GASP IV, and it is instructional to consider their listings of the FORTRAN routines which the language uses. As they point out, there are interesting problems to be solved when one considers the relative efficiencies of different ways of sorting through event-lists to determine the next event to occur.

The advantage of greater detail is that it is easier to incorporate features such as variance reduction, and also to provide additional analyses of the results. On the other hand, the disadvantage of greater detail is that the programming takes longer, and no doubt GPSS is popular on account of its extreme simplicity. There is also, associated with GPSS instructions, a symbolic representation, which is an aid to the use and description of the language. Tocher (1965) provided a review of simulation languages, while a more up-to-date comparison is to be found in Law and Kelton (1982, chapter 3), who also give programs for simulating the M/M/1 queue in FORTRAN, GASP IV, GPSS, and a further commonly used simulation language, SIMSCRIPT II.5. A further useful reference is Fishman (1978).

8.3 Markov models

8.3.1 Simulating Markov processes in continuous time

The Poisson process of Section 2.7 is one of the simplest Markov, or 'memoryless', processes. To take a process in time as an illustration, the probability rules determining the future development of a Markov process depend only upon the current state of the system, and are independent of past history. Generally speaking, Markov models tend to be easier to analyse than non-Markov models, and for this reason are popular models for real-life processes. In Section 9.1.2 we consider the simulation of a Markov chain, for which the time variable is discrete; here we shall consider Markov processes with a continuous time variable.

Kendall (1950) pointed out that we may simulate one such process, the linear birth-and-death process, in a different way from that of Section 8.1.1, though that approach may also of course be used. In the linear birth-and-death process the variable of interest is a random variable, $N(t)$, which denotes the number of live individuals in a population at time $t \geq 0$, and with $N(0) = n_0$, for some known constant n_0. Independently of all others, each individual divides into two identical 'daughter' individuals with birth-rate $\lambda > 0$, and similarly each individual has death-rate $\mu > 0$. For a full description, see Bailey (1964, p. 91). Thus at time t the event list of Section 8.1.1 contains $N(t)$ events (either births or deaths) as well as their times of occurrence. A simpler way of proceeding is to simulate the time to the next event as a random variable T with the exponential probability density function given by:

$$f_T(x) = (\lambda + \mu) N(t) \exp\left[-(\lambda + \mu) N(t) x\right] \qquad \text{for } x \geq 0$$

and then once the time to the next event has been chosen, select whether a birth or a death has occurred with respective probabilities $\lambda/(\lambda + \mu)$ and $\mu/(\lambda + \mu)$. The justification for this procedure comes in part from realizing that when there are $N(t)$ individuals alive at time t, then for each there is a Poisson process of parameter λ until a birth, and a Poisson process of parameter μ until a death, resulting in a Poisson process of rate $(\lambda + \mu)$ until one of these events, so that overall there is a Poisson process of rate $(\lambda + \mu) N(t)$ until the next event for the entire population. It is this Poisson process which results in the exponential density above. Conditional upon an event occurring, the probabilities of it being a birth or a death, given above, are intuitively correct (for full detail see Kendall, 1950). In this simulation the time to the next event is a single random variable from a single distribution, and it is not read from an event list. In this case there is a different kind of list which contains just two possibilities, i.e. a birth or a death, and now associated with these are relative probabilities of occurrence. If we now return to the BASIC program of Exercise 2.27 we see that this approach is used there for simulating an M/M/1 queue. This approach may also be used for quite general Markov processes in continuous time (see Grimmett and Stirzaker, 1982, pp. 151–152).

One feature of the M/M/1 queue is the different qualitative behaviour between a queue for which $\lambda \geq \mu$, and a queue for which $\lambda < \mu$, where λ and μ in that case refer to the arrival and service rates for the queue, respectively (see Exercise 2.27). A similar result holds for the linear birth-and-death process, simulations of which are given in Fig. 8.2.

In the case of the birth-and-death process, where λ and μ now refer respectively to the birth and death rates, if $\lambda \leq \mu$ then with probability 1 the population ultimately dies out, and then, unlike the case of queues, which start up again with the arrival of the next customer, in the absence of immigration that is the end of that population. However, if $\lambda > \mu$ then the probability of extinction of the population is $(\mu/\lambda)^{n_0}$. Simulations such as that of Fig. 8.2 are useful in demonstrating these two qualitatively different types of behaviour. We also find such differences in more complex Markov models such as models

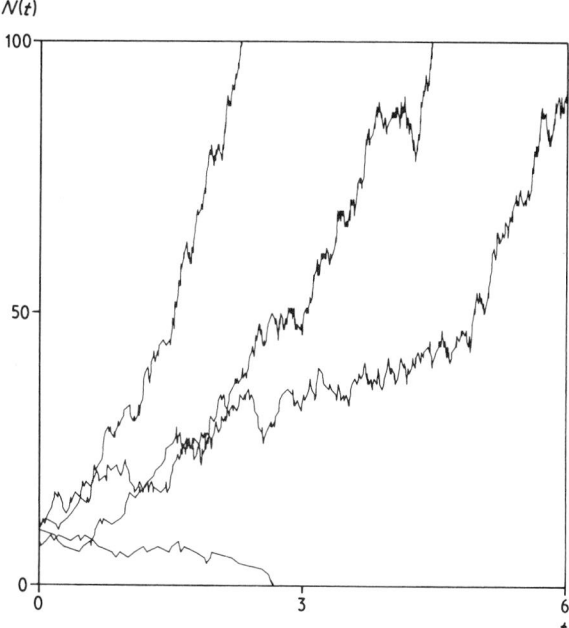

Figure 8.2   Four simulations of the linear birth-and-death process, for $\lambda = 1.5$, $\mu = 1$ and $n_0 = 10$.

for the spread of epidemics in a population (see e.g. Kendall, 1965), and also when we investigate far more complicated models for the development of populations.

The linear birth-and-death process could be used as a model for the development of only the simplest of populations, such as bacteria, though even here the model may be too simple (Kendall, 1949). McArthur *et al.* (1976) simulated a much more complicated model, for the development of a monogamous human population, with death and fertility rates varying with age, with rules for deciding on who was to marry whom, with (or without) incest taboos, and so forth. One set of their results is illustrated in Fig. 8.3. Population F is heading for extinction (in fact all of its 6 members after 135 years were male) while populations A, B and C appear to be developing successfully, and the future for populations D and E is unclear at this stage.

Quite often, in examples of this kind, simulation is used to investigate the qualitative behaviour of a model. As a further example, consider the elaboration of the linear birth-and-death process which occurs when some of the individuals carry an inanimate marker, which may not reproduce. Thus these markers are only affected by the death rate μ, if we suppose that when a cell divides the marker moves without loss into one of the daughter cells.

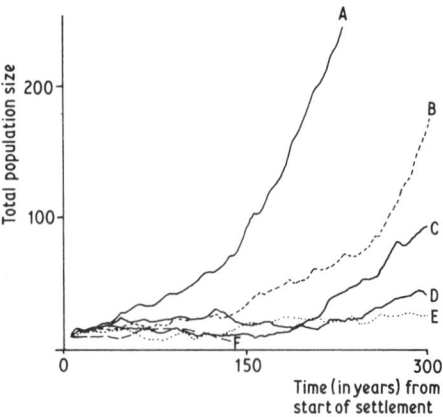

Figure 8.3   Taken from McArthur *et al.* (1976), this figure charts the development of six simulated human populations, under the same conditions of mortality, mating and fertility. Each population started from three married couples, with wives aged 21, 22 and 23 years and respective husbands aged 2 years older.

Meynell (1959) described experiments on cells, in which the marker was either a particular gene, or a bacteriophage. The aim, in his experiments, was to try and use the development of the population of markers to assist in the separate estimation of the cell birth and death rates, λ and μ, respectively. A mathematical analysis of the corresponding birth-and-death process model is given in Morgan (1974). If $N_1(t)$ and $N_0(t)$ denote the numbers of marked and unmarked cells, respectively, at time t, then

$$\mathscr{E}\left[N_1(t)\right] = N_1(0)e^{-\mu t}$$

$$\mathscr{E}\left[N_1(t) + N_0(t)\right] = (N_1(0) + N_0(0))e^{(\lambda - \mu)t}$$

and $$\mathscr{E}\left[N_1(t)\right]/\mathscr{E}\left[N_1(t) + N_0(t)\right] \propto e^{-\lambda t}$$

an observation which prompted Meynell's experiments. It is therefore of interest to see how the ratio $N_1(t)/(N_1(t) + N_0(t))$ develops with time, and this is readily done by simulations such as those of Fig. 8.4.

8.3.2   Complications

Markov processes are named after the Russian probabilist, A. E. Markov, and the 'M' of an M/M/1 queue stands for 'Markov'. As mentioned above, Markov processes may be far too simple even for models of lifetimes of bacteria, as they are also for the lifetimes of higher organisms such as nematodes (see Schuh and Tweedie, 1979; Read and Ashford, 1968). A further complication, that rates may vary with time, is the subject of the paper by Lewis and Shedler (1976),

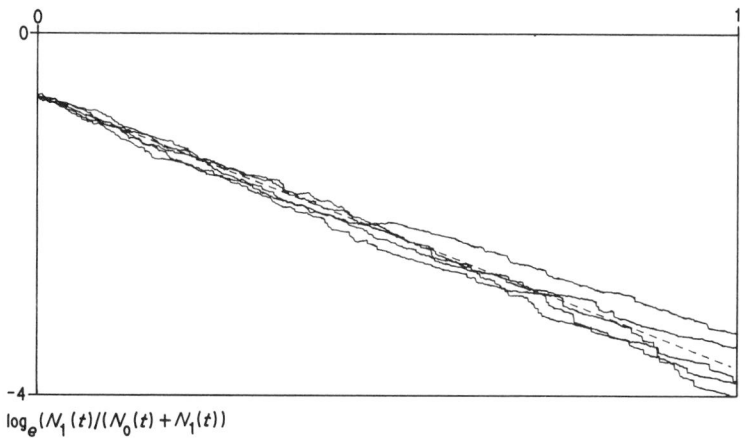

$$\log_e(N_1(t)/(N_0(t)+N_1(t)))$$

Figure 8.4   Simulations of a marked birth-and-death process. $N_1(t)$ denotes the number of marked cells at time t and $N_0(t)$ denotes the number of unmarked cells at time t. $N_1(0) = N_0(0) = 50$, $\lambda = 3$ and $\mu = 1$. It was of interest to compare the simulations with the dashed line, which is a graph of $\log_e (\mathscr{E}[N_1(t)]/\mathscr{E}[N_0(t) + N_1(t)])$ vs. t. Here, and in related simulations, the time t and the parameters λ and μ are measured in the same units.

who discuss the simulation of a non-homogeneous Poisson process; the IMSL library routine GGNPP of Section A1.1 allows one to simulate such a process. McArthur *et al.* (1976) emphasize that an unrealistic feature of their simulation study is that it assumes that the probability rules do not change with time. When lifetimes can be seen as a succession of stages in sequence, then each stage may be described by a Markov process, though the entire lifetime will no longer be Markov. For the case of equal rates of progression through each stage, the resulting lifetime will have a gamma distribution (see Section 2.10). Because service-times in queues can be thought to consist of a number of different stages, the gamma distribution is often used in modelling queues. This is the case in the example which now follows, and which illustrates a further complication in the simulation of queues.

EXAMPLE 8.1   *Simulating a doctor's surgery*

During the 1977 Christmas vacation, Miss P. Carlisle, a third-year mathematics undergraduate at the University of Kent, collected data on consultation and waiting times from the surgery of a medical practice in a small provincial town in Northern Brittany. From eight separate sessions, overall mean consultation and waiting times were, respectively, 21.3 minutes and 43.3 minutes. The surgery did not operate an appointment system, and it was decided to investigate the effect of such a system on waiting times by means of a simulation model. It was assumed that consultation times for different patients were independent, and that inter-arrival times were independent, according to an appointment system with superimposed early/late values simulated from an empirical distribution in Bevan and Draper (1967). Patients were seen in the order in which they arrived and patient non-attendance was taken as 10% (see Gilchrist, 1976; Blanco White and Pike, 1964). For some simulations, consultation times were simulated from an empirical distribution (using a histogram to give the c.d.f.), while in others they were simulated from a fitted $\Gamma(9.13, 0.43)$ distribution, using the algorithm of Cheng (see Exercise 5.22). Simulation allows one to experiment with different length appointment systems, and some results are given in Fig. 8.5.

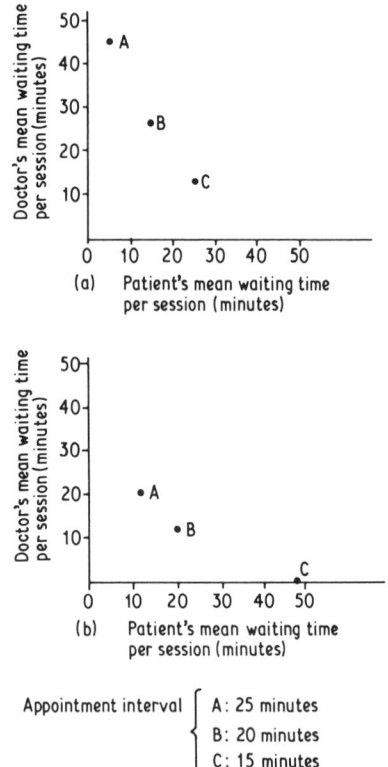

Figure 8.5 (from Carlisle, 1978) Mean waiting times from simulating a doctor's surgery with an appointment system: (a) simulating from the empirical distribution of consultation times, and (b) simulating from a gamma distribution fitted to the consultation times (see Exercise 8.17).

These results are qualitatively as one would expect, but simulation allows one to gauge effects in a quantitative manner. As described by Bevan and Draper (1967), appointment systems can be difficult to establish, and studies of this kind are needed so that the gains can be fully appreciated. This study was only tentative, provided no estimates of error (the subject of the next section) and did not employ any variance-reduction technique. Comparisons between the use of different appointment intervals could be improved by the use of common random numbers (see Section 7.3.1). More detailed studies of this kind are described by Bailey (1952) and Jackson *et al.* (1964), relating to British surgeries, and related studies can be found in Duncan and Curnow (1978), *ABC*, p. 159, and Gilchrist (1976) (cf. also Exercises 1.6 and 1.7).

As can be seen from Nelsen and Williams (1968), the analytic approach to queues with delayed appointment systems is mathematically complex, and so simulation here can play an important rôle. Furthermore, analytic solution of queues is often directed at equilibrium modes, while in the case of the above surgery, over the eight sessions of observation, the number of individuals attending varied from 6 to 13, reflecting a quite different situation from that investigated for the M/M/1 queue in Section 7.3.3, in which average waiting-times were computed for groups of size 100 customers. In practice we shall often want to investigate distinctly non-equilibrium conditions for models, and even for the simple M/M/1 queue mathematical solution is then difficult. However, simple models have a particular utility in preparing us for different possible modes of behaviour, as can be appreciated from a comparison of Figs 8.2 and 8.3.

*8.4   Measurement of error

A frequent end-product of simulation modelling is the point estimate of a population parameter, such as a mean waiting-time, but point estimates are of little use unless they are accompanied by measures of their accuracy. This is a subject that we touched upon in Chapter 7, and here we consider in greater detail the approach to be adopted when, as is frequently the case, the point estimate is a function of a sum of *dependent* random variables.

Let us suppose that we observe a process in equilibrium, such as the M/M/1 queue for $\lambda < \mu$, that we take observations $X_1, \ldots, X_n$ on some random variable, and we want to estimate $E[X_i] = \theta$, $1 \le i \le n$. The usual estimator would be:

$$\bar{X} = \frac{\sum_{i=1}^{n} X_i}{n}$$

Let us now suppose that $\mathrm{Var}(X_i) = \sigma^2$, $1 \le i \le n$, and $\mathrm{Corr}(X_i, X_{i+s}) = \rho_s$, dependent only upon the separation of X_i and X_{i+s}, and not upon i. Then it can be shown (see Exercise 8.8) that

$$\mathrm{Var}(\bar{X}) = \frac{\sigma^2}{n} \left\{ 1 + 2 \sum_{s=1}^{n-1} \left(1 - \frac{s}{n} \right) \rho_s \right\} \tag{8.1}$$

Ignoring the effect of the $\{\rho_s\}$ can result in too small an estimate of $\text{Var}(\bar{X})$; for instance, Conway (1963) reported a factor of 8, found from a particular empirical study. Law and Kelton (1982, p. 146) point out that the three main simulation languages, GPSS, GASP IV and SIMSCRIPT II.5, make the serious mistake of estimating $\text{Var}(\bar{X})$ by σ^2/n alone. In practice one does not know σ^2 or $\{\rho_s\}$ and so they have to be estimated from the data. The simplistic approach of a method of moments estimation of each ρ_s (see Exercise 8.15) is not in general to be recommended, as seen from Law and Kelton (1982, p. 147).

Moran (1975) has compared a variety of approaches. One way to proceed, suggested by Hannan (1957) is to estimate $\text{Var}(\bar{X})$ by:

$$\hat{V}_1 = \frac{n}{(n-k)(n-k+1)} \left\{ \sum_{j=-(k-1)}^{k-1} \left(1 - \frac{|j|}{n}\right)(c_j - \bar{X}^2) \right\}$$

where
$$c_j = (n-j)^{-1} \sum_{s=1}^{n-j} X_s X_{s+j} \quad \text{for } 0 \leq j \leq k-1 \qquad (8.2)$$

and k is chosen large enough so that

$$\rho_j \approx 0 \qquad \text{for } j \geq k$$

As shown by Moran (1975), the bias in $\hat{V}_1$ involves only terms incorporating ρ_j for $j \geq k$. Jowett (1955) has considered the variance of $\hat{V}_1$. We see from (8.2) that computation of $\hat{V}_1$ could involve the calculation of a large number of cross-product terms.

An alternative approach, suggested by Blackman and Tukey (1958, p. 136), and which we have in fact already used in Section 7.3.3, is to divide the sequence of $X_1, \ldots, X_n$ into blocks, and to compute the sample mean of each block. The idea here is that for reasonable size blocks, the block means, which will each estimate the parameter of interest, θ, will be effectively independent and then we can estimate $\text{Var}(\bar{X})$ by:

$$\hat{V}_2 = \left(\sum_{i=1}^{b} \bar{X}_i^2 - b\bar{X}^2 \right) / (b(b-1)) \qquad (8.3)$$

where b denotes the number of blocks, and $\bar{X}_i$ is the block mean of the ith block, $1 \leq i \leq b$. Moran (1975) considers the case $\rho_s = \rho^s$ and demonstrates that in this case it is possible that appreciable bias can result from the use of (8.3). We shall examine this case in the following example.

EXAMPLE 8.2   (*Moran, 1975*)
An investigation of the bias resulting from using block means as a device for estimating the variance of an estimator which is a function of dependent random variables.

Suppose $n = bk$, that $\sigma^2 = 1$,

and $$\rho_s = \rho^s \quad \text{for} \ -1 < \rho < +1$$

for $1 \leq s \leq n$, while for $\hat{V}_2$ we assume additionally that $\rho_s = 0$ for $s > k$.

Here,

$$\text{Var}\,(\bar{X}) = n^{-1}\left(1 + \frac{2\rho}{1-\rho} - \frac{2\rho}{(1-\rho)^2}(1-\rho^n)\right)$$

and

$$\mathscr{E}[\hat{V}_2] = \frac{1}{(b-1)}\left\{\frac{1}{k}\left(1 + \frac{2\rho}{(1-\rho)} - \frac{2\rho}{(1-\rho)^2}(1-\rho^k)\right)\right.$$
$$\left. - \frac{1}{n}\left(1 + \frac{2\rho}{(1-\rho)} - \frac{2\rho}{(1-\rho)^2}(1-\rho^n)\right)\right\}$$

Whence, the bias in $\hat{V}_2$ is

$$\mathscr{E}[\hat{V}_2] - \text{Var}\,(\bar{X}) = \frac{2\rho}{k(b-1)(1-\rho)^2}\left\{\frac{(1-\rho^n)}{n} - \frac{(1-\rho^k)}{k}\right\}$$

Examples are given in Table 8.1.

Table 8.1   Values of the bias in $\hat{V}_2$, given by Equation (8.3)

(a) for $n = 1000$, $k = 50$, $b = 20$

ρ	0.2	0.4	0.5	0.6	0.8	0.95	0.99
Bias	−0.000 012 5	−0.000 044 4	−0.000 08	−0.000 15	−0.000 80	−0.013 97	−0.143 8

(b) for $n = 100$, $k = 5$, $b = 20$

ρ	0.2	0.4	0.5	0.6	0.8	0.95	0.99
Bias	−0.001 25	−0.004 40	−0.007 74	−0.013 8	−0.052 4	−0.282 4	−0.721 6

The values for $\rho = 0.5, 0.95$ and 0.99 were given by Moran (1975); however, we now see that for smaller values of ρ, which might be expected in practice (see Daley, 1968), the bias is far less serious, especially if, as is likely, experiments of size approximating (a), rather than (b), are employed.

In practice a difficulty lies in the choice of b, and further empirical results are given by Law and Kelton (1982, p. 301).

8.5   Experimental design

Gross and Harris (1974, p. 425) describe the simulation of a toll booth, where the question of interest was whether or not to change a manned booth to an automatic booth, which could only be used by car-drivers with the correct change. From observations on the inter-arrival times of cars, the times for service at a manned booth, and also a reasonable guess at the service time at an automatic booth (including the possibility of drivers sometimes dropping their change) there remained, effectively, just the one parameter to vary, viz. the ratio of manned : automatic booths; even then the constraints of the problem meant that only a small number of possibilities could be tried.

We have seen from Example 8.1 that a similar situation arises when one investigates the optimum time-interval for patient arrival at a waiting-room (such as a doctor's surgery) with an appointment system.

However, more complicated systems certainly arise, in which one is interested in the effect of changing a large number of different factors. In ecology, for instance, quite complex models have been proposed for the behaviour of animals such as mice and sheep, and these models can contain as many as 30 parameters. In such cases one needs to simulate the model a large number of different times, each corresponding to a different combination of parameter values. A simple instance of this is given in the next example, from the field of medicine.

EXAMPLE 8.3

Schruben and Margolin (1978) present the results of a pilot study, carried out to investigate the provision of beds in a special hospital unit which was to be set up in a community hospital for patients entering with heart problems. On entry to the unit, patients were assessed and passed either to an intensive-care ward, or a coronary-care ward. Whichever ward a patient entered, after a period of time the patient was either transferred to an intermediate-care ward, prior to release from the hospital, or released from the hospital, possibly due to death. It was thought that the hospital could cope with numbers of beds in each ward of roughly the following magnitudes:

$$\left.\begin{array}{lr} \text{intensive care} & \text{14 beds} \\ \text{coronary care} & \text{5 beds} \\ \text{intermediate care} & \text{16 beds} \end{array}\right\} \tag{8.4}$$

and presumably these numbers also reflected the hospital experience and skills in dealing with patients with heart problems. Such experience also enabled the system to be modelled as follows:

(a) A proportion 0.2 of patients leave the hospital via intensive care, without spending time in the intermediate care beds.
(b) A proportion 0.55 of patients pass through intensive and intermediate care wards.
(c) A proportion 0.2 of patients pass through coronary and intermediate care wards.
(d) The final proportion, 0.05 of patients leave via the coronary care wards, without spending time in the intermediate care beds.

Stays in all wards were taken to have a log-normal distribution (see Exercise 4.21) with the following parameters (time is measured in days):

	μ	σ
Intensive care	3.4	3.5
Coronary care	3.8	1.6
Intermediate care for intensive care patients	15.0	7.0
Intermediate care for coronary care patients	17.0	3.0

Arrivals at the unit were taken to be Poisson with parameter 3.3. We thus have here a more complicated example of a network of queues than in Exercise 7.26. The problems arise when queues begin to form, for instance of individuals who have completed their stay in coronary care and are due for a period in intermediate care, yet find that no intermediate-care beds are available. The effect of such congestion can be a lack of free beds in the intensive-care and coronary-care wards for new patients, and the response Y, say, of the model was the number of patients per month who were not admitted to the unit because of a lack of suitable beds. Without simulation of the system, no one knew the effect on the response of the distribution of beds between wards.

In the simulation it was therefore decided to consider small variations in the bed distribution of (8.4). The number of beds in each ward was taken as a separate factor, and each factor was taken at one of the two levels: the value of (8.4) ± 1, resulting in 8 different, separate simulations. Each simulation was run for a simulated time of 50 months, after a 10-month initialization period (cf. Exercise 8.16). The results obtained are given in Table 8.2.

Table 8.2

Simulation experiment	Number of intensive care beds	Number of coronary care beds	Number of intermediate care beds	Mean response $\bar{Y}$	Estimated* variance of $\bar{Y}$
1	13	4	15	54.0	1.75
2	13	4	17	47.9	1.59
3	13	6	15	50.2	1.81
4	13	6	17	44.4	1.65
5	15	4	15	55.2	1.35
6	15	4	17	48.9	1.76
7	15	6	15	50.5	1.26
8	15	6	17	44.1	1.62

* These values were obtained by means of a blocking approach, as described in the previous section.

Comparison of the results of experiments 1 and 8 reveals what one might expect, viz., an increase in numbers of beds in all wards reduces the mean response. However, more detailed comparisons are more revealing, and we see that the change in the number of intensive-care beds does not seem to have much effect on the response. The information in Table 8.2 can itself be modelled and suggestions such as this can then be formally tested within the model structure. When this was done by Schruben and Margolin, they concluded with the model:

$$\mathscr{E}[\bar{Y}] = 51.1 - 2.1(X_1 - 5) - 3.1(X_2 - 16) \tag{8.5}$$

where X_1 and X_2 respectively denote the number of beds in the coronary-care and intermediate-care wards. In fact the model of (8.5) arose from a slightly different set of experiments from those of Table 8.2, and we shall explain the difference shortly.

This example was only a pilot study, but although (8.5) only holds for the range of factor levels considered, it provides an indication for further investigation. In order to try and reduce variability, common random numbers (see Section 7.3.1) were used for all of the experiments in Table 8.2. A novel feature of the study by Scruben and Margolin (1978) was the repeat of half of the experiments of Table 8.2, using antithetic random number streams, rather than uniformly common random number streams (see Exercise 8.13 for the results using the antithetic simulations). Both of these variance-reduction techniques were shown to perform well, compared with the approach of using no variance reduction, and simply employing different streams of random numbers for each experiment. (See Schruben and Margolin, 1978, for further discussion and for a comparison between the two different variance-reduction techniques.) Duncan and Curnow (1978) discuss the general problem of bed allocations in hospitals.

Consultant statisticians are accustomed to advising scientists on how to design their experiments. In the context of model simulation statisticians also adopt the rôle of experimenters, and this can unfortunately lead to a neglect of advice that would be given to other experimenters. The experimental design of Example 8.3 was that of a 2^3 factorial design, and the IMSL routine AFACN (see Section A1.4) provides one example of a computerized algorithm for the general analysis of a 2^k factorial design. When k is appreciably larger than 3, the number of separate simulation experiments may become prohibitively large for a full factorial design, and in such cases alternative designs such as fractional replicates may be used. For example, the experiments of Exercise 8.13 form a 2^2 fractional replicate of the experiments in Example 8.3. Standard analysis of standard designs allows one to investigate how the response variable for the model is affected by the factor levels adopted in the design, and the conclusion of the analysis may be described by a simple equation such as that of (8.5), for the ranges of factor levels considered. The exploratory approach of varying factor levels with the intention of trying to *optimize* the response is termed 'response-surface methodology', and is described by Myers (1971); we shall return to this topic in Section 9.1.3.

8.6   Discussion

We have seen from this chapter that model simulations may range from elementary explorations of the qualitative predictions of a model, to elaborate explorations of a realistic model described by a large number of parameters, all of which may be varied. Whatever the simulation, the approach would make use of the material in Chapters 3–7, simulating random variables from a variety of distributions and if possible using variance-reduction techniques.

An interesting feature of the use of common random numbers for variance reduction, and of the use of antithetic variates, is that they preclude using rejection methods for the generation of derived random variables unless the uniform random variables used in one simulation run are stored before being used in a comparison run. However, Atkinson and Pearce (1976) show how Forsythe's method, a rejection method described in Exercise 5.41, may be used to produce approximately antithetic variates. Thus, for example, if such variance-reduction methods were likely to be used, and normal random variables were required, one might well prefer to use the Box–Müller method, rather than the Polar Marsaglia method, even though the latter may be faster, as the former does not involve rejection (see Section 4.2). For an explanation of how to use common random numbers in GPSS, see O'Donovan (1979, chapter 8).

The analysis of the results of simulation experiments is a subject in itself, and in Sections 8.4 and 8.5 we have only touched briefly upon certain aspects of this analysis. For further discussion see Hunter and Naylor (1970) and Kleijnen (1975, 1977). Schruben and Margolin (1978) used a simulation language for the simulation described in Example 8.3. Many large-scale model simulations are made with the aid of simulation languages, but as we have seen, these may not always provide the required analyses.

The topic of this chapter has been the simulation of models, and model simulation also occurs in Chapter 9. However, as we shall see, in that chapter the main interest is often in the evaluation of techniques, rather than in the models themselves.

8.7   Exercises and complements

8.1   Write a BASIC program to simulate an M/M/1 queue by the approach of Section 8.1.1, and compare the complexity of this program with that already given in Exercise 2.27. Modify your program to simulate M/D/1 and M/E$_2$/1 queues, in which service-times are, respectively, constant (D stands for 'deterministic') and $\Gamma(2, \mu)$ (E stands for 'Erlangian', a synonym for $\Gamma(n, \mu)$ distributions in which n is a positive integer – named after A. K. Erlang).

8.2 Modify the program of the last exercise to admit 'discouragement', i.e. let the arrival parameter be a decreasing function of n; specifically, let $\lambda = \lambda/(n+1)$. It can be shown theoretically that such a queue with $\lambda = 2$ and $\mu = 1$ has the same equilibrium mean size as the M/M/1 queue with parameters $\lambda = 2$ and $\mu = 3$. Verify this by simulation, and compare the variability in queue size for the two different queues. cf. Exercise 2.28.

†8.3 Repeat the simulation approach of Fig. 8.1, but using the antithetic block data from Table 7.2.

8.4 If the birth-rate λ in the linear birth-and-death process is set to zero, we obtain a linear death process. Show that for this process, if the death-rate $\mu = 1$ then the time to extinction of the population starting with n individuals, is given by the random variables Y and Z of Exercise 2.11.

8.5 Simulate a linear birth-and-death process, and consider how to incorporate the complexity of marked individuals as in the simulation of Fig. 8.4.

8.6 A further bivariate birth-and-death process is presented in Cox and Miller (1965, p. 189). In this model individuals are either male or female, and only females may reproduce. Any female thus has two birth-rates, λ_1 (for female offspring) and λ_2 (for male offspring). Death-rates for males and females are, respectively, μ_1 and μ_2. It can be shown analytically that if $\lambda_1 > \mu_1$, then as time $t \to \infty$, the ratio of the expected number of females to the expected number of males is $(\lambda_1 - \mu_1 + \mu_2)/\lambda_2$. Discuss this model, and how you would investigate this result using simulation.

*8.7 The Galton–Watson branching process is described in Cox and Miller (1965, p. 102). In this process a population of identical individuals is considered in successive generations, and any individual in any generation independently gives rise to a family of i descendants in the next generation with distribution $\{p_i, i \geq 0\}$. It can be shown analytically that if this distribution has probability generating function, $G(z) = \Sigma_{i=0}^{\infty} z^i p_i$, and mean $\mu = \Sigma_{i=0}^{\infty} i p_i$, then the probability of ultimate extinction of the population is given by the smaller root, η, of the equation $x = G(x)$. Furthermore, if $\mu \leq 1$ then $\eta = 1$, but if $\mu > 1$ then $\eta < 1$. From a study of white American males in 1920, Lotka (1931) suggested the following form for $G(x)$:

$$G(x) = (0.482 - 0.041\,x)/(1 - 0.559\,x)$$

In this example the individuals in the population are white males, and the Galton–Watson branching process may be used in a rudimentary fashion to determine the probability of survival of a family surname. Use simulation to investigate the probability of extinction for the population with this $G(x)$.

*8.8   Prove the variance formula of Equation (8.1).

8.9   Use the early/late data of the histogram of Fig. 8.6 to repeat the simulation of a doctor's waiting room, described in Example 8.1, based on the fitted gamma distribution. Use common random numbers to compare the effect of different appointment intervals and discuss your findings (cf. Exercise 1.6).

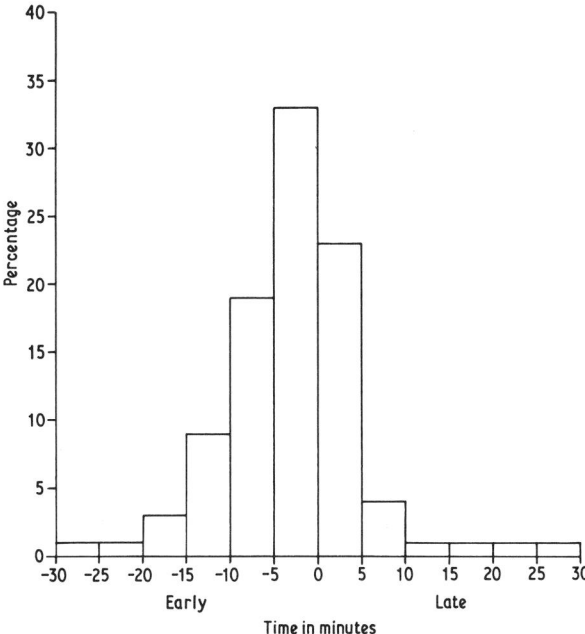

Figure 8.6   Histogram showing the distribution of patients' arrival times, relative to appointment times, based on data collected by Bevan and Draper (1967).

8.10   Devise a rejection method for simulating exponential random variables and then use this method to repeat the antithetic variate investigations of Table 7.3. Comment upon your results.

†**8.11** Produce a flow diagram for the toll-booth simulation mentioned in Section 8.5. Discuss how you might validate results arising from simulating the toll-booth process.

8.12 Give a flow diagram for the simulation reported in Example 8.3. In connection with this example, comment upon the proportion of individuals turned away.

*8.13 The antithetic simulations of Schruben and Margolin (1978) resulted in the following data:

Simulation experiment	Number of intensive care beds	Number of coronary care beds	Number of intermediate care beds	Mean response $\overline{Y}$	Estimated variance of $\overline{Y}$
9	13	4	17	51.3	1.73
10	13	6	15	53.9	1.49
11	15	4	15	58.5	1.83
12	15	6	17	47.8	1.56

Use the analysis of variance to verify the fitted model of Equation (8.5), resulting from these data, together with the data from experiments 1, 4, 6, 7 in Table 8.2.

8.14 (Moran, 1975)   Verify the form given for $\mathrm{Var}(\overline{X})$ in Example 8.2.

*8.15 Use your program from Exercise 8.1 to simulate an M/M/1 queue and estimate the mean delay experienced by a customer when the queue is in equilibrium. Use Equation (8.1) and a method-of-moments estimation of ρ_s to estimate the variance of your estimator of mean delay. Daley (1968) provides the theoretical values of ρ_s which may be used to gauge the performance of Equation (8.1).

8.16 Exercises such as Exercise 8.15 presuppose that one is able to judge when a queue has reached equilibrium. Schruben and Margolin (1978), for example, ran their simulation, described in Example 8.3, for a simulated time of 50 months, but first ran it for an initialization period of 10 months. Consider how you would approach this problem. For further discussion, see Gross and Harris (1974, p. 419).

8.17 Discuss the differences between Figs 8.5 (a) and (b), and comment on the relative merits of simulating from an empirical distribution and a fitted theoretical distribution.

9

FURTHER EXAMPLES
AND APPLICATIONS

Statisticians are likely to employ simulation in a wide variety of situations. Some of these uses of simulation were mentioned briefly in Section 1.5, and in this chapter we shall consider these examples in greater detail, together with additional applications.

9.1 Simulation in teaching

We have already encountered examples of the use of simulation in teaching, in the comparison of the histograms of such as those of Figs 2.8–2.11, in the pebble-sampling experiment of Exercise 3.5, and in the Markov process simulations of Figs 8.2 and 8.4. Further illustrations are considered here.

9.1.1 Simulation in MINITAB

The MINITAB computer package is widely used for a range of elementary statistical analyses, described by Ryan, Joiner and Ryan (1976). This in itself gives rise to an invaluable teaching aid. However, a bonus provided by MINITAB is that it incorporates within the package a number of simulation facilities, described in chapter 3 of Ryan *et al.* (1976). Thus, for example, the student is encouraged to simulate convolutions such as that of Exercise 4.8, in addition to using imagination and mathematical analysis (Ryan *et al.*, 1976, p. 80). In MINITAB it is possible to simulate directly from the $U(0, 1)$, $N(\mu, \sigma^2)$, Poisson and binomial distributions, as well as simulate uniform random integers and any finite-range discrete distribution. Instructions for simulating from exponential and Cauchy distributions are given in Fig. 9.1. Note that in MINITAB it is possible to operate directly on column vectors of data, such as C1, and 'LOGE OF C1, PUT IN C2' will take the natural logarithms of each element of the column C1, and enter them in the same order in column vector C2.

```
NOPRINT
URANDOM 50 OBSERVATIONS, PUT IN C1
LOGE OF C1, PUT IN C2
DIVIDE C2 BY -2, PUT IN C3
PRINT C3


NOPRINT
NRANDOM 50 OBSERVATIONS, WITH MU=0.0, SIGMA=1.0, PUT IN C1
NRANDOM 50 OBSERVATIONS, WITH MU=0.0, SIGMA=1.0, PUT IN C2
DIVIDE C1 BY C2, PUT IN C3
PRINT C3
```

Figure 9.1   MINITAB programs for (a) simulating 50 variables with an exponential distribution, parameter $\lambda = 2$, using the method of Equation (5.5); (b) simulating 50 variables with a standard Cauchy distribution, using the method of Exercise 2.15(b). The 'NOPRINT' commands ensure that the individual uniform and normal variates are not printed following their generation.

Such a simulation facility, combined with a simple histogram command, 'HISTO C3', for example, enables a ready investigation of central limit theorems (Section 2.9) and a consideration of how rate of approach to normality of the distribution of a sum of random variables as the number of terms in the sum increases, can depend on the form of the distribution of the component variables comprising the sum. Further examples of this use of MINITAB are given in Exercise 9.1.

MINITAB contains a number of programs for carrying out statistical tests. A graphical teaching aid results when such programs are combined with in-built simulation mechanisms, as can be seen from the MINITAB programs of Example 9.1. The statistical terminology of this example is given in *ABC*, chapters 16 and 17, for example.

EXAMPLE 9.1   *Investigation of (a) confidence-intervals and (b) type I error, using MINITAB.*

```
NOPRINT
  STORE
  NRANDOM 10 OBSERVATIONS, WITH MU=69, SIGMA=3, PUT IN C1
  ZINTERVAL 90 PERCENT CONFIDENCE, SIGMA=3, DATA IN C1
  END
EXECUTE 30 TIMES


NOPRINT
  STORE
  NRANDOM 10 OBSERVATIONS, WITH MU=0.0, SIGMA=1.0, PUT IN C1
  ZTEST MU=0, SIGMA=1, DATA IN C1
  END
EXECUTE 30 TIMES
```

The example of (a) is suggested by Ryan *et al.* (1976, p. 122), the normal distribution adopted being suggested as a possible distribution for male adult human heights (cf. Fig. 5.8). The 'EXECUTE' statement executes the commands between 'STORE' and 'END' the number of times specified after the EXECUTE statement. Thus the program (a) produces 30, 90 % confidence intervals for μ, assuming $\sigma = 3$, each interval being based on a separate random sample of size 10 from an $N(69, 9)$ distribution. The theory predicts that roughly 90 % of these intervals will contain the value $\mu = 69$, and students can now compare expectation with reality.

Program (b) proceeds in a similar way, producing 30 two-sided significance tests of the null hypothesis that $\mu = 0$, each test at the 5 % significance level. When, as here, the null hypothesis is true, then the null hypothesis will be rejected (type I error) roughly 5 % of the time, and again it is possible to verify this using the simulations. Other computer packages, such as S, mentioned in Section A1.4, and BMDP also contain simulation facilities.

9.1.2 Simulating a finite Markov chain

Feller (1957, p. 347) and Bailey (1964, pp. 53–56) consider mathematical analyses of a 'brother–sister' mating problem, arising in genetics. In stochastic process terms the model considered is a 6-state Markov chain (see Grimmett and Stirzaker, 1982, chapter 6), with states E_1 and E_5 absorbing, i.e., once entered they are never left, like the random-walk barriers of Section 7.3.4. The transition-matrix is given below and the BASIC simulation program is given in Fig. 9.2.

Probability transition matrix for the 'brother–sister' mating problem

		Following state					
		E_1	E_2	E_3	E_4	E_5	E_6
	E_1	1	0	0	0	0	0
	E_2	0.25	0.5	0.25	0	0	0
Preceding	E_3	0.0625	0.25	0.25	0.25	0.0625	0.125
state	E_4	0	0	0.25	0.5	0.25	0
	E_5	0	0	0	0	1	0
	E_6	0	0	1	0	0	0

If we denote this matrix as $\{p_{ij}, 1 \le i, j \le 6\}$ then

$$p_{ij} = \text{Pr(next state} = j \,|\, \text{last state} = i)$$

Thus each row of the matrix constitutes a probability distribution over the integers 1–6, and as we can see from the following BASIC program, movement to the next state can be determined by, say, the table-look-up method applied to the distribution determined by the current state.

Results from a single run of the above program, which simulates the chain 10

```
10   RANDOMIZE
20   DIM A(6,6),B(6)
30   REM SIMULATION OF A SIMPLE 6-STATE
40   REM MARKOV CHAIN,STARTING IN STATE N
50   REM INPUT TRANSITION MATRIX
60   FOR I = 1 TO 6
70    FOR J = 1 TO 6
80     READ A(I,J)
90    NEXT J
100  NEXT I
110  LET K = 1
120  REM CALCULATE CUMULATIVE SUMS OF PROBABILITIES FOR
130  REM EACH ROW, FOR TABLE-LOOK-UP SIMULATION
140  FOR I = 1 TO 6
150   FOR J = 2 TO 6
160    LET A(I,J) = A(I,J)+A(I,J-1)
170   NEXT J
180  NEXT I
190  FOR K1 = 1 TO 10
200   PRINT
210   LET N = 3
220   PRINT "   1    2    3";
230   PRINT "   4    5    6"
240   FOR I = 1 TO 6
250    LET B(I) = 0
260   NEXT I
270   LET B(N) = 1
280   FOR I = 1 TO 6
290    IF B(I) = 0 THEN 320
300    PRINT "    *";
310    GOTO 340
320    PRINT "     ";
330   NEXT I
340   IF N = 1 THEN 430
350   IF N = 5 THEN 430
360   LET U = RND
370   FOR I = 1 TO 6
380    IF U < A(N,I) THEN 400
390   NEXT I
400   LET N = I
410   PRINT
420   GOTO 240
430  NEXT K1
440  DATA 1,0,0,0,0,0
450  DATA 0.25,0.5,0.25,0,0,0
460  DATA 0.0625,0.25,0.25,0.25,0.0625,0.125
470  DATA 0,0,0.25,0.5,0.25,0
480  DATA 0,0,0,0,1,0
490  DATA 0,0,1,0,0,0
500  END
```

Figure 9.2   BASIC simulation program for 6-state Markov chain.

times, each time starting in state 3, are shown below:

3 → 1

3 → 1

3 → 3 → 2 → 2 → 3 → 4 → 5

3 → 4 → 4 → 4 → 5

3 → 6 → 3 → 2 → 2 → 1

3 → 4 → 4 → 3 → 3 → 4 → 4 → 5

3 → 5

3 → 5

3 → 5

3 → 2 → 2 → 3 → 4 → 5

Thus we can estimate Pr(end in absorbing state 1|start in state 3) by 0.3 in this case. Further discussion of this example is given in Exercises 9.2 and 9.3; the latter also provides the theoretical solution to the probabilities of ultimately ending in state 1 or state 5.

9.1.3 Teaching games

Simulation can play an important rôle in teaching games. One illustration is provided by Farlie and Keen (1967), where the technique being taught is an optimization procedure. A variety of further examples are provided by Mead and Stern (1973), one being a capture–recapture study for the estimation of a population size already known to the simulator, and we shall return to this topic in the next section. In Section 8.5 we mentioned the exploratory side of experimental analysis and design, and this is a feature which is well suited to teaching by means of a game. In the game the teacher establishes a known model for response to factor levels, together with conventions for possible plot and block effects; the aim of the student is to try to optimize the response using experimental designs within special constraints. Mead and Freeman (1973) describe such a game for a fictitious crop yield, which was expressed as a known function of the levels of six nutrients, together with an additive plot effect. Also described are the performances of a number of students, some of whom used rotatable response surface designs. Clearly a similar approach could be adopted for a situation such as that of Example 8.3, replacing Equation (8.5) by a more complicated model outside the region of the parameter space considered in Example 8.3. White *et al.* (1982) make extensive use of simulation examples in the teaching of capture–recapture methodology, one of the topics of the next section.

9.2 Estimation and robustness in ecology

International agreements to limit the exploitation of certain species such as whales are made in the light of estimates of the sizes of mobile animal populations. Naturally, such estimates result from sampling of some kind, as in transect sampling, where an observer traverses an area of habitat and records the individuals seen (see Burnham *et al.*, 1980).

In the case of birds, observations can be taken during migration (see Darby, 1984), or by the censusing of particular areas, as occurs in the Christmas Bird Census of North America, and the Common Bird Census of the United Kingdom (see Mountford, 1982; Upton and Lampitt, 1981). The approach we shall consider here is one of repeated sampling of an area, in which captured individuals are marked in some way prior to release. For full discussion of this topic, see Ralph and Scott (1981), and Morgan and North (1984).

Suppose the population size to be estimated is *n* and one captures and marks

m of these and then releases them. The distribution of marked animals in any subsequent sample is hypergeometric (see, e.g. *ABC*, p. 147) if one assumes that the marking process does not affect recaptures. Typically, n will be very much larger than m, and the hypergeometric distribution can be approximated by a more tractable binomial or Poisson distribution. Suppose that in a further sample of size $\tilde{n}$ the number of marked individuals is M, then a natural estimate of n is

$$\hat{n} = \frac{m\tilde{n}}{M}$$

Attributed to Petersen (1896), this is the maximum-likelihood estimate of n under the various parametric models that may be proposed (see Cormack, 1968). Using a normal approximation to the distribution of M,

$$M \approx N\left(\tilde{n}\frac{m}{n}, \tilde{n}\frac{m}{n}\left(1 - \frac{m}{n}\right)\right)$$

enables us to write down approximate confidence intervals for n if we approximate m/n by $M/\tilde{n}$. For example, if $n = 10\,000$, $m = 500$ and $\tilde{n} = 500$, respective 95% and 90% confidence intervals for n, given $M = 26$, are: (6 997, 15 366) and (7 317, 14 018). We explain below why we have taken $M = 26$ to obtain these intervals. It is disappointing that the width of the confidence intervals is of the same order of magnitude as n itself. This conclusion follows also from simulation of this capture–marking–recapture experiment, conducted by Clements (1978) and illustrated in Fig. 9.3. The value of $M = 26$ chosen for illustration above was the modal value of the appropriate set of simulations.

More precise estimates of n result if the capture–recapture cycle is repeated many times (see Exercise 9.4). The modes of analysis that may then be used are described by Cormack (1973), and Bishop and Sheppard (1973) compared the performance of different approaches by means of a simulation study: the simulated population was sampled either 10 or 20 times (at equal time-intervals), the probability of survival from sample to sample was taken to be 0.5 or 0.9, the population size n was taken as either 200, 1000 or 3000, and the proportion of the population sampled was taken as either 0.05, 0.09, or 0.12. In all there were therefore 36 separate simulations. Simulation is clearly a most useful procedure in such investigations as it enables us to try out an estimation technique in a situation where all the parameters are known.

Models such as these rely on assumptions such as 'equal catchability', i.e. marked individuals returned to the population are assumed to be no more or less likely to be caught when the next sample is taken. Evidence exists that this assumption may well be violated (Lack, 1965, regularly found a particular robin in his trapping nets) and simulation may then be used to investigate the *robustness* of a methodology with respect to departures from assumptions. A simple illustration is provided by the data of Table 9.1, taken from Bishop and Bradley (1972). Here the population size of interest was the number of taxi-

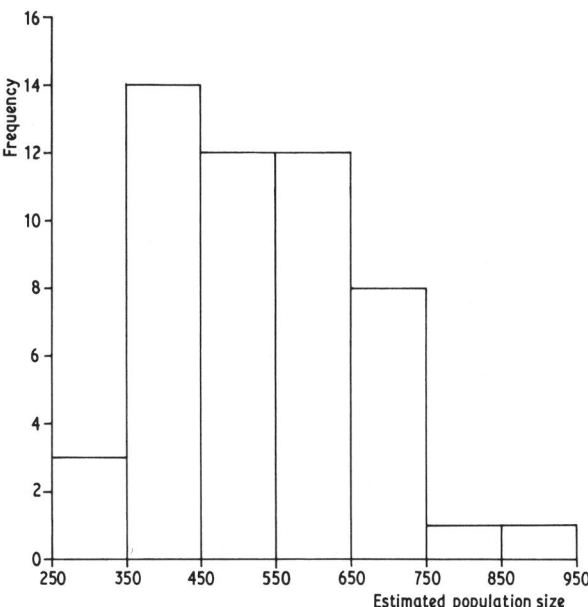

Figure 9.3   (From Clements, 1978) Results of a simulation study of the Petersen estimate $\hat{n} = m\tilde{n}/M$. Here the known value of $n = 500$; $m = 100$, $\tilde{n} = 50$, and 50 simulations were taken.

cabs in Liverpool (known at that time to be 300). Only 62 of these taxis operated regularly from Lime Street Station, and so repeated samples, taken between 14.30 and 15.30 hrs at that station were inappropriately analysed by a model which assumed equal catchability.

Table 9.1   Estimates of the known number (300) of taxi-cabs in Liverpool (source: Bishop and Bradley, 1972), using the capture–recapture model of Jolly (1965). Data were collected from two different sampling points: Lime Street, and Lime Street Station.

Day of sampling	Lime Street	Lime Street Station
Wednesday	208.4	97.1
Thursday	336.3	98.0
Friday	352.8	130.4
Saturday	286.9	178.8
Sunday	213.9	267.2
Monday	230.7	90.5
Tuesday	240.7	59.6
Mean	267.1	131.6

A computer simulation study is reported by Buckland (1982): 13 successive samples were taken on a cohort of 500 animals, each of which was given the probability 0.95 of surviving between samples. It was shown that significantly large biases arose in the estimation of the probability of survival if the simulated animals were given different catchabilities, not accounted for by the model used. An elementary theoretical approach to unequal catchability is to be found in Exercise 4.19, where we can consider the $\Gamma(n, \theta)$ distribution for the Poisson parameter λ as describing unequal catchability. If this model holds then numbers of individuals caught will be described by a negative-binomial distribution, rather than a Poisson distribution in the case of equal catchability. For far more detailed theoretical studies, see Burnham and Overton (1978) and Cormack (1966).

Here we have been considering robustness in the context of particular ecological models. A more standard study of robustness, using simulation, is the work of Andrews *et al.* (1972).

9.3   Simulation in multivariate analysis

9.3.1   Linear discriminant analysis

Discrimination and classification are natural human activities—for instance, we classify individuals as men or women, and we can also devise rules for discriminating between them. In statistics the traditional way of discriminating between two populations is to use the linear discriminant function: $d = \mathbf{x}'\alpha$, in which α is a previously derived vector of coefficients and $\mathbf{x}$ is a vector of measures taken on an individual. If $d > c$, for some additional derived constant c, then the individual is assigned to population 1, but if $d \leq c$, the individual is assigned to population 2. The construction of α and c from $\mathbf{x}$ values on already classified individuals is described by Lachenbruch (1975, p. 9) and other texts on multivariate statistics. Suppose we have vector means $\bar{\mathbf{x}}_1$ and $\bar{\mathbf{x}}_2$ from two samples, one from each population, and let $\mathbf{S}$ denote the pooled sample variance–covariance matrix, estimated from both of the samples. The linear discriminant function is then (with an additional term)

$$d = \{\mathbf{x} - \tfrac{1}{2}(\bar{\mathbf{x}}_1 + \bar{\mathbf{x}}_2)\}' \mathbf{S}^{-1} (\bar{\mathbf{x}}_1 - \bar{\mathbf{x}}_2)$$

If it is equally likely that the individual with measures $\mathbf{x}$ comes from either of the two populations, and if the costs of the two possible incorrect decisions can be taken as equal, then the rule which minimizes the expected cost of misclassification is that the individual is classified into population 1 or 2 according as $d > 0$ or $d \leq 0$, respectively. (For an example of the data that may arise, see Exercise 9.5.) Of course, in classifying further individuals one may incorrectly assign individuals; this occurs for one of the women in Exercise 9.5, if one applies the discrimination rule back to the data that produced it. The likelihood of misclassifying individuals should depend upon the amount of separation between the two populations, and indeed, $\Phi(-\hat{\delta}/2)$ is used as an estimate of the probability of misclassification, where $\hat{\delta}$ is the sample Mahalanobis distance between the two samples, given by:

$$\hat{\delta}^2 = (\bar{\mathbf{x}}_1 - \bar{\mathbf{x}}_2)' \mathbf{S}^{-1} (\bar{\mathbf{x}}_1 - \bar{\mathbf{x}}_2)$$

Unfortunately, the sampling distribution of $\Phi(-\hat{\delta}/2)$ is unknown, and Dunn and Varady (1966) have investigated it using simulation.

Let $\tilde{p}$ denote the true probability of correct classification when the linear discriminant function is used, let the cost of misclassifying an individual be the same, from whichever population the individual comes, and suppose that individuals are equally likely to come from either population. $\tilde{p}$ is estimated by $\Phi(\hat{\delta}/2)$, and in a simulation $\tilde{p}$ is also calculated explicitly from the known parameters of the simulation. Dunn and Varady (1966) estimated percentiles of the sampling distribution of $\{\tilde{p}/\Phi(\hat{\delta}/2)\}$, using 1000 pairs of random samples, each pair being of n variables of dimension k. Table 9.2 gives their estimates of α such that $\Pr(\tilde{p}/\Phi(\hat{\delta}/2) \leq \alpha) = \theta$, for a range of θ values. The two populations were taken with population means separated by a distance δ, and each population was multivariate normal, with the identity matrix as the variance–covariance matrix. Standard normal random variables were obtained from mixed congruential generators followed by the Box–Müller transformation, an approach employed and tested by Kronmal (1964) (see Exercise 6.25).

Dunn and Varady (1966) also use their results to obtain confidence intervals for $\tilde{p}$, given $\Phi(\hat{\delta}/2)$. Clearly these results are a function of the particular simulation parameters employed, but they allow one to appreciate the influence of values of n and k in a known context. One might query the number of simulations used in this study, and the use of three decimal places in presenting the results of Table 9.2, without any discussion of accuracy of the estimates of the percentiles, and we shall discuss these matters in Section 9.4.2. A device for improving efficiency is suggested in Exercise 9.6.

We shall return to this use of simulation in Section 9.4. For further discussion, and alternative discriminatory techniques, see Lachenbruch and Goldstein (1979) and Titterington *et al.* (1981).

9.3.2 Principal component analysis

The measures taken on individuals are unlikely to be independent, and a useful and commonly used multivariate technique consists of finding the 'principal components', $\mathbf{y} = \mathbf{A}\mathbf{x}$, for which the elements of $\mathbf{y}$ are uncorrelated, and have progressively smaller variances (see, for example, Morrison, 1976, p. 267). A difficult but important question concerns the number of principal components that will suffice to provide a reasonable description of the original data set, and

Table 9.2 (Source: Dunn and Varady, 1966 reproduced with permission from The Biometric Society) An illustration of using simulation to determine the sampling distribution of $\{\tilde{p}/\Phi(\hat{\delta}/2)\}$ in linear discriminant analysis. The table gives estimates, from 1000 simulations in each case, of α, such that $\Pr(\tilde{p} \leq \alpha\Phi(\hat{\delta}/2)) = \theta$, for given values of θ, δ, k and n, explained in the text.

		δ					
		2			6		
n	k	$\theta = 0.05$	$\theta = 0.50$	$\theta = 0.95$	$\theta = 0.05$	$\theta = 0.50$	$\theta = 0.95$
25	2	0.911	0.984	1.080	0.998	0.999	1.006
	10	0.825	0.899	0.972	0.991	0.997	1.000
500	2	0.982	0.999	1.019	0.999	1.000	1.001
	10	0.975	0.995	1.013	0.999	1.000	1.000

Jeffers (1967) provides a 'rule-of-thumb', resulting from experience. Simulation provides a useful way of augmenting experience, and Jolliffe (1972a) derived such rules by analysing simulated data sets of known structure. He suggested retaining those principal components corresponding to eigenvalues greater than ≈ 0.7, when the components follow from finding the eigenvalues and eigenvectors of the sample correlation matrix of the original variables.

Principal components retained may be many fewer in number than the original variables, but each principal component is still a linear combination of the original variables, and a further question is whether one might dispense with some of the original variables without the loss of too much information. Jolliffe (1972a) considered this problem for a number of structured data sets, one of which involved the measurement of variables $\{X_i, 1 \leq i \leq 6\}$ on each individual, where $X_i = N_i$, for $i = 1,\ 2,\ 3,\ X_4 = N_1 + 0.5N_4$, $X_5 = N_2 + 0.7N_5$, $X_6 = N_2 + N_6$, and the N_i, $1 \leq i \leq 6$, were independent standard normal random variables, constructed by means of the generalized rejection method of Butcher (1960) (see Exercises 5.29 and 9.7). Further discussion of this example is given in Exercise 9.8. Applying rejection methods to real data sets, Jolliffe (1972b) found that the pictures produced by principal component analyses did not differ appreciably if a substantial fraction of the original variables were removed in an appropriate manner.

9.3.3 Non-metric multidimensional scaling

A useful feature of a principal component analysis can be an *ordination* of individuals, i.e. a scatter-plot in which proximity of individuals is related to their similarity. Ordination is the aim of techniques such as non-metric multidimensional scaling, described by Mardia *et al.* (1979, p. 413), Gordon (1981, Section 5.3) and Kruskal and Wish (1978), with additional illustrations found in Morgan (1981) and Morgan, Woodhead and Webster (1976). Non-metric multidimensional scaling tries to position points representing individuals in some r-dimensional space so that the more similar individuals are then the closer together are the corresponding points. 'Similarity' here can be gauged from the distance separating individuals in the original n ($> r$) dimensional space resulting if n variables are measured on each individual, but it is often calculated in other ways. The performance of non-metric multidimensional scaling in r dimensions can be judged by a measure called the 'stress', which is usually expressed as a percentage. Large stress values might suggest that a larger value of r is necessary in order to obtain a reasonable representation of the similarities between the individuals. Kruskal (1964) gave guidelines as to what are acceptably small stress values, but his guidelines were not calibrated with respect to the number of individuals concerned, viz. $0\% =$ perfect fit, $2.5\% =$ excellent fit, $5\% =$ good fit, $10\% =$ fair fit, 20% $=$ poor fit. Intuitively, one would expect such a calibration of stress to change with respect to the number of individuals involved, and by the end of the 1960s simulation studies of the distribution of stress were undertaken by Stenson and Knoll (1969), Klahr (1969) and Wagenaar and Padmos (1971). The results of Table 9.3 are taken from Klahr (1969).

Table 9.3 (From Klahr, 1969) Stress values (given as percentages) when non-metric multidimensional scaling is applied to uniform random similarities between n individuals. For $n = 8$, 100 separate runs were used, while for $n = 12$ and $n = 16$, 50 separate runs were used. Each of the $\binom{n}{2}$ similarity values was obtained by selecting an independent $U(0, 1)$ random variable.

Number of individuals		Number of dimensions (r)		
(n)		2	3	4
8	Average	16.0	6.5	1.6
	standard deviation	3.4	2.7	1.8
12	Average	24..0	14.4	8.8
	standard deviation	1.7	1.6	1.6
16	Average	27.9	18.5	13.0
	standard deviation	1.4	1.0	1.1

The pattern of Table 9.3 thus provides a quantification of one's intuition in connection with this technique: 'Good' stress values were often obtained for 8 individuals in 3 dimensions, but for $n > 10$ no 'good' stress values were obtained when the similarities were randomly generated.

Further details are provided in the source papers. While Arabie (1973) points out a possible flaw in the scaling procedure used, the value of the basic approach is quite clear.

9.3.4   Cluster analysis

A logical preliminary to the discrimination discussed in Section 9.3.1 is classification (also called cluster analysis), described by Everitt (1979) as 'probably one of the oldest scientific pursuits undertaken by man'. Even so, it is certainly true that the area of classification can be something of a minefield. This is because there are many different methods of cluster analysis, as well as possibly different algorithms for any particular method. In addition, for any one method a variety of different 'stopping-rules' may be proposed, which suggest the appropriate number of clusters to describe the data (see Everitt, 1980, Section 3.4). Unfortunately these rules may suggest structure when none is present, and alternatively they may miss known structure. For illustration, see Day (1969), and Everitt (1980, p. 74). An obvious way of assessing and comparing different methods is by simulating populations of known structure, and then applying the methods to such populations. The following example is taken from Gardner (1978).

EXAMPLE 9.2
Figure 9.4 illustrates three random samples, each of size 70, one from each of three bivariate normal distributions, with respective means $(-2, 0)$, $(2, 0)$, $(0, 3.46)$ and unit variance–covariance matrix. The population means of these samples therefore lie at the corners of an equilateral triangle with side equal to 4 standard deviations. Also shown on Fig. 9.4 is the result of the three-cluster solution of a cluster analysis using the method of Beale (1969), an algorithm for which is provided by Sparks (1973, 1975). We can see that the cluster analysis has captured most of the known structure in the simulated data, with only 9 of the 210 points misclassified. Calinski and Harabasz (1974) propose a measure, C, for indicating an appropriate number of clusters for describing a set of data, and it is interesting to evaluate the performance of such a measure on simulated data. For Beale's method and the data of Fig. 9.4, we find:

Number of clusters sought (k)	C
6	239
5	254
4	279
3	341
2	133

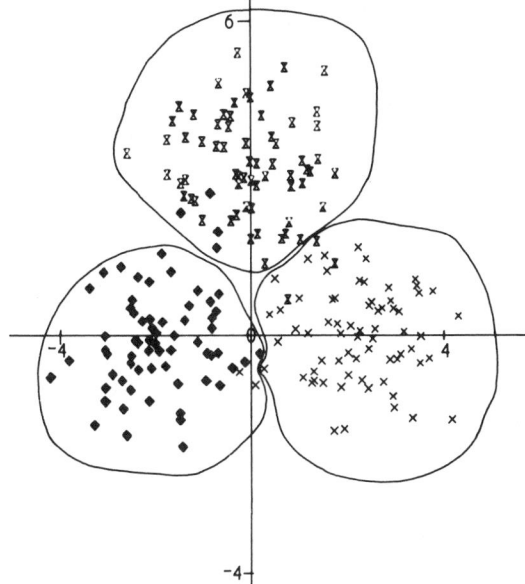

Figure 9.4   (From Gardner, 1978) Three bivariate normal samples together with a three-cluster solution using the method of Beale (1969). Objects within each contour form a cluster, and the different symbols indicate the different sample memberships.

A local maximum in the graph of *C* vs. *k* is an indication of a good description of the data for that value of *k*, and we see that for these data this measure is performing as anticipated.

Many more examples of the uses of simulation in classification are referenced by Gordon (1981), and further discussion of Example 9.2, as well as additional illustrations, are found in Exercise 9.9.

9.4   Estimating sampling distributions

The use of simulation to investigate sampling distributions arising in multivariate analysis was encountered in the last section, and in Section 1.5 we encountered the early example of this use by W. S. Gossett. Two further examples are given here.

9.4.1   The chi-square goodness-of-fit test

This test was described in Section 2.14, and employed in Chapter 6. As emphasized already, the chi-square test is only an approximate test when expected values are 'small' (say less than 5), and one way to examine how and when the approximation breaks down is by means of simulation. An illustration is provided by Hay (1967), who considered both the October mean pressure at the point 63°N, 20°W in the Atlantic, and a measure of subsequent winter temperature in Central England, for 45 specified years (see Table 9.4). In this case the chi-square goodness-of-fit statistic is computed as explained in *ABC*, p. 277, resulting in a value of $X^2 = 20.2$ with 8 degrees of freedom. This value is just significant at the 1 % level using the standard chi-square test, but in Table 9.4, under the assumption of independent row and column classifications, *all* the expected cell values are less than 5, and so one would question the validity of the standard chi-square test, while at the same time drawing some comfort from the extreme significance of the result obtained. Craddock and Flood (1970) used simulation to show that the standard chi-square test is in fact conservative for this application. For each of a variety of table sizes, and grand totals of table entries, they simulated 10 000 values of X^2 with independent row and column cassifications and uniform marginal distributions, and drew up histograms of the resulting sampling distributions. The relevant estimated percentiles for the above application are given in Table 9.5.

Table 9.4   Relationship between monthly mean pressure in October at a location 63°N, 20°W and a measure of subsequent winter temperature in Central England for 45 years. (From Hay, 1967). The figures in the body of the table are frequencies of years.

October mean pressure (terciles)	Winter temperature (quintiles)					Total
	1	2	3	4	5	
Upper	8	3	1	2	2	16
Middle	1	2	2	4	5	14
Lower	0	4	6	3	2	15
Total	9	9	9	9	9	45

Table 9.5   Value of z such that $p = \Pr(X \leq z)$. (From Craddock and Flood, 1970)

Grand total of (5 × 3) table	$100\,p$									
	1	2	5	10	50	90	95	98	99	99.9
40	1.84	2.25	2.98	3.80	7.7	13.3	15.2	17.5	19.2	24.9
50	1.80	2.19	2.91	3.70	7.6	13.3	15.3	17.7	19.4	25.2
∞	1.65	2.03	2.73	3.49	7.3	13.4	15.5	18.2	20.1	26.1

Estimates of error are also given by Craddock and Flood (1970), who suggest that, apart from the 99 and 99.9 percentile values, which are subject to larger errors, the estimated percentiles should differ from the true values by less than 0.2 (see Kendall and Stuart, 1969, p. 236). We shall return to error estimation in sampling distributions in Section 9.4.2.

It is interesting to note that Craddock and Flood (1970) were unfortunate in their choice of random number generator, discarding a discredited generator of Kuehn (1961) and replacing it by:

$$x_{n+1} = (kx_n + x_{n-1}) \mod (2^{48})$$

where $k = 5^{20}$, which, at a later stage, was found to fail the runs up-and-down fact conservative for this application. For each of a variety of table sizes, and grand totals of table entries, they simulated 10 000 values of X^2 with Exercise 9.13. Note that no variance-reduction method was used, as is true also in a more recent study by Cox and Plackett (1980), who use simulation to evaluate the distribution of X^2 for small samples in three-way as well as two-way tables. Hutchinson (1979) provides a bibliography of related research. By contrast, the first use of control variates in simulation was by Fieller and Hartley (1954), whose approach we shall describe in the next section.

*9.4.2   Precision and the use of control variables

As we have seen, if one has a random variable of unknown distribution then one approach is to simulate the variable many times and then draw a histogram to visualize the resulting sample. This procedure is illustrated by Fieller and Hartley (1954) for the range of a sample of size 5 from an $N(0, 1)$ distribution. In this case of course the distribution of the range is given by Equation (6.2), and Fieller and Hartley were able to produce the following table:

	Value of the range						
	0–1	1–2	2–3	3–4	4–5	5–∞	Total
Observed values	56.0	330.5	414.0	163.0	33.5	3.0	1000
Expected values	45.0	336.6	407.5	173.2	34.0	3.7	1000

If p_i is the probability that the range is in the ith category above, for $1 \leq i \leq 6$, then we can estimate p_i by

$$\hat{p}_i = \frac{n_i}{n} \quad \text{and} \quad \text{Var}(\hat{p}_i) \approx n_i(n - n_i)/n^3$$

when n_i of the n values are found to lie in the ith category, $1 \le i \le 6$. Thus we obtain the estimated distribution given in Table 9.6.

Table 9.6

	\multicolumn{6}{c}{Value of the range}
	0–1	1–2	2–3	3–4	4–5	5–∞
Estimated probability of range being in this interval	0.056	0.331	0.414	0.163	0.034	0.003
$n \times$ variance of estimated probability	0.053	0.221	0.243	0.136	0.032	0.003

The error for any interval is $(N_i - np_i)$, and from Table 9.6 we see that while the variance of the error is greatest for $p_i \approx 0.5$, the variance of $(\hat{p}_i/p_i)$ is greatest for $p_i \approx 0$ or 1. In deciding upon the value of n, it is sensible therefore to consider the *relative* error,

$$X = \left(\frac{N_i - np_i}{np_i} \right)$$

for which

$$\mathcal{E}[X] = 0 \quad \text{and} \quad \text{Var}(X) = (1 - p_i)/(n_i p_i)$$

Suppose we want to choose n so that

$$\text{Pr}\left\{ \left| \frac{N_i - np_i}{np_i} \right| \le \delta \right\} = \theta \tag{9.1}$$

for some small $\delta > 0$ and $1 \ge \theta \ge 0$. N_i has a $B(n, p_i)$ distribution, and making a normal approximation to the binomial distribution and employing a continuity correction (see, e.g., *ABC*, p. 227) we can approximate to (9.1) by

$$2\Phi\left(\frac{\delta - \frac{1}{2}}{\sqrt{(q_i/np_i)}} \right) = \theta \quad \text{where} \quad q_i = (1 - p_i)$$

i.e.,

$$n = \left\{ \frac{\Phi^{-1}(\theta/2)}{\delta - \frac{1}{2}} \right\}^2 \left(\frac{q_i}{p_i} \right) \tag{9.2}$$

We see from (9.2) that if we want to choose n so that (9.1) is satisfied for all i, then the size of n is determined by $\min_i \{p_i\}$. Thus we may obtain initial estimates of the $\{p_i\}$ by means of a pilot simulation, and then select n according to (9.2) and the smallest value of $\hat{p}_i$, to correspond to predetermined values of θ and δ. There was no discussion of precision in the studies reported in Tables 9.2 and 9.3, for which the choice of the number of simulation runs taken was not explained.

As we have seen in Chapter 7, for a given precision the size of a simulation experiment can be reduced by using a variance-reduction technique, and we shall now consider the use of a control variate for estimating the sampling distribution of the range described above. The idea of using a control variate is the same as in Section 7.3.1, namely, we want to estimate the distribution of X using simulated random variables, and we introduce a correlated random variable Y of known distribution. We use the same random variables to estimate the distribution of Y and hope to increase the precision for X by making use of the correlation between X and Y and the known distribution of Y. The details which follow come from Fieller and Hartley (1954) and while they are qualitatively different from the analysis of Section 7.3.1, they utilize the same underlying idea. Y will be taken as the sample corrected sum-of-squares, $(n-1)s^2$, and X will be the sample range. Each simulated sample of size 5 from an $N(0, 1)$ distribution therefore gives rise to a value of X and a value of Y. In this particular example we would certainly expect a high positive correlation between X and Y. If n such random samples are taken in all, and if the range of X is divided into n_x categories, while the range of Y is divided into n_y categories, then

$$n = \sum_{i=1}^{n_x} \sum_{j=1}^{n_y} n_{ij}$$

where n_{ij} samples give rise to an X value in the ith X category and a Y value in the jth Y category. For $n = 1000$, Fieller and Hartley (1954) presented the data reproduced in Table 9.7.

Table 9.7   An illustration of the use of a control variate in the investigation of a sampling distribution, (From Fieller and Hartley, 1954).

Categories for the range, X	Categories for the corrected sum-of-squares, Y						Totals
	0–0.71	0.71–2.75	2.75–4.88	4.88–9.49	9.49–14.86	14.86–∞	
0–1	49.5	6.5					56
1–2	5.0	305.0	20.5				330.5
2–3		39.0	306.0	69.0			414
3–4			2.5	142.0	18.5		163
4–5				10.0	19.5	4.0	33.5
5–∞						3.0	3.0
Totals	54.5	350.5	329.0	221.0	38.0	7.0	1000

Let $p_{ij} = \Pr(X \in i\text{th } X\text{-category and } Y \in j\text{th } Y\text{-category})$.

Thus $\sum_i \sum_j p_{ij} = 1$.

For simplicity, let us also use the summation notation:

$$n_{i.} = \sum_j n_{ij}, \qquad p_{.j} = \sum_i p_{ij}, \text{ etc.}$$

We want to estimate $\{p_i\}$. We can introduce the effect of the control variate by writing:

$$p_{i.} = \sum_j p_{ij} = \sum_j \left(\frac{p_{ij}}{p_{.j}} p_{.j} \right)$$

and substituting the known values for $\{p_{.j}\}$ while estimating $(p_{ij}/p_{.j})$ from the data. Thus we can form:

$$\hat{p}_{i.} = \sum_j \left(\frac{n_{ij}}{n_{.j}} \right) p_{.j} \qquad \text{as long as } n_{.j} > 0.$$

If $n_{.j} = 0$ for some j it is suggested that $(n_{ij}/n_{.j})$ be replaced by $(n_{i.}/n)$, but we are unlikely to have $n_{.j} = 0$ in a well-designed sampling experiment (see below). It is shown by Fieller and Hartley (1954) that

$$\mathscr{E}[\hat{p}_{i.}] = p_{i.} - \sum_j (1 - p_{.j})^{n-1} (p_{ij} - p_{i.}p_{.j})$$

and $\qquad \mathrm{Var}(\hat{p}_{i.}) \approx p_{i.}(1 - p_{i.})/n - \sum_j (p_{ij} - p_{i.}p_{.j})^2/(np_{.j})$

and they also present the exact formula for $\mathrm{Var}(\hat{p}_{i.})$. We see that if X and Y are independent, $p_{ij} = p_{i.}p_{.j}$, and so the use of the variate Y accomplishes nothing. However, for $p_{ij} \neq p_{i.}p_{.j}$, the variance of $(\hat{p}_{i.})$ is reduced from its value when Y is not employed. The price to pay for this variance reduction is the bias in $\hat{p}_{i.}$, but this will be small as long as the (known) $p_{.j}$ are not too small (Fieller and Hartley suggest choosing the intervals for Y such that $np_{.j} \geq 8$).

For the data of Table 9.7 we obtain the values:

							Total
$n\hat{p}_{i.}$	51.8	327.9	396.0	184.8	37.3	2.1	999.9

and the comparison of the variances is shown in Table 9.8.

Table 9.8 Evaluating the effect of the use of the control variate in Table 9.7. Here $\hat{p}_{i.} = (n_{i.}/n)$.

Categories for the range, X	Estimated variance		$\dfrac{\text{Var}(\tilde{p}_{i.})}{\text{Var}(\hat{p}_{i.})}$
	$\text{Var}(\tilde{p}_{i.})$	$\text{Var}(\hat{p}_{i.})$	
0–1	0.053	0.011	4.8
1–2	0.221	0.063	3.5
2–3	0.243	0.104	2.3
3–4	0.136	0.063	2.2
4–5	0.032	0.021	1.6
5–∞	0.003	0.002	1.5

Tocher (1975, p. 92) presents the results of a simulation study with $n = 819$, for which X is a sample median and Y is a sample mean. Further discussion, involving the choice of the number of categories to use, is given by Fieller and Hartley (1954).

*9.5 Additional topics

In this section we take a brief look at two relatively new developments in statistics, each of which involves simulation.

9.5.1 Monte Carlo testing

In Section 9.4.1 simulation was used to estimate the sampling distribution of the X^2 statistic for a particular type of two-way table, and this enabled us to judge the significance of one X^2 value for such a table. This example reveals two features: when estimating distributions using simulation in this way one is frequently interested in the tails of the distribution, and we also saw that the tails are estimated with least precision, with much of the simulation effort being of little relevance to the tails. An alternative approach was suggested by Barnard (1963), resulting in what is now called Monte Carlo testing. We shall illustrate what is involved here by reference to an example considered by Besag and Diggle (1977), which is relevant to questions posed in Chapter 6 regarding the randomness or otherwise of two-dimensional scatters of points.

One approach to scatters of points in two dimensions is to compute the distance from each point to its nearest neighbour and then form the sum, s_1 say, of these nearest-neighbour distances. For regular patterns of points, small nearest-neighbour distances will not occur, and so s_1 will tend to be large, while for patterns of points that appear to be clustered, as in Fig. 9.4, s_1 will tend to be small. For a given scatter of n points we can compute s_1 and then we need to be able to assess its significance. One approach is to simulate many uniformly random scatters of k points, for the area in question, possibly using the rejection approach of Hsuan (1979). For each such random scatter s_1 can be calculated and ultimately the sampling distribution can be estimated, using ideas such as those of Section 9.4.2. An alternative approach involves $(n-1)$ such simulations, for some n, resulting in the sums of nearest-neighbour distances $\{s_2, s_3, \ldots, s_n\}$. The Monte Carlo test approach is simply to form the ordered sample $s_{(1)}, s_{(2)}, \ldots, s_{(n)}$; the significance level of s_1 is then obtained from the rank of s_1 among the n order-statistics. Investigations of the power of such a test have been carried out by Hope (1968) and Marriott (1979). It is of interest to apply a Monte Carlo test to the problem of Craddock and Flood (1970) considered earlier, and that is done in the following example.

EXAMPLE 9.3

Suppose we want to perform a one-tail significance test of size α on some given null hypothesis. In general we choose an integer n and carry out $(n-1)$ simulations under the given null hypothesis. The value of the test statistic is calculated for each simulation and also for the observed data. In addition, we choose an integer m such that $m/n = \alpha$, and the null hypothesis is rejected if the value of the test statistic for the data is among the m largest of the n values of the statistic. Marriot recommends that we ensure $m \geq 5$. Two-way tables of the same dimension as Table 9.4 were simulated with column totals of 9 and with each row given probability $1/3$ of occupation by each datum. For $\alpha = 1\%$ the values of $m = 5$, $n = 500$ were chosen, and for three separate runs the following values for the rank, r, of the value $X^2 = 20.2$ were obtained: $r = 12$, $r = 14, r = 12$, each resulting in a significant result at the 3 % level, in reasonable agreement with the finding of Craddock and Flood (1970) from Table 9.5, though the discrepancy merits further investigation.

9.5.2   The bootstrap

Associated with the parameter estimates of Section 9.2 were estimates of error, obtained from the curvature of the likelihood surface and making use of an asymptotic approximation to a sampling distribution. Suppose that in a general problem we have a random sample of data $(x_1, \ldots, x_n)$ which is used to estimate a parameter θ, resulting in the estimate $\hat{\theta}$. As discussed in Section 8.4, $\hat{\theta}$ by itself is of little use without some indication of its precision, which is a property of the sampling distribution of $\hat{\theta}$. The usual approach is to make an asymptotic approximation, but this could be misleading if confidence intervals are required, and repeated sampling to produce further sets of real data may be impractical. The idea of the bootstrap is to sample repeatedly, uniformly at random, and with replacement, from the one sample of real data $(x_1, \ldots, x_n)$.

Successive bootstrap samples will therefore contain the elements $x_1, \ldots, x_n$, but some may be repeated several times for any sample, and then others will be missing. For each bootstrap sample the estimate $\hat{\theta}$ of θ may be calculated, enabling a form of sampling distribution to be estimated, which may then be used to gauge the precision of $\hat{\theta}$, as well as a wealth of further detail. Due to Efron (1979a), this technique is well illustrated in Efron (1979b), with further discussion in Efron (1981). The bootstrap relies upon multiple simulations of a discrete uniform random variable, which could be done by means of the NAG routines GO5DYF or GO5EBF, or the IMSL routine GGUD, as well as within MINITAB (see Appendix 1). Except in cases when theoretical analysis is possible (see Reid, 1981, for example), the bootstrap procedure would have to utilize a computer, and is therefore a truly modern statistical technique, as implied by the title of Efron's (1979b) paper.

9.6 Discussion and further reading

The examples we have considered here reveal the power and versatility of simulation methods. At the same time we can see the limitations of some of these methods – as when variance–covariance matrices are chosen arbitrarily to be the identity matrix, for example.

An alternative approach to the simulation of spatial patterns of points is given by Ripley (1979). Forests provide real-life examples of spatial arrangements, in which the points are trees; when foresters want to estimate the density of forests they may use quadrat sampling, which for long and narrow 'belt' quadrats, or transects, is analogous to the transect method mentioned in Section 9.2, which was investigated using simulation by Gates (1969). Quadrat sampling can be extremely time-consuming and laborious, and recently there has been much interest in the use of methods which involve measuring distances from points to trees and from trees to trees; see Ripley (1981, p. 131) for an introduction. Here again the various methods proposed have been evaluated using simulation; see Diggle (1975) and Mawson (1968).

Simulation continues to be used in multivariate analysis, as can be seen from the work of Krzanowski (1982) who used simulation to help in the comparison of different sets of principal components. Just as simulated data are useful for demonstrating and investigating methods of cluster analysis, such data may be similarly used in discriminant analysis, for example, and are also used in the evaluation of 'imputation' procedures for estimating missing values from sample surveys. Principal components and discriminant analysis use linear combinations of the original variables, which may be thought to be somewhat restrictive (for this reason, Cormack, 1971, favoured non-metric multidimensional scaling). In discriminant analysis an alternative approach is to employ a quadratic discriminant function, and Marks and Dunn (1974) have used simulation to compare the performance of linear and quadratic discriminant functions.

In this chapter we have concentrated on examples of the use of simulation in a small number of areas, and even here it has been necessary to select only a very small subset of the possible examples. There are many more applications of simulation. Rubinstein (1981, chapters 5–7) describes some of these, showing, for instance, that the solution to certain integral equations is given in terms of the mean time to absorption in particular Markov chains with an absorbing state. In this application an aspect of a completely deterministic problem is estimated by simulation of the related Markov chain (cf. Sections 9.1.2 and 7.3.4). There are also important uses of simulation in statistical physics.

9.7   Exercises and complements

(Some of the following are best appreciated if one has a prior knowledge of MINITAB and multivariate analysis.)

†9.1   Discuss the following MINITAB program for simulating chi-square random variables on 10 degrees of freedom.

```
NOPRINT
 STORE
 NRANDOM 10 OBSERVATIONS, WITH MU=0.0, SIGMA=1.0, PUT IN C1
 LET C2=C1*C1
 SUM C2, PUT IN K1
 JOIN K1 TO C3, PUT IN C3
 END
 EXECUTE 100 TIMES
 HISTO C3
```

Suggest how you might write MINITAB programs to simulate logistic and beta random variables.

9.2   Modify the program of Fig. 9.2 to produce estimates of probabilities of absorption, and suggest how you might employ the variance-reduction approach of Section 7.3.4.

*9.3   Feller (1957, p. 395) provides the following analytical solution to the brother/sister mating problem. Starting in the jth state, for $j = 2, 3, 4, 6$, the probability of being absorbed in the first state after n steps is:

$$p_{j2}^{(n-1)}/4 + p_{j3}^{(n-1)}/16$$

and the corresponding probability for the fifth state is:

$$p_{j3}^{(n-1)}/16 + p_{j4}^{(n-1)}/4$$

where

$$p_{jk}^{(n-1)} = \sum_{r=1}^{4} \frac{\theta_{jk}^{(r)}}{s_r^{n-1}}$$

$$s_1 = 2, s_2 = 4, s_3 = \sqrt{5} - 1 \quad \text{and} \quad s_4 = -(\sqrt{5} + 1),$$

$$\theta_{jk}^{(r)} = c_r x_j^{(r)} y_k^{(r)}$$

where $c_1 = \frac{1}{2}$, $c_2 = 1/5$, $c_3 = (1 + \sqrt{5})^2/40$, $c_4 = (\sqrt{5} - 1)^2/40$,

and $$\{x_j^{(1)}\} = (1, 0, -1, 0) = \{y_k^{(1)}\}$$

$\{x_j^{(2)}\} = (1, -0, 1, -4)$

$\{x_j^{(3)}\} = (1, (\sqrt{5} - 1), 1, (\sqrt{5} - 1)^2)$

$\{x_j^{(4)}\} = (1, -1 - \sqrt{5}, 1, (1 + \sqrt{5})^2)$,

$\{y_k^{(2)}\} = (1, -1, 1, -0.5)$

$\{y_k^{(3)}\} = (1, \sqrt{5} - 1, 1, (\sqrt{5} - 1)^2/8)$

$\{y_k^{(4)}\} = (1, -1 - \sqrt{5}, 1, (1 + \sqrt{5})^2/8)$.

Use these expressions to calculate the probabilities of ultimate absorption in states 1 and 5, and compare your results with the estimates of Exercise 9.2. Comment on the relative merits of the simulation and analytic approaches to this problem.

*9.4 Alternative approaches to the estimation of the size of mobile animal populations include multiple marking, following a succession of recaptures (see, e.g., Darroch, 1958), and inverse sampling, in which following marking, the second-stage sampling continues until a prescribed number of marked individuals are caught. Identify the distribution of the number of animals that will be taken at the second-stage sampling. Find the mean of this distribution, and suggest a new estimator of the population size (see Chapman, 1952).

*9.5 The following data provide measurements, in inches, on members of a final-year undergraduate statistics class at the University of Kent.

			Men			
Chest	Waist	Wrist	Hand	Head	Height	Forearm
37.5	31.0	5.5	8.0	23.0	67.0	18.0
34.5	30.0	6.0	8.5	22.5	69.0	18.0
41.0	32.0	7.0	8.0	24.0	73.0	19.5
39.0	34.0	6.5	8.5	23.0	75.0	20.0
35.0	29.0	6.4	8.0	22.5	66.0	16.0
34.0	28.0	6.2	8.1	19.7	66.0	17.3
34.0	29.0	6.5	7.5	22.0	66.0	18.0
36.0	31.0	7.0	8.0	23.0	73.0	18.0
40.0	36.0	7.5	8.5	25.0	74.0	19.0
38.0	30.0	7.5	9.0	22.0	67.0	17.0
38.0	33.0	6.5	8.5	22.3	70.0	19.2
34.5	30.0	6.0	7.8	21.8	68.0	19.3
36.0	29.0	6.0	8.0	22.0	70.5	18.0

Women

Chest	Waist	Wrist	Hand	Head	Height	Forearm
35.0	25.0	6.0	7.5	22.0	63.0	16.0
35.0	25.0	5.8	7.3	21.5	67.5	18.0
34.0	26.0	6.0	7.0	21.0	62.8	17.0
36.0	27.0	6.0	7.0	23.0	68.0	17.5
34.0	26.0	6.0	7.0	21.0	65.0	18.0
35.0	26.0	5.0	7.0	22.0	65.0	16.0
36.0	30.0	6.5	7.5	24.0	70.0	17.0
34.0	24.0	6.5	7.0	22.0	63.0	16.5
36.0	26.0	6.5	7.5	22.0	67.5	16.5
30.0	23.0	6.0	7.0	21.0	59.0	14.5
32.0	24.0	5.8	7.0	22.5	60.0	15.0
35.5	28.0	6.0	7.0	22.0	69.5	17.5
34.0	22.0	5.0	6.0	22.0	66.0	15.0
36.0	27.0	6.0	7.0	22.0	66.5	17.5
38.0	29.0	6.0	7.5	22.0	64.5	17.5

(i) Compute the principal components, using the correlation matrix, for men and women separately. Can you suggest variables which may be omitted?

(ii) Calculate the linear discriminant function and discuss the probability of misclassification using the linear discriminant function for these data.

*9.6 Discuss the use of $\hat{\delta}$ as a control variate in the sampling investigation summarized in Table 9.2.

9.7 As mentioned in Section 9.3.2, Jolliffe (1972b) used a method of Butcher (1960) for simulating standard normal random variables. The method employed the generalized rejection method of Exercise 5.29 for the half-normal density, writing

$$\sqrt{\left(\frac{2}{\pi}\right)}e^{-x^2/2} = cg(x)r(x) \qquad \text{for } x \geq 0$$

in which

$$g(x) = \lambda e^{-\lambda x}$$

and

$$r(x) = \exp\left(\lambda x - x^2/2\right)$$

Deduce c, and show that the resulting algorithm is that already considered in Example 5.5.

9.8 For the structured data set of Section 9.3.2, if ρ_{ij} denotes the correlation between X_i and X_j, show that $\rho_{14} = 0.894$, $\rho_{25} = 0.819$, $\rho_{26} = 0.707$, and $\rho_{56} = 0.579$. If it is decided to reject three out of the six variables, explain why the best subsets of variables to retain are: $\{X_1, X_2, X_3\}$ and

$\{X_2, X_3, X_4\}$, while the next best ('good') subsets are: $\{X_1, X_3, X_5\}$, $\{X_1, X_3, X_6\}$, $\{X_3, X_4, X_5\}$, $\{X_3, X_4, X_6\}$. Two of the variable-rejection rules considered by Jolliffe (1972a, b) were: B2—reject one variable to correspond to each of the last three principal components; and B4—retain one variable to correspond to each of the first three principal components. For B4 it was found that for 11 out of 96 simulated data sets, the best subsets were retained, while for B2 it was found that the subsets retained were always good, but never best.

†9.9 Discuss the results given below, from Gardner (1978). Two further populations, A and B, were simulated as in Example 9.2, with the differences that for A the three group means were $(-1.75, 0), (1.75, 0), (0, 3.03)$, and for B there was no group structure, all the values being obtained from a bivariate normal distribution with mean $(0, 0)$. Samples of size 60 were taken from the populations, and the results from two of the samples from A are given below for the 3-cluster solution produced by Beale's (1969) method:

	Cluster no.	Cluster size	Cluster	centre	No. of mis-classifications
	1	15	2.3	0.0	1
1st sample	2	27	−1.7	0.2	4
from A	3	18	−0.2	3.2	1
	1	20	1.7	0.5	3
2nd sample	2	17	−2.2	0.3	1
from A	3	23	−0.2	2.9	2

In addition, the values of the C measure for 4 samples of size 60 are as follows:

No. of clusters	Population of Example 9.2	Population A, 1st sample	Population A, 2nd sample	Population B
6	69.6	60.1	68.9	48.2
5	78.5	64.6	72.8	48.6
4	91.3	69.5	73.5	49.4
3	101.0	77.5	66.7	47.3
2	38.2	39.0	31.1	39.1

9.10 Morgan, Chambers and Morton (1973) applied non-metric multidimensional scaling to measures of similarity between pairs of the digits

1–9. 'Similarity' here was derived from experimental results summarizing how human subjects acoustically confused the digits. A two-dimensional representation is given in Fig. 9.5 in which the six largest similarities are also linked. Comment on the stress value of 6.44%, in the light of the results of Table 9.3 (cf. the discussion of digit confusion in Section 6.7).

Figure 9.5 Two-dimensional non-metric multidimensional scaling representation of the acoustic confusion of digits.

9.11 Use the result of Exercise 4.14 to simulate bivariate normal samples as in Exercise 9.9, but with a different dispersion matrix.

9.12 Everitt (1980, chapter 4) simulated a bivariate sample of size 50, 25 elements from the bivariate normal:

$$N\left(\begin{pmatrix}0\\0\end{pmatrix},\begin{pmatrix}16.0,1.5\\1.5,0.25\end{pmatrix}\right)$$

and the remaining 25 from the bivariate normal:

$$N\left(\begin{pmatrix}4\\4\end{pmatrix},\begin{pmatrix}16.0, & 1.5\\1.5, & 0.25\end{pmatrix}\right)$$

Using single-link cluster analysis he obtained the following 5-cluster solution (numbers 1–25 refer to the first population, and 26–50 to the second):

Cluster	Objects
1	1–13, 15, 16, 18–25
2	26–41, 43–50
3	14
4	17
5	42

Comment on these results. Single-link cluster analysis is described by Everitt (1980, p. 9). He simulated his population using the Box–Müller method followed by the transformation of Exercise 4.14.

†9.13 Combine categories in Table 9.4 in such a way as to enable a standard chi-square test to be justified, and compare the results of this test with that for which there is no pooling. Suggest a possible control variate for use in the sampling distribution investigation of Section 9.4.1.

9.14 Repeat the study of Section 9.4.2 for a different pair of correlated variables.

*9.15 Write a computer program to provide a Monte Carlo test of whether the pattern of points in Fig. 9.4, or a similar pattern, generated as in Exercise 9.11, is random.

9.16 Investigate further the discrepancy observed in Example 9.3. If a BASIC program is written for the Monte Carlo test, an ordering subroutine can be obtained from the variety provided by Cooke, Craven and Clarke (1982, chapter 3).

9.17 Comment upon the relationship between the estimated probability and the variance of the estimated probability in Table 9.6.

9.18 Consider how you would use MINITAB to simulate Gossett's simulation mentioned in Section 1.5.

9.19 How would you use MINITAB to investigate central limit theorems? (cf. Mead and Stern, 1973, p. 194).

*9.20 Any hierarchical method of cluster analysis, such as single-link analysis mentioned in Exercise 9.12, can be represented by a set of 'ultrametric' distances between the objects (see Johnson, 1967). Consider how you might compare such ultrametric distances with the original between-object similarities, and suggest how simulation may be used to investigate the behaviour of your measure of comparison (see Gower and Banfield, 1975; Jenkinson, 1973).

APPENDIX 1

COMPUTER ALGORITHMS FOR GENERATION, TESTING AND DESIGN

The main aim of this section is to document some of the more convenient computer algorithms for the generation and testing of pseudo-random numbers. We shall also indicate the availability of programs for simple experimental designs.

A useful starting point in a search for such algorithms is the GAMS (1981) index, which indicates the IMSL and NAG libraries as the main sources of the computer algorithms of interest to us. IMSL is a commercial library containing about 500 FORTRAN subroutines. The library is available from IMSL, Inc., Sixth Floor, NBC Building, 7500 Bellaire Boulevard, Houston, Texas 77036, USA. NAG is a commercial library containing about 600 subroutines in FORTRAN and ALGOL, and NAG is available from the Numerical Algorithms Group Ltd, Mayfield House, 256 Banbury Road, Oxford OX2 7DE, UK, and 1131 Warren Avenue, Downer's Grove, Illinois 60515, USA. Other sources of algorithms exist, and we shall discuss some of these later, but first we shall describe what is available in NAG and IMSL. It is important to realize that this appendix is not intended as a reference manual for these libraries, and it will be necessary to consult the manuals for detailed running instructions. The information below is obtained from Edition 8 of the IMSL library, and mark 10 of the NAG library, and we only list routines which are user-callable. For both libraries, names of routines are strings of letters and numbers, as seen below.

A1.1 Generation routines available in NAG and IMSL

Beta

IMSL:GGBTR

In the notation of Section 2.11, the method used depends on the size of α and β:

(i) $\alpha < 1$ and $\beta < 1$, the method of Jöhnk (1964) is used.

(ii) $\alpha < 1$ and $\beta > 1$, or $\alpha > 1$ and $\beta < 1$, the method of Atkinson (1979b) is used.

(iii) $\alpha > 1$ and $\beta > 1$ and fewer than 4 variates required, then the algorithm BB of Cheng (1978) is used.

(iv) $\alpha > 1$ and $\beta > 1$ and ≥ 4 variates required, the algorithm B4PE of Schmeiser and Baliu (1980) is used.

(v) $\alpha = 1$ or $\beta = 1$ then the inversion method is used (cf. Exercise 5.33(b)).

NAG : GO5DLF

Uses the method of Exercise 2.14, obtaining the component gamma variates from GO5DGF.

GO5DMF

Generates a variate from a beta distribution of the *second* kind, defined as (and generated by) X/Y in the notation of Exercise 2.14.

Binomial

IMSL : GGBN

Simulates $B(n, p)$ variates. If $n < 35$, a simple counting procedure is employed with re-use of uniform variates to improve efficiency. If $n \geq 35$, the method of Relles (1972) (see Exercise 4.17 and Section 4.4.1) is used.

NAG : GO5DZF

Simulates a $B(1, p)$ variate, with responses 'True' and 'False'.

GO5EDF

Uses the table-look-up method; used in conjunction with GO5EYF.

Cauchy

IMSL : GGCAY

Uses the inversion method of Exercise 5.8.

NAG : GO5DFF

Uses the rejection method of Exercise 5.27.

Chi-square

IMSL:GGCHS

Uses the method of Section 4.3, involving convolutions of exponentials, and possibly the further sum of a squared $N(0, 1)$ variate if necessary.

NAG:GO5DHF

Uses GO5DGF for gamma variates.

Exponential

IMSL:GGEXN

Uses the inversion method of Section 4.3.

GGEXT

Simulates from a mixture of two exponential densities.

NAG:GO5DBF

Also uses the method of Section 4.3.

F

NAG:GO5DKF

Sets $F = (nY)/(mZ)$, where Y, Z are independent χ_m^2 and χ_n^2 variables, respectively, obtained as $\Gamma(m/2, 2)$ and $\Gamma(n/2, 2)$, from GO5DGF, described below.

Gamma

IMSL:GGAMR

Simulates $\Gamma(a, 1)$ variates, the method used depending on the value of a. If $a < 1$ and $a \neq 0.5$, a rejection method described in Ahrens and Dieter (1974) is used. If $a = 0.5$, squared and halved standard normal variates are used. If $a = 1$, exponential deviates are used, while if $a > 1$ a ten-region rejection procedure of Schmeiser and Lal (1979) is employed.

NAG: GO5DGF

Simulates $\Gamma(n, \lambda)$ where $2n$ is a non-negative integer, using the approach of Section 4.3.

General distributions

IMSL: GGDA

Simulates from any given discrete distribution using the alias method of Exercise 5.42.

GGDT

Simulates any given discrete distribution using the table-look-up method, using bisection in the early stages.

GGVCR

Simulates any continuous random variable, using the inversion method. The user may provide only partial information on the distribution function, and the method of Akima (1970) is then used to interpolate, using piecewise cubics. The method, together with descriptions of accuracy, is described in Guerra *et al.* (1972).

NAG: GO5EXF

Used in conjunction with GO5EYF, for any given discrete distribution. This routine forms the cumulative sums needed for the table-look-up method.

GO5EYF

Performs an indexed search through tables of cumulative sums of probabilities.

Geometric

IMSL: GGEOT

Simulates a geometric random variable using the definition of such a variable in Section 2.6, i.e. $U(0, 1)$ variates are chosen sequentially until one is less than p, the geometric success probability.

Hypergeometric

IMSL : GGHPR

Simulates hypergeometric random variables using the definition of the hypergeometric distribution (see for example *ABC*, p. 147).

NAG : GO5EFF

Employs the table-look-up method. Used in conjunction with GO5EYF.

Pseudo-random integers

IMSL : GGUD

Generates n uniform random integers over a specified range, using GGUBFS and truncation.

NAG : GO5DYF

Generates a uniform random integer over a specified range, using GO5CAF and truncation.

GO5EBF

The table-look-up approach for uniform random integers; used in conjunction with GO5EYF.

Logistic

NAG : GO5DCF

Uses the table-look-up method (see Exercise 5.13(a)).

Log-normal (See Exercise 4.21)

IMSL : GGNLG

Returns the value $\exp(\alpha + \beta N)$, where N is an $N(\mu, \sigma^2)$ random variable obtained from GGNPM.

NAG : GO5DEF

Returns the value $\exp(N)$, where N is a suitable normal variate obtained from GO5DDF.

Multivariate

IMSL: GGMTN

Simulates multinomial random variables. In the notation of Section 2.15, first a $B(n, p_1)$ distribution is simulated, and then successive conditional binomial simulations are used, simulating X_j from
$B(n - x_1 - x_2 \ldots - x_{j-1}, p_j/(1 - p_1 - p_2 - \ldots - p_{j-1}))$, for
$2 \leq j \leq m$, with $X_i = x_i$, for $1 \leq i \leq j - 1$.

GGNSM

Simulates multivariate normal random variables with zero means when the dispersion matrix, $\boldsymbol{\Sigma}$, is specified. The subroutine LUDECP provides the required triangular factorization of $\boldsymbol{\Sigma} = \mathbf{A}\mathbf{A}'$ (see the solution to Exercise 4.14).

GGSPH

Simulates points uniformly at random from the surface of the unit sphere in 3 or 4 dimensions (see Marsaglia, 1972a). In three dimensions, for example, if V_1 and V_2 are independent and $U(-1, 1)$, then conditional upon $S = V_1^2 + V_2^2 < 1$, set $Z_1 = 2V_1 \sqrt{(1 - S)}$, $Z_2 = 2V_2 \sqrt{(1 - S)}$ and $Z_3 = 1 - 2S$. The $\{Z_i\}$ are then the Cartesian co-ordinates of the required points.

ZSRCH

Generates points in n-dimensional space, for use with nonlinear optimization routines that require starting points. Such a procedure is needed for non-metric multidimensional scaling (see Section 9.3.3).

NAG: G05EAF

Sets up a reference vector which is used in conjunction with G05EZF to generate multivariate normal random variables. The same approach is used as with GGNSM.

Negative binomial (cf. Exercise 4.19)

IMSL: GGBNR

Employs the table-look-up method.

NAG: GO5EEF

Employs the table-look-up method. Used in conjunction with GO5EYF.

Normal

IMSL: GGNML

Uses the inversion method, the inverse normal c.d.f. function being provided by the IMSL routine MDNRIS: n variates are generated.

GGNPM

Uses the Polar Marsaglia method of Section 4.2.2.

GGNQF

The version of GGNML that is used if only a single $N(0, 1)$ variate is required at each call.

NAG: GO5DDF

Uses Brent's (1974) GRAND algorithm, which employs Forsythe's method (see Exercise 5.41).

Order statistics

IMSL: GGNO

Generates a set of order statistics from the normal distribution. Initially GGUO is used, and then the ordered uniform variates are transformed, using the inversion method as in GGNML.

GGUO

Generates a set of order statistics from the $U(0, 1)$ distribution. If a full set is required, spacings between successive order statistics are obtained as exponential variables. If only a subset is required, a beta variable is used for one of the order statistics and then the others are generated conditionally. For general discussion of the generation of order statistics, see Gerontidis and Smith (1982).

Permutations

IMSL : GGPER

Generates a random permutation of the integers 1 to k, using an exchange sort procedure, described by Cooke *et al.* (1982, p. 15).

NAG : GO5EHF

Method used is only possible for $k < 20$.

Poisson

IMSL : GGNPP

Simulates a non-homogeneous Poisson process (see Lewis and Shedler, 1976; and Section 8.3.2).

GGPON

For use when the Poisson parameter λ changes from call to call, as may occur, e.g., if one was using the Poisson and gamma distributions to simulate a negative-binomial random variable (see also Kemp, 1982; and Exercise 4.18). Uses the method of Fig. 4.4 for $\lambda \le 50$. For $\lambda > 50$ a normal approximation is used.

GGPOS

For use when the Poisson parameter does not change often. For $\lambda \le 50$ the standard table-look-up method is used. For $\lambda > 50$ a normal approximation is employed.

NAG : GO5ECF

Calculates probabilities for use with GO5EYF and the table-look-up method.

Sampling from a finite population without replacement

IMSL : GGSRS

Given a population of size n and a sample of size $m < n$, this routine selects this sample sequentially without replacement, so that when $j < m$ items have been selected, the ith object is selected with probability $(m-j)/(n+1-i)$ (see

Bebbington, 1975, for an explanation, and McLeod and Bellhouse, 1983, for a new approach).

Stable distribution

IMSL:GGSTA

Simulates from a range of stable distributions indexed by two parameters, using the method of Chambers *et al.* (1976). Possible distributions include the Cauchy and the normal.

t

NAG:GO5DJF

Calculates $W = Y \sqrt{(n/Z)}$, in which Y has a $N(0, 1)$ distribution, provided by GO5DDF, and Z has an independent χ_n^2 distribution, obtained as $\Gamma(n/2, 2)$ from GO5DGF.

Time series

NAG:GO5EGF

Sets up a reference vector for an auto-regressive moving-average time series model with normally distributed errors. The series is initialized to a stationary position using the method of Tunnicliffe–Wilson (1979).

GO5EWF

Simulates the next term from an auto-regressive moving-average time series model, using the reference vector set up in GO5EGF.

Triangular

IMSL:GGTRA

Simulates random variables from the symmetric triangular density of Exercise 5.5, but over the range (0, 1). The inversion method is used, as in Exercise 5.5(a).

Uniform

IMSL:GGUBFS

Uses the multiplicative congruential generator $(7^5, 0; 2^{31} - 1)$. Note that for simplicity, $U(0, 1)$ pseudo-random deviates are obtained from normalizing by 2^{31}, and not $2^{31} - 1$. The seed may be specified by the user.

GGUBS

Generates n pseudo-random variates using GGUBFS.

GGUBT

Uses the multiplicative congruential generator $(397\,204\,094, 0; 2^{31} - 1)$. The algorithm is slower than GGUBFS, due to portable coding. The seed may be specified by the user.

GGUW

Employs 'shuffling', with $g = 128$ (see Section 3.6) applied to GGUBFS.

NAG: GO5CAF

Uses the multiplicative congruential generator $(13^{13}, 0; 2^{59})$ with initial value, $x_0 = 123\,456\,789 \times (2^{32} + 1)$ unless GO5CBF or GO5CCF is also used (see Best and Winstanley, 1978, for comments).

GO5CBF(I)

Sets the seed for GO5CAF as $x_0 = 2I + 1$.

GO5CCF

Generates a random seed for GO5CAF, using the real-time clock.

GO5CFF

Records the current value in a sequence from GO5CAF. The sequence can then be continued at a later stage by calling the routine GO5CGF.

GO5CGF

Restarts a sequence from GO5CAF, following a previous call to GO5CFF.

GO5DAF

Simply transforms a $U(0, 1)$ pseudo-random variable from GO5CAF, to give a $U(a, b)$ variate.

Weibull

IMSL : GGWIB

Uses the inversion method of Exercise 5.13 (b).

NAG : G05DPF

Also uses the inversion method of Exercise 5.13 (b).

A1.2   Testing routines available in NAG and IMSL

IMSL : GFIT

Chi-square goodness-of-fit test.

GTCN

Used prior to a chi-square goodness-of-fit test, to determine suitable number of class-intervals (see Mann and Wald, 1942).

GTDDU

Computes the squared distance between successive pairs of pseudo-random $U(0, 1)$ variates and then forms an appropriate tally which is used in GTD2T.

GTD2T

Performs a test of the tallies produced by GTDDU (see Gruenberger and Mark, 1951).

GTNOR

Uses a chi-square goodness-of-fit test for $N(0, 1)$ variates.

GTPBC

Performs a count of the number of zero bits in a specified subset of a real word in binary form; tests are then applied to the resulting counts.

GTPKP

Used in the preparation of expected values for the poker test; this routine evaluates $B(n, 0.5)$ probabilities.

GTPL

Produces tallies for use in the poker test, performed by GTPOK.

GTPOK

Performs the poker test, using statistics computed by GTPL.

GTPR

Produces a tally of pairs (or lagged pairs) in a sequence of pseudo-random numbers; used in GTPST.

GTPST

Performs a test on pairs produced by GTPR.

GTRTN

Computes numbers of runs up and down in a sequence of variates.

GTRN

Tests the runs produced by GTRTN. Note that this test does *not* use the appropriate covariance terms (cf. Equation (6.1)) (see Kennedy and Gentle, 1980, pp. 171–173).

GTTRT

Produces a tally of triplets in a sequence of pseudo-random numbers; used in GTTT.

GTTT

Tests the randomness of triplets produced by a pseudo-random number generator, and previously tallied by GTTRT.

NKS1

Performs the Kolmogorov–Smirnov one-sample test, with the user supplying the theoretical distribution via a FORTRAN subroutine.

NKS2

Performs the Kolmogorov–Smirnov two-sample test.

NAG : GO8CAF

Performs the Kolmogorov–Smirnov one-sample test. The parameter 'null' provides alternative forms for the theoretical distribution. These are: $U(a, b)$, $N(\mu, \sigma^2)$, Poisson and exponential. This routine was used in connection with Exercise 6.21.

A1.3   Features of the NAG and IMSL algorithms

Because of inertia, library subroutines are unlikely to keep pace with the latest developments in the generation of random variables; while some of the routines of Section A1.1 are remarkably up to date, using methods of Chapter 5, some are clearly less so. Comparisons of efficiency for certain algorithms for beta, gamma and normal variables are given by Atkinson and Pearce (1976). It is interesting to contrast the different emphases used in these two different libraries. Thus IMSL places much importance on the testing of random variables, as can be seen from the last section. An attractive feature of certain of the routine descriptions is that they specify the empirical tests satisfied by the generated numbers, and provide the results of such tests. A number of the IMSL tests proceed in two stages, the first stage forming a tally which is then used by the second stage. A similar two-stage procedure is employed in NAG for simulating discrete random variables. The table-look-up method is employed through the GO5EYF routine which performs an indexed search through tables of cumulative sums of probabilities, set up individually by a separate routine for each distribution.

A1.4   Further algorithms

Published computer algorithms for simulation can be found in the journals: *Communications of the Association for Computing Machinery, Computer Bulletin, Computer Physics Communications, Computer Journal, Computing, The Journal of the Association for Computing Machinery*, and *Applied Statistics*. All of the algorithms in *Applied Statistics* up to the end of 1981 are indexed in *Applied Statistics* (1981), **30** (3), 358–373, while simulation algorithms from the other journals mentioned here are indexed in *Collected Algorithms* from ACM. *Applied Statistics* algorithms which complement those of Section A1.1 are:

Ripley (1979), on simulating spatial patterns (cf. Section 9.5.1);

Patefield (1981), on simulating random $r \times c$ tables with given row and column totals (cf. Section 9.4.1);

Smith and Hocking (1972), on simulating Wishart variates (cf. Exercise 4.15);

Wichmann and Hill (1982b) provide a FORTRAN program for their portable random generator, which combines the multiplicative congruential generators (171, 0; 30 269), (172, 0; 30 307) and (170, 0; 30 323) in the way explained in Exercise 3.18.

Golder (1976a, b) provides the spectral test, mentioned in Section 6.1.

This last test has recently been revised and a FORTRAN program for the new algorithm is provided by Hopkins (1983a). Hopkins (1983b) provides a FORTRAN algorithm for the collision test mentioned in Section 6.3. As mentioned in Chapter 6, the package of Cugini *et al.* (1980) provides a range of empirical test programs, written in BASIC.

Computerized algorithms for the analysis of experiments can be found in standard statistical packages such as GENSTAT, BMDP and SPSS. In addition, the IMSL routine AFACN provides an analysis for a full factorial experiment (see Section 8.5).

Other statistical packages provide simulation facilities. Some of those to be found in MINITAB have already been described, and we have seen how these may be extended by the use of the MINITAB language (see Exercise 9.1). The S language, developed at Bell Laboratories, also provides a wide range of simulation possibilities, and the opportunity arises within BMDP3D for using simulation for simple tests of robustness. See: C17, 'Quick and Dirty Monte Carlo'—p. 865 of the 1976 (not 1981) BMDP handbook.

Simulation programs in BASIC can also be found in Cooke *et al.* (1982, chapters 7 and 9) and in Wetherill (1982, Appendix 1).

APPENDIX 2
TABLES OF UNIFORM RANDOM DIGITS, AND EXPONENTIAL AND NORMAL VARIATES

These tables provide pseudo-random, rather than random, variates, but they should be adequate for small-scale investigations.

Table A2.1   Pseudo-random digits

Each digit was obtained from the first decimal place of pseudo-random $U(0, 1)$ variables resulting from the NAG routine GO5CAF. The digits have been grouped as shown for convenience of reading. The digits may be used to provide $U(0, 1)$ variates to whatever accuracy is required, by placing a decimal point in front of groups of digits. Thus for 5 decimal places we would obtain, from the first row of the table: 0.36166, 0.15217, etc.

36166	15217	88906	60493	36211	02862	68789	35346	83423	38001
48734	61061	82801	32055	99587	51156	61919	85682	16253	06162
29474	76062	40096	88802	52678	92156	61784	75192	39087	90198
24632	13950	13723	02027	29179	28792	75928	03507	74295	27971
84842	84503	07919	72887	77041	57728	05468	08203	97672	00856
63337	02467	90923	50023	69684	01854	68186	17018	31268	51312
61917	09485	90672	72283	97043	20066	92073	93723	60124	67424
88162	29658	70156	63238	55560	00192	09480	91738	99359	04009
19864	59125	89677	42774	93979	39397	78970	42590	49189	26010
57823	07638	75417	25906	85532	80853	47920	32719	79086	76277
95883	03872	88357	22660	41639	51747	20188	33676	37997	58689
86817	63621	51718	85194	56953	49026	53298	05380	23302	72846
75071	05936	63958	80809	09052	43912	31379	24941	82897	55358
19385	05924	17643	37034	81099	86478	66570	32685	81290	47747
50653	95687	02929	88847	59817	97697	52342	77772	41516	21306
62108	71354	38481	64582	68355	98234	89441	50133	33179	46922
38756	20124	18911	68285	40299	69862	57529	00433	57503	13604
86402	23441	04471	92961	50458	12385	24719	54217	17412	56950
21521	24909	18469	94693	58424	50233	87267	77388	66093	36902
77782	26088	39561	99008	29704	73404	42854	72034	35340	65979

36999	40918	12940	42293	81239	70291	36004	13710	74061	82940
03841	08765	62184	81399	07316	64641	94691	65074	61898	04083
00807	78726	59416	63247	45718	88664	85898	13795	87046	85866
02917	82605	09600	29041	81189	18604	19172	93031	05855	69612
6460i	93871	48040	57314	34586	32937	16346	31772	30045	14411
56595	13487	12824	01773	73622	45794	65307	27776	66889	20934
16032	78673	69922	10028	83325	45572	77482	15638	25912	65162
14456	95942	95841	77315	60149	09003	24361	99812	55686	26936
67222	14182	76751	05780	02212	58651	77991	62466	96086	90989
10247	27376	83657	45033	35590	09304	23567	22613	76589	37302
21422	63373	21711	38058	75287	34529	69725	93297	65289	93188
82388	94499	77254	24158	55167	26300	93631	15096	33918	18141
72684	89170	04762	90070	98818	03719	65486	42392	45248	15136
28699	19170	73753	23241	68546	25212	02970	16920	45402	02343
71021	32190	86137	89319	80906	99572	41952	72071	42417	91892
83980	22152	44200	73564	83758	25394	64887	07239	94281	16310
75389	30623	64567	55612	45316	88660	67815	73256	32369	29460
62047	58958	59585	76603	86878	21102	16943	11897	90282	52079
75247	99939	86152	58634	35428	30692	14171	85562	77202	25123
13137	81688	37850	72674	83095	32217	75930	41517	99201	61852
96445	95977	07199	22046	11044	36433	02260	26243	32479	88557
13460	82626	45375	01319	55155	49979	47690	96011	43922	02048
50803	46870	12341	45281	52413	65600	72523	82302	69144	72124
10865	30925	15863	84504	64695	92529	00309	63797	65493	48420
02274	91320	96809	78475	72529	79329	61460	46642	99924	53647
18542	64085	02190	75560	31551	75421	65491	74278	54110	24016
81494	88532	44757	52378	26088	46991	83883	02387	84262	07839
55678	51663	00409	51342	30827	34336	46256	29043	71083	65480
18828	06215	71236	38371	82643	12269	85808	80589	52286	94826
71579	84319	57755	52609	00554	58061	63462	45746	31033	34576
92243	23481	71940	43545	68539	71537	86147	99940	53937	07833
47839	75693	93902	88982	35549	59077	16083	44916	02950	24889
11381	70221	12843	72933	31494	01026	33125	93650	73428	56666
43407	45189	23534	91948	16877	50161	56625	66547	82253	64260
21605	99899	13588	22150	95336	50900	33526	97706	19356	22391
71671	20240	28871	14852	81695	92408	68110	43747	44635	98671
68905	40211	21246	11299	29360	37829	32326	15975	17952	70767
90950	72574	79462	16444	76097	89310	72939	47577	71549	17270
24440	50126	62637	99076	65063	54117	89122	00806	96591	44700
77885	12787	57930	48813	36474	93627	85568	44294	86627	56531
18957	98793	64674	03885	09056	02257	50615	30698	27609	20132
33692	50936	18026	14555	14991	92073	34368	98843	68804	69235
86728	19108	69750	35394	87272	32514	69272	71707	77148	51250
99061	08714	57491	55037	60722	17467	21814	82460	77249	11848
14932	41756	57749	77093	29400	64792	21228	55738	31278	34764
84678	98353	15743	07920	45989	04466	38259	07986	44432	12614
00911	64580	95310	62810	42932	54119	08522	03883	38260	30253
93849	32655	56899	37396	39997	75403	31062	95681	39325	89946
81791	87778	46219	27048	61745	72576	73076	43531	08671	19885
93070	82481	36492	76023	61538	46864	41831	14911	20292	16995
94374	47133	71545	13578	80441	49114	99380	87654	15887	55035
46498	96367	09824	04516	24591	18289	72452	03971	77274	71729
73585	61883	00121	94029	65071	07975	99806	92880	88828	62018
24062	65158	44854	58600	91996	67859	34081	99534	64688	38482
82957	29903	57694	06955	92314	88307	94859	30320	50689	84158

(contd.)

Table A2.1 (contd.)

58116	77716	50045	86995	75517	66352	52300	55676	52761	20817
43105	70000	74113	92189	07764	73417	37965	98864	74921	71121
48642	29111	60116	96879	19642	70697	48955	94622	03892	98759
57806	93529	78862	30676	75643	23121	42213	85314	07570	23608
69391	03367	32175	76497	73839	97718	25877	24022	22924	28968
80830	78624	85966	95513	90641	21868	87699	00841	77126	30649
30771	33366	52247	67479	90247	46755	60993	49041	63318	69732
67996	79930	84564	70705	57831	20618	86605	94286	90522	52203
31856	84574	93822	15991	08263	01539	87868	49187	56313	67487
70795	88068	55503	75541	85589	70521	00714	94507	81677	63946
10220	57484	30795	06334	03875	52013	30290	04465	50883	81493
69375	77419	22914	76692	06106	71121	58998	89485	14909	07367
26587	58120	81086	28041	98928	80003	67575	93899	92768	21919
19494	51024	11371	27222	60832	54625	55611	74639	01664	47290
47191	40142	35681	85465	88982	92657	22040	64126	50301	98272
93090	51159	41773	20704	89490	23192	45756	10748	22563	01230
09344	48492	90624	04082	26159	79189	65601	97914	97136	25494
02312	84666	63794	94848	53765	80070	73313	74398	10206	92343
68661	01193	11425	42226	95654	60885	29321	89929	07239	39568
41379	50393	65240	89433	51113	86395	78271	71631	81527	73933
67084	13061	22106	11072	26118	97197	34766	47556	71441	40461
31358	26123	18891	76728	34605	12075	46392	04324	28985	99182
00767	05805	76506	28204	79129	46979	98472	63428	03755	53764
29777	38405	57199	09465	10119	73633	12131	16098	41159	34470
63711	36059	27403	19035	97961	79169	29648	36408	62714	85352
75928	21317	70128	73744	17432	81679	27578	69882	99225	76110
33986	07916	77054	49838	53563	23448	38564	55787	45299	67177
00097	14652	78153	41081	56950	80972	83711	32490	42143	97116
43287	72228	88220	54054	69210	69473	99599	61404	15400	40170
25488	90700	20844	30501	53500	60058	10689	11854	93029	26010
49015	81077	73106	84215	59146	74846	47773	92276	27200	31617
07679	61012	85057	21966	96622	44041	34440	88431	93782	34566
13983	76295	46221	90843	30436	94432	50801	05014	57670	35168
25092	44402	37977	89575	84117	87202	57610	11668	52681	12062
54144	47384	51232	84698	14973	87909	78415	88714	65015	58385
94659	48073	92404	40755	34113	04453	34126	71323	03381	21602
81582	98238	04333	76282	45360	92503	14556	45931	77687	31809
02828	99064	35230	35302	48969	20868	61852	48547	53949	61402
68395	72433	73080	63041	20266	37790	34964	90692	00908	41666
47855	78200	05131	36223	09609	01310	59600	59347	93379	09306
45200	59476	58331	58983	04752	42041	04034	86689	79104	10358
01918	91945	76546	29141	76238	83569	53021	17257	82111	82930
61540	28614	16849	03525	76376	00034	19491	08069	79091	29986
21784	77463	00629	59748	31781	72763	35753	06669	25366	58467
76309	28173	66598	78866	18983	55363	51882	53612	18179	90367
64868	96259	94895	31664	72553	46067	79129	41317	43249	94019
54222	89245	18609	81706	89002	10318	43117	56097	16846	46140
00670	64654	56324	63234	20928	88497	86712	71340	19249	46024
89738	90350	88020	14263	68442	65743	08167	97330	71354	76866
78724	47107	75096	75954	55046	05828	12946	08225	68254	19754
17253	08757	72241	73101	48783	75679	08801	67848	69759	60701
01695	03347	76198	21391	45264	54639	94383	29324	61533	69315
18179	57886	72056	10430	10098	34826	01970	42570	29802	97836
93284	26630	00436	89979	18825	59778	80234	18065	92327	38938
89674	64578	56497	00294	35061	66174	56981	70068	37099	92404

70758	58852	44908	12902	80367	20544	55501	61804	25625	53932
22158	02610	03366	21233	70854	47126	67408	41123	46509	15397
73754	19177	90691	47458	46097	91095	77467	16462	18557	93615
24481	96281	13584	33060	64401	43071	32464	40241	07161	80686
91059	04211	83450	30025	72293	74025	73148	45185	63359	83061
27916	02822	96352	09633	58924	75279	99967	17782	58272	73690
92001	18234	97267	12545	80438	75775	87711	63518	22947	87339
70714	33659	47341	51517	65875	84552	18719	87924	92700	81501
70980	02613	90570	91853	35434	01132	03392	56199	22093	15606
99346	17545	97398	39715	56819	04725	97472	60312	52951	23979
81870	76792	41803	58485	23160	92630	56568	62125	26546	66474
37833	26472	60993	61249	21816	51996	86750	52771	19772	38716
19729	57397	70534	51836	25213	26797	09106	47203	28940	99930
82538	67161	66611	82661	11668	10649	18831	70045	47794	83740
82615	42264	16995	56773	31603	07879	23724	62811	34053	20721

Table A2.2   Pseudo-random standard normal variates, i.e. with zero mean and unit variance, resulting from the NAG routine G05DDF. If a random variable N has an $N(0, 1)$ distribution, then we may obtain a random variable X with an $N(\mu, \sigma^2)$ distribution by setting $X = \mu + \sigma N$.

-0.3817	1.7309	0.1020	0.1781	-2.1976	0.2756	0.8807	-0.0815	0.9479	0.8028
-1.4627	-0.0787	0.9825	-0.9018	0.1992	1.2393	-0.8749	-0.6252	-0.1124	0.5100
0.3425	-2.5194	0.7917	-0.6435	-0.8305	0.5140	1.0899	1.1004	-0.3905	-0.0751
0.2134	0.2637	0.0484	0.4427	0.7435	0.1946	1.1059	-0.7671	1.6726	-2.3685
1.9877	-0.3445	0.2755	0.3305	-1.2995	0.5384	0.8722	-1.0991	0.5924	-1.0997
-0.8110	1.5247	0.6916	-0.2860	-0.0646	0.9838	-0.8654	0.2936	-0.6669	-2.7949
1.2422	3.2607	1.8091	-0.1199	-0.9951	-1.2879	1.2767	-0.5209	-0.4755	-0.4369
1.7016	-0.8645	-0.5770	0.6983	0.9749	1.9517	0.5771	-0.8968	1.0238	-1.8262
-2.2935	-0.2379	0.4626	-0.5964	-0.8931	0.7290	-1.7184	-0.8512	-1.0430	-0.5693
0.7397	0.0578	-0.8564	0.0379	1.1788	-1.5122	-1.2791	0.3306	0.9229	0.5169
0.2863	-0.2170	0.7516	-0.6194	0.7882	-1.8403	1.1957	-0.0081	-0.3016	0.2164
-0.9733	-1.2938	-1.1274	-0.8729	0.6601	-1.2445	1.2535	0.1542	0.3109	-0.2190
1.0996	-0.4117	-0.3625	-0.9987	-1.0937	0.6487	2.0155	0.2707	1.5944	-0.9749
-0.0864	-0.9278	-0.3483	-1.1420	0.0262	1.7378	-0.0644	1.4343	0.9960	0.3269
0.2443	0.5603	1.9258	0.5536	-1.0949	0.6032	-1.4501	0.4264	-0.6752	0.0338
-0.4326	-1.0698	-1.6900	0.3524	0.5440	-0.4991	-0.2465	2.2318	-1.3405	-0.5392
1.6935	1.1556	0.2844	0.3268	0.1115	-0.8292	1.0769	-1.4394	0.4225	-0.1589
0.2875	0.1739	0.3121	-0.4311	0.3326	-0.0997	0.1986	0.1609	0.4360	1.2305
0.3443	1.5596	0.3378	-1.7689	2.3660	-0.5133	-0.6426	0.3270	-0.6778	0.7603
0.5787	0.4870	0.4032	0.6988	-0.3843	0.2841	1.5372	-0.6841	0.7115	-0.7106
0.7738	0.5707	-0.0395	0.3851	-0.1621	-1.9569	1.6740	0.8431	1.2604	-0.6320
0.9049	-1.3804	0.7392	0.6756	-0.4579	-1.3956	0.3706	-1.0403	0.8779	-0.4107
-0.4386	1.7108	-0.4503	-0.6238	0.5754	1.6640	0.0940	0.2241	-0.4473	0.6381
0.7867	-2.5127	-0.3218	2.3873	-0.6108	1.1309	1.8718	0.8949	0.0993	-0.4548
-1.2941	-0.0209	-0.6836	2.0827	0.6614	1.9945	-0.7858	0.0739	-2.1420	0.4447
-0.2455	0.0987	-2.5699	0.0936	-0.1062	0.3717	1.0177	0.3336	-0.6937	-1.7130
-1.4142	0.9285	-0.8045	-0.3002	-1.1250	1.2432	-1.5822	-0.2261	0.9079	-0.4599
-0.6906	-0.8680	0.1212	-1.0759	0.3097	0.0314	-0.2841	-1.5271	-0.7704	0.1679
-1.3015	-0.1933	-1.0627	-0.4595	-1.2822	-1.4311	1.7182	-0.1883	0.4226	-0.2443
-1.0915	0.3373	-0.8352	-0.7284	-0.8283	0.2353	-0.0307	1.3551	0.6767	1.1028

(contd.)

Table A2.2 (*contd.*)

-0.7447	-0.8772	0.0127	0.7654	-1.7962	-0.6207	-1.6598	-0.3508	-0.0217	-0.8522
-0.1038	-0.5699	-0.0175	0.8108	0.1930	0.4096	-0.3671	-0.3834	-0.1277	-0.1389
-0.1214	0.0438	-0.2015	0.5480	0.4599	-0.0736	0.0520	1.5282	0.3362	1.1986
0.6330	-0.1110	0.2029	0.2977	-0.2666	-0.2293	0.0287	0.3880	-0.5429	1.4054
0.5491	-0.4730	0.1286	0.2657	0.0265	-1.2633	-1.8893	0.4530	0.7296	0.7892
0.2180	0.2324	-0.1088	-0.1978	-0.4070	0.3857	1.3599	0.0081	-0.1153	-0.7821
0.1982	-1.0065	0.2581	1.0381	0.7913	-0.7412	1.1229	0.6890	-1.3086	-0.9308
-0.7594	1.4248	0.3888	0.0650	-2.2091	-1.2866	1.1597	-2.1434	-0.5391	1.0803
-1.6500	-0.5480	-0.4918	1.2628	-0.3704	0.5616	-0.2671	0.0672	-0.2623	-0.9198
1.5428	-0.1753	-0.5112	-0.6321	-0.0249	0.0315	-1.2220	0.8575	1.2922	-0.6323
1.4909	-0.5305	0.4103	1.7201	0.7324	-1.1984	0.0065	-0.2614	1.8041	-0.6362
-1.0791	-0.6958	-0.1386	-2.1384	-1.4571	0.1021	-0.6138	-1.1380	-0.7032	0.5105
-1.4939	0.4900	0.8626	-2.1653	-1.8681	-0.0380	-1.0705	-0.3707	0.8859	-0.9370
0.2543	-0.2685	0.9255	-0.2155	0.9099	-1.2442	-0.8209	0.7985	0.5065	1.1831
-0.1023	-0.0806	-0.2510	0.3311	0.3626	-0.9315	-1.6741	0.0759	-0.2548	0.1467
0.3196	-1.1455	-0.0005	-1.2296	1.3916	-0.9621	-0.4561	-0.7797	0.1499	0.9646
-1.4982	-0.4769	0.5693	-0.0364	-0.0107	0.2022	-0.9352	-0.5443	-0.7543	0.1192
0.1937	0.2299	-0.1068	0.2201	-0.3486	0.0246	-1.1463	1.1911	0.2864	0.2987
0.8412	0.7450	0.3409	-0.1437	-1.3503	-0.6071	0.4920	-0.6524	-0.1416	-0.6042
0.5355	0.3768	0.3261	-0.8482	1.1350	0.0969	0.4290	0.9840	0.0393	-0.1375
1.1153	0.4419	-1.1267	0.9633	1.7656	0.9445	0.1617	-1.1596	-3.3947	-0.0885
-1.4848	-0.8722	-0.5615	0.9176	-1.1794	-1.4326	1.9189	-0.4979	-0.1419	-1.6472
-0.3317	-2.0402	1.9038	-0.4714	1.2864	-0.2311	-1.5293	1.0073	-2.4238	0.2967
-0.1971	-2.4073	1.9169	-2.1454	-0.4002	0.4891	-0.8868	0.4606	-0.8800	-1.3264
-0.6606	0.6312	0.1003	-0.6720	0.8703	0.9839	-0.3304	0.1415	-0.1812	-0.5166
-1.2673	0.0231	1.2253	1.1969	-2.0353	-0.0101	-0.6852	-0.0361	-0.8582	1.9879
0.4946	0.2027	-1.3595	-0.6540	-0.2104	0.0999	1.5437	-0.1192	-0.0978	-1.1926
0.0113	0.5291	0.6703	-0.4035	0.4518	0.5733	-0.6426	-0.2681	0.9294	-0.5792
-0.7008	0.4030	0.9852	-0.4456	0.1074	-2.1391	1.2920	-1.2210	0.1845	-0.4000
0.3877	0.4916	-0.0256	0.7662	-0.9510	-1.2123	0.7984	0.4067	0.2765	0.0120
0.2176	-0.0410	1.1940	0.5138	0.0490	0.2083	-0.3644	-0.4395	0.6796	-2.1861
2.4568	1.1273	-1.8763	0.9266	-0.5361	0.4281	1.0042	-1.3309	0.7990	0.2737
-0.5189	-0.7305	0.1622	-0.3210	2.6591	-1.2294	-1.3681	-0.1728	-1.6764	-1.5392
0.4605	-1.5739	-2.0301	0.6998	-0.7805	-0.0328	-0.5359	1.3519	1.0376	0.5738
-0.2112	0.2712	-0.5583	-0.6672	-1.7556	1.5365	-0.9355	-2.1280	-0.5481	-0.4610
0.5879	0.3855	0.3576	0.6718	0.2949	0.8121	-1.1718	0.1422	-0.7861	1.3364
1.2061	-1.1127	1.1804	0.6032	-0.5295	1.3583	0.7758	0.3685	1.2931	0.4544
0.0002	0.4946	-0.6920	0.4599	2.2684	0.3157	1.2824	0.4818	0.0673	0.4632
-0.2253	0.3013	-0.2282	-0.4879	-0.8050	-0.8271	-0.3503	-0.8143	0.1021	-0.3118
0.5471	0.9078	0.2249	0.2213	-1.1060	0.4849	-0.7031	-0.0051	-2.0247	0.0957
0.1652	-0.2295	-0.1413	-0.2856	-0.1498	-0.2851	-0.4935	0.8605	0.7028	1.7356
-0.6826	0.1470	0.1734	0.1971	0.8672	0.6629	1.6643	0.5277	-0.2714	-0.1074
-0.1796	0.2998	1.2826	-1.2854	1.4688	-1.1570	-2.1032	-0.8779	0.8949	1.1306
-1.1163	0.3578	2.0840	-0.3008	-0.1733	0.6392	-2.5278	-0.6316	0.1519	0.7871
1.7477	0.6345	0.5110	-0.6359	0.4579	0.7412	-0.5617	0.5451	-1.1775	0.2286
1.0788	0.4764	-1.6086	0.4507	0.7641	1.0745	1.0140	0.6907	-0.5315	0.7717
-1.2137	0.7048	-0.3749	0.4213	-0.0404	-1.5166	-0.2283	0.1414	-1.2080	-0.5067
0.6453	0.5468	-0.0953	0.8074	-0.4192	1.2891	0.4142	-0.3544	0.2319	0.0946
0.6607	-0.6891	-0.1132	0.8950	0.0689	1.1429	0.0596	0.2354	-1.1305	0.0949
0.5430	-0.7035	0.0226	0.7450	-0.1272	0.1311	0.5547	0.1567	-0.4982	0.0276

0.3460	0.7198	-1.2223	1.2202	0.0050	-0.2626	-0.5388	-1.1806	0.2583	1.6386
-0.2589	-0.2162	1.1548	-2.6792	-1.1973	-0.5551	-0.1702	-1.5690	-1.6533	2.3711
-1.1040	0.0462	-1.3020	-0.0301	-1.0134	2.4415	0.3828	0.7004	-0.1860	-1.3155
0.9261	0.3282	1.1284	0.9049	1.1271	1.1463	-2.1408	-0.4295	-1.2538	1.0178
2.0035	0.7039	0.1641	-1.1972	-1.2732	-1.2802	-0.5247	-1.5819	0.4825	-2.5208
0.8375	-0.5823	-0.3391	0.7842	-0.2856	1.1753	-0.1701	0.2960	-0.1099	-2.2621
-0.2829	1.1759	0.6676	-0.4985	-0.1122	0.2444	-1.3184	-0.5538	2.1025	-0.9317
-0.1922	-0.0924	1.0498	1.2507	0.7364	-0.6094	0.3139	-0.8567	-2.3104	1.0240
0.2171	0.6905	0.2018	-0.9239	-0.4136	-0.1313	0.0318	0.2032	1.4773	0.7604
-1.3191	-1.6618	-0.0480	0.0743	0.2409	0.9758	-0.0492	-1.6672	-1.0366	-0.7924
0.5560	0.1231	1.1472	-2.0845	0.7485	-0.5174	-0.3905	0.0276	0.8491	1.1972
-1.5706	0.2128	0.7111	0.3511	-0.0728	-0.2712	-0.8729	0.2279	-0.1911	0.6673
-0.9456	0.6403	1.7273	0.6093	-1.9654	-0.5398	-0.0139	-0.5650	0.1421	0.5615
-0.1165	-1.5958	-0.9846	-0.2547	1.0011	1.1833	-1.0947	0.1415	-0.9593	-1.1241
-0.3379	0.6091	-0.1162	0.0141	0.5700	0.2966	0.0885	-1.3566	-0.0854	0.3437
0.7381	-1.2642	0.5075	0.4373	-1.2345	-0.4483	-0.5594	-0.4709	0.4174	0.2098
-0.8579	0.0646	0.9651	1.2438	0.4470	-0.9279	0.3644	0.3957	-0.6866	0.0309
-1.5837	0.3119	0.6497	-0.2339	0.4062	-0.2241	2.1615	1.5279	0.7134	0.9049
-0.3588	-1.9944	-0.6753	-0.3621	0.4242	-0.2129	-0.5683	0.0644	0.3488	-0.2284
0.6035	-0.2967	-0.8854	0.0490	-1.8515	0.4484	-1.1920	0.7712	-0.0391	2.4790
0.1209	0.9153	-0.6865	-0.2160	0.1633	-0.4249	-0.0635	0.4999	-1.4120	-0.6085
-0.0868	-0.0022	0.7671	-0.7399	1.0856	0.2305	-0.9537	-1.8478	-0.8942	-1.0270
-2.0612	0.6422	0.1293	-0.5036	0.8353	0.7489	0.0347	-1.0474	1.7749	-0.6057
0.6965	0.0510	-0.1628	-0.6597	-0.3132	-1.5787	2.1354	1.2608	1.4539	0.4451
-0.4552	0.5669	-0.6006	1.7116	0.3294	0.0586	-1.5486	0.5342	-0.0617	-0.5186
0.8342	0.2675	-0.6018	-0.6515	-0.4376	-0.8055	-0.3741	-1.1115	-0.3490	-0.5677
0.3083	0.4172	-0.7705	-0.5042	0.3185	1.0291	-0.9206	1.4008	1.3564	0.4660
-0.9023	-0.0349	-1.7964	-1.1040	0.7958	0.0920	0.6723	-0.9108	-0.8027	-1.1455
-0.0739	-1.0777	-0.9581	0.5148	-1.3367	0.5243	-1.0246	-0.3141	0.9019	-1.0278
0.9581	1.8532	-1.6620	1.4843	-1.2946	0.4273	-0.0871	0.2727	0.0847	-0.2205
-1.1082	0.0403	-0.9305	1.0516	0.2338	0.9036	-0.3409	0.2501	0.0306	0.7370
-0.3651	0.9010	0.1728	0.3939	0.3449	0.7790	2.3268	0.6927	-0.2406	-0.9163
-0.0878	1.4724	-0.5195	-0.1867	0.4123	-0.6673	-0.0279	-0.4646	0.5088	0.0256
1.3949	-0.4008	-0.2809	-0.2315	0.9563	0.1550	-1.3613	-2.3085	-0.1026	1.9934
-0.9507	-0.8173	-0.6481	-0.5517	0.3518	0.3320	-1.1557	0.6753	-2.0882	-0.5401
-0.2975	1.0719	0.2245	-0.6437	-0.5597	0.7993	0.9188	0.3066	-1.8867	0.3124
0.3503	-1.1069	-1.1431	-0.0498	-0.7034	-0.4407	-1.7980	0.2801	0.4698	1.1033
1.4696	-1.8542	0.3778	1.4505	-1.8632	-1.3904	0.6922	1.7818	-0.2424	-0.6377
-0.6435	-1.1988	0.8109	-0.0664	1.1235	-0.1914	-0.2877	-1.7789	-2.2507	-0.0532
0.9780	-2.4088	-0.9229	0.8373	1.0870	-1.0420	0.4932	-2.4639	-0.5129	-2.3132
-0.7672	-1.2232	-0.1937	0.1592	0.1617	-1.5393	0.9292	-1.1157	-0.1974	0.0498
-0.4779	0.0708	-0.3296	-0.3235	0.6747	-1.4498	2.7573	-1.2530	-0.3147	-1.2958
-2.8630	-1.3078	1.0689	0.0017	-1.0147	-1.1387	1.1743	0.3187	0.0707	0.0575
0.1218	0.0250	0.7535	-1.3361	0.2371	0.8131	0.7527	1.3997	1.4066	-0.8182
-0.5065	-0.4000	-0.9144	1.8379	0.6187	1.5437	-0.7353	0.4856	-0.3588	-0.4198
0.7468	0.8258	0.0800	-0.1586	-1.3957	-0.9638	-0.3441	-0.1508	0.8359	-1.6932
-0.6942	-0.0367	-1.4847	-0.4460	-0.2375	-0.6544	0.0349	0.2408	1.6872	0.4398
-0.1597	-0.8739	1.6248	-0.5259	1.0277	1.1312	1.4566	0.5325	-1.1900	0.4827
-0.6789	-0.5154	-2.3408	0.6902	1.1743	0.8016	0.1503	0.1682	0.2502	0.7707
0.1046	-1.9092	2.1039	-0.6266	-0.8288	-0.3829	-1.0186	-0.7065	-2.1309	0.3020

(contd.)

Table A2.2 (contd.)

-0.0202	-1.0896	-0.0822	-0.0484	-0.8855	-0.4400	0.4770	0.5804	0.3339	1.0071
-0.9952	0.3841	0.2855	1.0495	1.0266	0.2986	-0.1652	-1.1661	-2.0677	1.7818
-1.0542	-0.2909	-0.2539	1.1788	0.5722	0.2289	-0.8464	-0.0999	0.0906	0.0099
1.2289	-0.7314	-0.9808	0.7221	-2.2937	0.8828	-1.3563	-1.3484	1.4520	0.6500
1.1963	0.9358	-0.7070	-0.3253	0.8227	0.1428	-1.2667	0.1695	0.7293	-0.9262
2.4309	-2.1713	0.0093	1.1039	1.6404	0.7972	-0.1515	1.0183	0.4111	-1.0596
-0.1958	0.3087	0.2570	0.6028	-1.0269	-0.3193	1.5805	-0.0009	1.9738	1.1816
1.5457	-0.5316	0.0106	2.0898	-0.1731	-0.0113	0.3493	0.7890	-2.1586	-0.3330
0.2267	0.0184	-1.9819	-1.1457	-1.0812	0.5891	1.2352	-0.6801	-0.1868	0.0982
-0.2612	-1.4163	-0.8742	-0.2906	-0.7184	1.7338	-1.1651	0.3753	-1.0519	1.1569
-0.8688	-0.8280	-1.2184	-1.4862	1.1289	0.0146	-0.4793	-0.6081	1.5993	-0.8757
-0.6955	1.1250	1.0204	1.0408	-0.9633	-1.0514	-1.3220	-0.1590	1.6191	0.4754
-0.4519	-0.6287	0.6040	1.3683	0.0742	-0.0605	0.1352	-1.4517	0.8944	0.9633
-0.6602	0.6240	0.2646	-0.3819	-0.0084	1.9886	1.1334	0.1009	0.0250	-1.0182
-0.1344	-0.1182	-1.1789	1.2170	1.5683	-1.4281	-0.1510	0.5872	0.0847	-0.0274
1.2640	0.0703	1.0076	-0.3489	1.2867	-1.7079	-0.0700	-1.2367	-0.3906	-1.6084
-1.9019	-0.8625	0.5933	-0.0961	-0.8162	0.3065	0.4936	0.3531	-0.0541	-1.2877
1.2069	0.3296	-0.3806	0.6904	1.3698	1.1400	-0.3950	0.1592	1.2722	-0.6745
0.8973	1.1281	0.4162	-0.1413	0.1180	0.2408	-1.9449	-0.6814	-0.2933	-0.3693
0.8424	-0.1854	-1.0996	2.1092	1.6036	-0.2453	1.3750	-0.2262	0.3836	-0.1857
-0.1649	0.6406	0.2668	0.7307	0.2373	0.6176	-0.0584	-1.1246	1.0290	-1.0958
-1.5278	1.5875	-0.2620	2.2898	0.2245	-0.4571	0.1227	-0.2653	-0.7171	0.8032
0.2816	1.0437	0.0577	0.7579	-0.4012	-0.9804	-1.9196	0.0878	0.2968	-0.3963
-0.2777	1.1807	1.0090	1.1751	-0.3217	-0.7736	0.8179	-0.1823	-0.6675	0.8259
-0.7149	-1.0806	-0.9845	1.4536	1.1983	0.4366	1.3928	0.4531	1.3993	-2.8221
-0.6375	1.8113	0.0048	1.1811	-0.8531	-1.9459	1.0539	-0.6002	-0.0357	-0.4037
-0.6547	-0.6046	0.5760	-0.9200	-0.5664	1.4467	0.0387	-0.4900	0.2345	2.0943
-0.3402	-0.7560	1.3898	0.3334	-0.8621	-0.0175	0.3185	0.8733	0.2367	0.5469
1.7026	0.8892	0.1971	-0.7230	-0.6838	-0.3847	1.3806	-1.8337	-0.9801	-1.0126
-1.5093	-0.5261	-1.7557	-1.5625	1.5003	-0.2731	0.2979	-1.7462	-0.5448	2.1122
-0.9819	-0.0530	-1.6216	-0.3596	1.4475	-0.1451	0.8106	2.0099	-0.9352	-0.1305
-0.6849	0.5243	0.2511	-0.5422	-0.0144	0.1959	0.1298	0.2764	-0.5023	-0.3567
0.7862	0.1847	2.2407	-0.6112	1.0421	-0.5862	-0.0981	0.9954	1.1684	0.5852
-1.0366	1.4752	-0.6907	1.6814	1.6319	0.4324	0.0041	0.3231	0.8942	-0.6281
-0.0153	0.4514	-0.9402	1.7857	-0.8629	-0.5054	0.1679	-0.4131	-0.5420	-0.6098
0.5000	-0.9703	-0.4820	-0.0260	-0.0864	1.9893	-0.9936	-0.1658	0.1289	-0.8644
0.3434	-0.0508	0.8022	-0.1013	-1.5885	-0.2033	1.1547	-1.3096	-1.3238	-1.8837
1.6726	0.7494	0.0275	0.5911	0.1651	-1.2728	-0.7027	0.3364	0.4084	3.1030
-0.1636	0.5416	0.1107	-0.1274	0.5649	0.4839	1.4732	0.8408	-0.2285	0.1433
0.9941	-1.4626	-0.6725	-0.2028	0.2368	0.1373	-0.3076	-3.1783	-0.7637	-0.4868
1.0448	1.2089	3.3747	0.1110	0.6652	-2.1108	0.7605	0.0861	1.1947	-1.3761
0.4274	0.0991	0.0780	-0.6661	0.7846	0.4445	-0.9208	1.2264	-0.5873	-0.1421
0.9986	-1.0404	-0.0846	-2.4126	0.5175	1.9468	1.7186	0.4114	-1.1982	2.0531
3.0454	-0.5958	0.0123	-0.4782	-0.6522	0.7619	0.9982	-0.2226	2.2064	0.0642
-0.3035	1.1293	-0.4138	-1.6477	1.6793	0.2961	2.1618	0.2845	1.2680	-0.5597
-0.1821	-0.2756	1.8769	-0.2809	-0.9521	-0.3845	-0.0187	1.6745	1.2674	-1.9071
-1.9166	1.4852	-1.3189	0.4355	-0.1369	0.3874	-0.7681	-0.8912	0.3885	-1.0778
-0.2656	1.3838	-0.0729	-0.2775	0.7002	-0.0637	-0.1728	1.1653	-0.2935	-1.1375
-1.4067	-0.9086	0.1491	-0.4140	-1.6494	-0.4681	-0.2469	0.3487	0.6849	-1.1230
-2.0422	-0.6236	0.2436	-0.8592	-0.8945	-0.3363	-0.4052	0.1950	0.5988	-1.0363

-0.3704	1.5122	1.0163	0.7257	-1.6742	-0.7219	-0.8545	-0.7573	1.0481	-1.9817
2.3111	0.0824	-1.1191	0.3736	0.0403	-0.7051	0.1438	0.1740	0.1905	-0.2504
-1.2928	1.6848	-0.4283	2.2530	-2.2737	-0.8996	-0.8064	-0.9003	1.5859	1.6073
-0.0680	0.2739	1.7214	0.6838	1.1518	0.5303	-1.4901	3.0725	1.5888	0.0467
0.6079	2.9847	-1.2016	-0.9288	-0.6611	1.0695	0.1654	1.3661	0.3479	0.9196
0.2902	-0.0783	-0.5685	1.7392	-0.8415	-0.2260	1.1166	-1.4446	1.0936	0.4532
1.4997	-0.5749	-1.7208	1.4801	1.1530	1.3533	0.0609	0.4370	0.6746	-0.1997
0.1881	1.2962	-1.2836	-0.4941	-0.0138	-0.3771	-1.5323	-0.1888	-0.0857	-0.0968
-1.0821	-0.8844	0.1861	0.9883	-0.9521	0.6014	0.5074	0.0574	-0.6358	-0.0886
-0.7167	-0.5368	-0.6665	-0.5636	-0.0041	0.1911	-0.8054	0.1000	1.7903	-0.1474
0.3655	-0.2901	2.6336	-1.0341	-0.0606	-0.5633	1.0543	-1.9596	-0.0259	0.5221
0.5883	0.5970	-0.6511	-0.2552	0.2872	-0.1955	0.1831	-2.0900	0.9477	0.1326
-0.1159	0.4299	0.9338	-0.4526	1.4113	0.2747	1.7919	-0.7276	-0.1354	-0.3181
0.8231	-0.5494	0.7583	-1.1222	-1.3870	-0.6589	-0.5949	1.6004	-0.2463	-0.7921
-0.5248	0.4338	0.3673	-1.9040	0.8940	-1.4962	₊0.0825	2.4494	-0.2543	0.0029
-0.4950	1.0338	-1.9280	-0.8219	-0.8489	-0.7477	-0.3854	-0.8685	-0.3945	1.1731
-0.5616	-0.1437	-3.1074	0.8348	0.5676	-0.5008	-0.6084	0.8369	-1.1583	-0.9287
-2.4219	0.2816	-0.8823	-0.5755	-0.3019	0.8035	0.9731	1.5311	0.5951	0.8272
-1.2578	-0.4639	-1.0632	-1.6230	1.0035	-0.1920	-0.4635	-0.3706	-2.3201	-0.1589
0.5826	-0.5626	0.3296	0.2349	0.1979	-0.8413	-1.3038	0.1725	-1.0930	-0.6337

Table A2.3   Pseudo-random exponential variates with mean unity, resulting from setting $E = -\log_e U$, where the U values result from the NAG routine GO5CAF. If we set $X = \mu E$ then X will have the exponential p.d.f.,

$$f_X(x) = \frac{e^{-x/\mu}}{\mu} \quad \text{for } x \geq 0.$$

1.8572	3.4592	1.0743	3.6763	0.0957	0.8504	0.4698	0.1111	1.0375	1.7342
0.5251	0.4830	0.2855	1.6446	0.2093	0.0736	2.0515	0.8111	0.3192	0.0668
3.2347	0.1517	0.2225	0.7250	1.0593	1.5821	0.4814	0.5294	2.4075	0.1771
0.1458	0.5038	2.1378	0.5180	0.2783	5.0171	0.2582	1.8211	1.4403	0.1366
0.8193	1.7958	0.9195	0.1614	1.6518	3.2883	0.7644	5.6159	1.5269	0.9151
0.9144	0.5225	1.0022	0.0600	0.9241	0.1664	1.1494	1.3339	2.4800	0.6779
0.5722	0.2910	0.3653	3.0089	0.8273	2.9518	0.3656	1.4026	0.4348	0.6532
0.1169	0.1251	0.5084	0.5678	1.2514	0.4883	0.8996	0.9814	0.2377	0.2227
0.2150	1.1252	0.0134	0.9149	0.2773	0.4019	0.4207	0.0516	0.7958	0.4141
0.9534	0.2030	0.2731	0.5001	0.0300	1.2154	0.1014	0.6190	0.1772	0.5083
0.4205	0.3205	1.3315	1.7900	0.9252	1.2341	2.6929	2.7975	0.0547	0.4471
1.0300	1.6961	1.7440	0.0587	0.1239	0.6041	0.3901	0.5108	1.2378	0.2289
0.5837	0.1656	5.1307	1.2200	0.6698	0.0185	0.1918	0.9945	0.2738	0.1020
0.1853	0.5972	1.3244	0.5484	1.1069	1.5228	0.4044	1.1261	0.0252	0.3613
0.0885	0.2271	2.8642	0.4575	1.1305	0.3324	1.7452	0.9543	0.3530	0.6410
0.1712	1.5262	2.9107	0.0398	0.3061	0.5607	2.9709	0.1740	0.6842	1.3186
0.5914	2.3353	0.9906	0.1750	1.6806	2.7451	0.1197	2.6395	0.1174	2.1589
0.3200	0.3388	0.6579	0.9450	0.3465	0.4968	0.4769	1.1113	0.0247	0.2832
1.6550	0.6806	0.1719	0.3168	0.6605	2.8238	0.0597	0.3388	1.4157	1.5625
1.5747	0.7014	0.2086	0.0838	0.3878	0.3077	0.9526	0.0637	1.1553	0.5379

(contd.)

Table A2.3 (contd.)

0.4185	1.2300	0.8475	0.9256	0.3846	0.6269	1.1691	0.6065	3.7019	2.8047
1.2737	1.4079	0.3099	0.5729	0.2212	1.6878	0.4891	0.4054	0.8321	1.0964
1.1325	1.6126	2.4728	0.9263	0.5674	0.4978	0.9920	1.4322	0.5031	0.5459
2.2052	0.6565	1.0535	2.3828	0.4405	2.0129	0.4054	0.4980	1.2940	0.0433
0.0701	1.0057	0.0953	2.1642	0.1510	1.6404	1.6011	0.9120	0.6383	0.1711
0.1714	3.4097	1.9012	0.0833	1.2162	2.9613	0.1718	0.9876	0.0804	1.3418
0.1110	2.6290	0.1302	0.1271	2.2356	0.6415	2.7836	1.8770	0.3003	0.0654
2.8029	0.2799	0.1435	0.5791	1.2004	2.0409	0.5535	0.8929	0.3591	0.7373
2.3754	0.5073	0.7327	0.4555	2.5153	0.7893	1.8496	1.6322	1.0184	0.4702
0.0041	0.9784	0.1867	2.6548	0.5169	0.5486	0.7794	0.0859	1.0186	0.2387
2.8148	0.7120	0.1445	0.6907	1.1812	1.7549	0.0111	1.2587	0.8341	0.3339
0.0751	1.5227	0.0236	1.9825	0.1325	1.1399	0.3508	0.4137	0.6494	1.8879
0.1474	1.3915	2.7638	0.9405	1.8763	0.4607	2.4377	2.2926	0.0510	2.5343
0.0466	0.4257	2.3732	1.8579	3.4050	0.0480	2.1816	2.0132	0.2805	0.0573
0.0083	4.4788	0.9013	0.4316	0.8931	0.0744	1.6461	1.0149	0.2268	1.4858
0.3649	0.7327	0.6490	0.1008	0.1406	0.4946	0.2665	0.1807	0.2858	0.6055
1.0391	3.6150	0.8064	0.2634	0.0073	0.2048	2.8955	0.1219	2.3012	0.5159
0.5666	0.5058	0.5475	3.3509	1.9163	1.2568	3.3307	0.8367	0.0436	0.2315
6.2686	0.0182	1.3739	0.4793	2.2085	0.0251	0.9262	1.2752	2.6874	0.4613
0.2839	0.0818	2.2989	0.3025	0.2859	1.3330	0.7476	2.5947	2.1751	0.2562
0.7553	0.1958	0.6482	3.1597	0.7049	0.4731	0.2981	0.9889	0.4722	0.3513
0.5294	0.0283	0.1890	1.8211	1.0868	0.8588	0.4446	0.9306	2.0225	1.1114
0.2212	0.0344	0.0365	1.1418	0.3201	2.2238	0.0531	0.0442	0.1343	2.1465
1.3380	0.5349	0.1021	1.8462	1.7937	1.0681	0.1962	0.4485	0.2526	3.5403
0.7430	0.1428	0.9713	0.4437	0.3263	0.1411	3.5229	1.8129	0.9101	1.5802
0.5901	1.8289	0.4033	0.1098	2.2648	2.0006	1.4166	0.1625	2.6244	0.3690
1.5884	0.2744	0.7705	0.6616	1.3312	0.2171	0.4255	0.2302	1.6358	2.4900
0.5440	0.2728	0.6584	2.8711	0.3936	0.3822	2.6415	0.4134	0.4690	2.0769
1.0460	1.0403	0.1330	0.1673	2.1810	0.0238	0.4644	1.7693	0.7595	0.2740
1.7797	0.0356	2.3025	0.7837	0.2083	1.8412	0.0175	0.4230	0.9848	0.9767
0.0223	0.6693	0.8409	0.6856	1.8164	6.6056	1.7954	0.9148	2.0384	0.2238
0.3709	0.4719	3.7242	0.2236	1.4922	0.3131	0.2265	0.1156	2.7813	2.9407
0.7348	1.0829	2.8747	0.0057	0.2925	0.5631	0.8403	1.5795	1.1995	2.0519
0.2760	1.9640	0.8123	0.6752	0.7699	0.0591	1.7510	2.3686	1.2948	0.2045
0.0221	1.2288	0.0772	0.1937	1.5765	0.3312	0.5948	2.7803	0.9025	0.1465
0.5576	2.9216	0.7447	0.8725	0.9607	0.0060	0.0655	0.1110	0.4979	0.8917
0.3385	0.5854	1.9777	0.9338	0.4273	0.5296	0.4373	0.6929	0.1340	0.9231
0.2356	3.6138	0.2794	0.6032	4.4243	0.8540	0.2926	2.1885	0.3949	0.1318
0.5869	2.8752	0.3111	0.0049	0.7986	2.6738	0.4131	2.7132	0.5547	3.4642
0.1087	1.8766	3.7290	1.2497	0.4815	1.3412	0.1428	0.6227	1.0827	0.0069
1.7004	0.3988	3.7033	0.9512	0.6112	4.6257	0.3721	0.1397	4.5403	0.1618
4.1954	3.4717	0.3393	0.2818	0.6557	0.5678	1.4490	0.4012	2.0571	0.3174
3.1273	0.2123	3.2346	0.4919	0.4834	1.2086	2.3945	0.2160	0.7635	0.2176
3.3386	0.5137	2.8209	2.1816	0.1608	0.2346	1.0335	0.1672	0.7071	1.0351
0.4727	0.3168	0.3568	0.1580	2.1702	0.2393	0.2997	5.6316	0.6097	0.4229
1.0465	0.3252	7.1367	0.4669	0.4372	2.2244	0.5163	0.2632	0.4837	0.7102
0.8417	1.5129	2.9798	0.9627	2.2240	1.8782	1.8176	2.5195	0.1047	1.1865
0.4068	0.5325	1.4654	0.2384	0.2246	0.6089	0.1324	1.4154	1.5960	0.3706
0.1895	2.3601	0.3702	0.9370	1.3506	1.7658	0.2896	0.1202	0.4319	0.4869
0.6706	0.7942	0.4774	0.1276	3.9749	1.2749	1.8331	1.3803	0.9938	2.1865

0.2262	2.8267	0.1052	0.2362	1.7041	0.2734	0.4391	2.8258	0.6541	1.7958
3.2953	0.1135	0.1731	0.1587	0.2409	0.6744	0.5500	0.5134	0.8409	1.8484
0.6126	0.3490	0.4303	0.5224	1.8155	0.3143	1.6027	3.3937	0.1878	1.2754
0.3283	0.3141	1.8492	0.8709	0.3674	1.8772	2.1114	2.1362	1.3396	0.1503
1.0728	0.5917	0.2623	0.8331	0.0419	0.2985	0.8854	2.3474	1.3154	0.0851
0.1265	0.1137	1.0678	3.5661	0.5626	1.4893	0.8538	0.0404	0.3173	1.5016
2.0436	0.3795	1.2001	0.7641	0.7103	0.0398	1.5690	0.6858	0.4605	0.1801
2.5115	0.6820	2.0325	0.5374	0.2298	0.2618	0.9126	0.4163	1.8925	1.0032
0.3692	0.0972	1.1210	0.7469	0.5936	0.3059	2.1837	0.6422	1.9291	0.0381
4.1908	1.9557	0.9553	1.2938	0.2662	0.8478	2.6194	0.3188	0.3720	0.4913
1.7484	0.4912	1.4788	0.3386	0.1233	0.4302	2.5079	0.6642	0.4174	0.9704
0.8150	0.5232	0.2354	0.6321	0.8006	0.0826	0.8072	0.4663	0.7566	0.2717
0.8636	1.4733	0.3092	0.0898	0.2649	2.6429	0.1031	0.7353	0.0149	0.2657
0.5870	1.7315	0.3341	0.1121	0.2112	2.0134	1.2306	0.9863	3.4826	1.9196
0.6496	1.4335	0.3021	1.3425	0.0060	0.7569	0.0021	2.9950	2.2585	1.5043
0.5196	0.1848	0.8745	1.6484	0.1868	0.1593	0.0461	1.1629	1.1766	0.9316
1.4844	0.8120	0.7313	1.8527	0.6266	0.3292	0.2989	0.3474	0.1179	0.8653
0.5861	1.7172	1.3469	0.4608	0.5784	0.5551	0.1093	0.0462	0.8539	0.6057
0.4101	1.4035	2.1997	1.1978	0.3406	0.4611	1.5759	0.3700	0.7084	1.1784
0.5488	1.0422	1.0366	0.6334	0.1717	0.3166	0.1696	0.0643	0.2461	2.4901
0.1236	0.5649	0.5670	0.2612	0.5140	0.2137	0.8628	0.0793	1.3068	0.2376
0.6278	0.6140	0.6958	0.0740	0.0807	0.2824	0.9182	0.1132	0.1802	0.0479
0.5319	0.7430	0.7235	0.3241	1.2398	0.2892	2.5901	0.2241	0.0082	2.3213
0.5642	1.4134	0.1912	2.8957	1.8297	1.1298	1.2301	1.9680	0.3615	0.4673
2.3539	0.0207	1.5929	0.0600	1.8815	1.7101	0.0981	0.1547	1.9830	0.2007
1.9547	1.0600	1.6758	0.6544	0.1895	0.3309	4.3876	3.2156	0.5301	0.5060
2.2706	0.1829	0.5427	3.0325	0.9998	0.0633	0.0367	1.2302	1.5948	0.5088
0.4744	0.2108	0.9395	2.1109	0.8330	1.0813	0.0515	1.1784	1.8747	1.4503
0.7436	0.1659	0.3701	0.3925	1.1715	1.1459	0.3943	1.9392·	3.6668	0.7452
4.2728	0.5936	1.6973	1.4687	2.7646	2.4024	0.6428	0.2875	1.6920	0.0018
0.0957	0.1430	0.2623	0.0217	0.2225	1.5412	3.3999	0.0035	0.9985	0.5645
0.9998	0.2178	0.0795	0.7800	1.3048	0.8586	0.3137	0.3108	0.0850	2.6314
0.5791	1.1190	0.3662	2.0195	0.2196	1.1779	2.4838	0.1701	0.0454	0.9289
3.1114	0.0944	3.9372	3.1869	0.8534	0.8903	0.0448	1.2545	0.8944	5.1982
1.0353	0.6959	1.8216	0.1180	1.0511	0.3947	0.4461	0.2814	1.4110	1.7693
1.9735	0.1029	0.4293	0.7401	0.8278	1.3691	0.4680	1.1357	3.6728	0.3248
0.1142	0.1513	0.8532	0.7742	0.7439	0.4131	0.7486	1.8577	0.2026	0.2623
2.5522	1.0665	0.0970	0.6111	2.3773	3.3147	0.9087	4.3171	1.2309	1.1514
0.1614	1.0738	0.0914	0.1081	1.1710	0.0784	0.3549	2.2707	1.8493	1.3236
0.0816	1.1954	0.6541	0.0907	0.4436	2.3404	1.4163	1.4363	0.7996	3.2709
1.2960	1.3858	0.5328	1.6771	1.0142	1.7405	0.8354	2.2597	0.2152	0.1080
0.2821	0.1510	1.0522	1.1828	0.1748	0.0983	1.7560	0.2089	0.7639	1.8624
1.5536	0.5723	1.7027	0.5789	0.3960	0.4795	0.8997	0.1382	0.0579	1.4898
0.4561	0.3852	0.7866	0.2477	0.3633	0.8102	0.0418	1.1972	0.2539	0.2425
0.7597	2.0661	3.5845	1.1805	0.0867	0.6872	0.4028	0.4954	2.3350	0.0021
1.5397	0.4409	0.2984	0.6214	0.9617	0.4515	0.1254	2.3894	1.6741	0.6003
2.5211	0.0564	2.3985	1.5490	1.2817	0.4700	0.2423	1.9756	1.0043	1.0038
0.3574	0.9720	0.6930	0.3806	0.0809	0.6515	0.8627	2.3706	3.3770	1.3854
1.0960	0.1730	0.6055	0.9324	0.2532	0.0956	0.1002	1.4651	0.0865	0.5684
1.4266	0.2631	0.4505	0.5244	0.2732	0.0784	0.0839	0.1109	0.5341	0.1828

(contd.)

Table A2.3 (contd.)

0.0020	1.5506	0.1412	1.3845	1.0397	0.3017	1.8554	0.4013	1.2064	0.5483
0.8056	0.8211	4.3835	0.7733	2.6632	0.5644	1.0729	1.6415	1.0340	0.9698
2.3716	0.7611	0.2588	0.7346	0.0218	2.3108	0.2370	0.2574	3.5895	1.7510
0.8240	0.0189	2.9614	0.9967	3.6090	3.9467	0.9286	1.0109	1.5118	0.3809
0.0273	1.3570	1.2247	1.4077	0.2343	0.9078	1.0065	0.7209	0.9502	0.7829
0.2267	0.0247	0.3038	0.2400	0.2400	0.7751	2.6584	1.3811	0.8652	1.3459
0.5598	0.1680	0.8589	0.4105	1.0735	2.7258	0.6697	1.8197	0.9354	2.8150
2.0794	0.0497	0.0814	0.2887	0.2973	0.9280	0.3838	1.5025	0.0980	1.2042
0.1951	0.5824	0.0336	3.2740	1.0851	0.3307	1.2155	0.9797	1.6466	2.8243
0.0868	0.1754	0.9027	0.7336	0.1776	0.2053	2.1264	0.1210	6.2721	1.0742
0.7221	0.1250	1.2740	5.3879	0.4954	0.9579	0.5828	2.5294	1.9390	1.8485
0.0204	0.0423	0.0501	0.4905	0.2521	1.1115	4.3949	2.2700	0.3386	0.8384
0.8472	3.1575	2.8907	0.4862	0.3371	1.4351	0.5208	0.3927	1.6363	0.4697
0.9920	0.9528	1.0351	0.6715	0.1306	0.5560	0.4613	0.0649	0.0304	0.9037
0.3503	0.4055	0.2853	0.7234	0.4398	3.6355	0.6676	0.5404	2.4822	0.1294
1.0599	1.0631	1.0945	0.3827	0.0704	0.2922	0.4514	0.1156	1.2851	0.2050
0.2141	0.6771	0.4807	0.7462	0.8734	1.2791	1.2755	3.4165	0.6051	0.2491
0.1566	1.4162	0.7977	1.4290	0.2720	1.4363	0.2324	2.2534	1.2951	1.6273
0.4487	1.0669	2.4955	0.0868	1.0839	0.0637	0.1374	0.5649	0.6746	0.3855
0.3972	0.3519	0.2145	0.5495	0.0113	0.3364	4.7241	0.2439	3.8726	0.1497
5.5211	1.9647	0.4753	0.2350	0.1076	0.6676	1.4092	0.8066	0.4367	0.1922
1.1163	0.1145	2.2686	0.3810	0.8976	1.6584	0.6624	1.0529	0.8873	2.4664
1.3387	0.1553	0.2712	0.8206	0.1716	0.4042	0.6780	0.0934	0.4501	1.0862
1.6541	0.9755	0.1364	0.1942	1.1767	0.2006	0.1681	1.0499	0.2130	0.0662
0.2194	3.8295	1.7801	0.7952	0.0226	0.2201	2.2478	0.2974	0.6153	0.1645
0.8518	2.1984	0.4149	0.9990	1.3600	0.1521	0.9890	4.0700	1.2248	0.0405
0.6637	2.1369	2.9399	0.4496	1.4639	1.1258	0.6278	1.7280	0.0522	1.8045
2.0584	0.5447	0.3175	0.1708	1.5913	1.2572	0.0060	1.7244	2.9564	1.1627
0.3382	2.5738	0.5194	0.0429	1.6915	0.2239	0.1227	0.1405	0.8199	1.6706
2.0646	0.0767	0.3615	0.6232	0.1053	5.5540	1.6225	0.0757	2.7629	0.2297
0.1164	0.9222	0.8609	0.4352	0.2178	0.0403	2.2668	1.7038	2.2501	3.7455
0.4037	0.5324	2.2477	0.5693	0.6259	1.6627	0.7400	1.3183	0.2363	0.6818
1.2446	0.5208	0.1301	0.7437	1.2288	0.0627	1.2719	0.6786	0.3839	0.0121
0.1328	0.0227	2.1513	0.3881	0.4592	0.1841	3.0301	0.4392	0.0048	1.3458
0.2047	0.3365	0.3302	0.5236	0.0138	0.1869	2.8403	0.5162	0.8853	0.4528
0.7170	0.2295	0.0153	0.5862	0.6261	0.0188	1.1097	0.9166	0.1952	0.2451
1.2079	1.4569	0.2900	0.6899	1.2776	0.1906	1.1876	0.9581	0.2968	0.8214
0.2446	0.3464	0.7005	3.2946	1.3201	0.2762	0.2149	0.0184	2.1297	1.7916
0.0430	5.8433	0.0135	1.8503	0.8965	0.8207	0.3920	1.4928	0.0278	0.0293
3.0440	0.0599	0.2378	0.6456	0.7093	0.5758	0.5150	0.7358	0.9877	0.9797
0.9938	0.2199	3.7546	0.5525	0.7431	0.7999	0.3385	1.0272	1.8057	1.7799
1.8878	0.8093	2.0029	2.6923	0.2624	0.3475	5.4931	3.2558	0.9715	1.3978
0.0627	0.9066	0.4246	0.9075	0.9085	1.7473	2.2634	1.1941	0.6448	0.2894
2.7327	0.2536	1.0699	0.3799	0.1146	0.7085	0.2637	1.8162	0.0179	0.4552
1.2805	0.3679	0.4568	1.1520	0.6807	0.5302	0.6891	0.8711	0.1018	4.9238
0.4077	0.6182	2.7196	0.0453	0.0994	1.1690	0.1256	0.6729	1.5018	1.3534
1.6524	0.0578	0.9401	0.4844	1.7384	0.4012	0.0896	0.2522	0.7065	0.7156
3.0129	0.5579	0.4581	1.0525	0.2073	0.1652	0.3050	0.2480	0.8802	0.2568
2.5483	0.2625	0.3315	1.0044	2.2402	0.6137	0.1468	0.2303	0.3303	3.8178
0.9173	0.3956	1.3693	4.4865	0.3643	1.8122	0.1770	0.0925	1.3591	2.4642

2.2040	0.6658	2.3207	0.1306	0.5352	2.0896	0.3820	0.7302	0.0935	2.1066
0.7918	2.3115	1.0047	0.0639	1.5118	0.4024	0.0268	2.3029	1.0775	0.0326
0.5436	3.1605	0.1936	1.2563	0.1289	0.0199	0.0419	0.4355	1.7336	0.3374
0.7121	1.1316	0.2907	0.2430	0.2678	1.0157	0.1725	5.8671	0.9236	0.7738
0.9205	0.4097	0.8732	0.7612	1.2238	0.0666	1.1736	0.5410	0.2813	0.3833
0.5685	1.8259	0.1638	0.6100	1.3283	0.2501	0.3153	1.1460	0.4860	0.3625
0.8819	1.3797	0.1310	1.6174	3.8816	1.1261	4.3691	0.5803	0.1833	1.1955
0.7753	3.6829	1.5551	0.5724	0.8455	0.4765	0.2525	2.3464	0.3067	1.3141
0.0397	0.9742	0.3634	0.6421	0.9689	1.3898	2.1468	0.2756	0.0560	0.3212
0.2504	1.2670	0.1704	0.6064	1.0659	0.8960	0.1546	0.0858	1.1147	4.2771
1.7969	0.6427	2.0124	0.6597	0.1216	1.6819	1.6728	3.2960	0.1293	1.6918
0.0224	1.0260	0.5243	0.1674	0.9266	0.2525	0.3436	0.1084	0.5761	0.3710
1.4744	0.6261	0.0737	0.5459	0.0148	0.8895	0.8107	2.7103	0.0233	1.5065
2.9960	1.1638	0.8562	0.9402	0.4523	1.6983	2.5498	0.3047	0.4901	0.1664
0.0432	0.1314	1.8979	0.0683	0.3236	0.2117	0.1061	0.8164	2.9188	2.2930
0.0121	0.1066	1.2953	0.7185	1.6484	0.2372	0.5439	0.2761	0.3360	0.4578
0.2233	0.1084	0.2725	0.5746	0.3096	1.2195	0.1693	2.8480	1.7535	3.1994
0.5355	0.0281	0.4577	0.9886	0.3570	0.4880	0.7886	0.6464	6.9621	0.4395
0.5545	0.1149	0.0821	1.0438	0.0970	1.1984	0.6901	0.5444	0.0791	1.5138
0.0163	1.6258	0.4038	0.5542	0.2152	0.4569	0.4728	2.1605	1.1753	0.0358
1.1129	0.4219	0.4613	0.7902	1.9451	0.0682	0.0067	0.0892	1.7962	0.1651
0.4483	3.2982	0.6813	0.4375	0.1428	0.6159	0.8452	2.8102	3.2282	0.7926
1.7602	0.3388	1.2218	0.4790	0.8864	2.4507	2.1559	2.3118	0.1544	0.1709
0.6097	1.2873	0.6693	1.6206	0.6881	0.5629	0.2347	0.2352	3.8297	0.3592
1.7148	1.8480	0.2305	1.8525	0.1288	0.3194	0.5693	0.8343	0.2692	2.5546
0.7045	0.0797	0.9792	1.7046	1.4487	0.4401	0.9002	3.3004	0.4273	0.7470
5.1411	1.5252	0.4152	0.3399	1.5862	1.1284	0.2603	0.0219	1.2524	0.8409
0.5345	2.0084	3.7378	0.0091	0.5290	2.5321	0.3915	0.2326	1.1834	2.2947
1.8014	0.6379	0.2535	2.4549	2.1129	0.9765	0.5183	0.3234	0.9804	0.9316
0.0965	0.0863	0.3828	2.2234	0.6006	2.5960	0.5562	0.7388	0.4843	0.3086
0.0696	0.0164	0.8179	1.4533	0.4580	0.0882	0.5530	0.1249	0.9188	1.0442
2.9240	1.6763	1.0844	0.2994	0.0216	0.3540	0.2473	1.9228	0.9097	0.0706
1.0356	0.8787	4.1542	1.4730	0.7638	3.1557	0.0749	0.8124	3.2439	2.1204
1.7099	0.4228	0.3146	0.0427	4.9240	0.4535	0.0519	0.2886	1.0888	2.9278
1.9067	0.2603	1.2458	2.1826	0.0330	0.3630	0.7221	0.4798	0.6748	1.0014
0.7160	0.0519	1.1099	0.1611	1.8574	0.4757	0.3693	0.1072	0.3277	0.0193
1.2309	0.2504	0.0629	1.9590	0.5589	1.2354	0.9823	0.0538	0.6819	0.1222
0.4223	0.5136	0.8320	0.5826	0.5883	0.8406	0.2686	0.1345	1.3802	0.0766
1.4697	1.4338	0.1059	0.2931	1.7057	0.0158	2.6117	1.3460	0.9927	0.9198
1.1331	2.1323	0.0701	0.1078	0.3921	1.7266	0.1386	1.4767	0.7636	0.5444
0.3738	2.3786	0.4930	1.4381	2.6188	0.9457	0.1261	0.8035	1.2941	1.3657
0.8953	0.1155	5.6212	1.2060	0.5895	1.6094	0.3086	0.3622	0.4592	1.0345
0.3922	0.3893	0.7624	0.6278	3.3524	2.5566	0.2322	0.1354	3.1449	1.4659
1.0988	0.4790	0.6383	0.2182	2.2340	0.3634	0.4640	0.0696	1.2084	0.2201
0.2146	0.1600	0.7314	1.0836	1.2925	1.8054	0.3488	0.6463	0.2516	0.9840
0.7489	2.6789	0.6826	0.2122	0.6610	0.0736	2.6766	0.0830	0.2461	0.6852
0.8515	0.3768	0.3656	0.0757	0.0308	0.6919	0.7830	0.3295	3.6192	1.5458
0.1258	1.0159	0.0182	0.1844	0.4960	0.2649	0.4626	0.5512	0.1497	0.0108
2.3288	2.3537	0.0419	1.5865	0.2152	0.2077	2.1020	1.3048	0.3295	1.0944
0.1882	0.3909	1.1431	0.1479	0.0602	1.3079	0.3594	0.7733	0.2706	2.4657

(*contd.*)

Table A2.3 (contd.)

3.5202	0.2246	0.2192	0.9789	0.6949	0.1418	1.1849	2.1674	0.1957	3.9806
0.0287	2.6889	0.4187	1.8645	0.3404	1.3758	0.8467	1.2666	1.1033	0.2709
1.2280	0.2721	0.4442	0.1851	0.0453	1.9203	0.0124	0.0920	0.3002	0.2983
0.0297	0.7480	0.8055	0.2170	3.1127	0.7236	0.1325	4.1433	0.8884	0.3438
0.0591	0.2924	0.2276	0.3067	1.5974	0.4718	1.7170	0.0248	0.8436	0.9206
1.0641	0.8135	0.6843	0.8648	0.3294	0.0637	0.9801	1.4394	0.8895	0.3319
1.7783	0.1189	0.5233	0.3184	3.1181	1.3062	0.6397	0.2234	4.1162	4.5862
4.5601	0.5416	0.3030	0.3294	0.7134	1.1931	0.3981	1.3931	0.2265	1.0007
0.1466	0.3228	1.4144	0.0524	0.5553	1.2189	0.9171	0.1416	1.4753	0.5104
0.6415	0.7804	1.0163	0.6759	1.5126	0.4745	1.3837	2.1061	0.5932	0.1320
0.5116	0.6517	1.4896	0.0848	1.5269	0.5994	0.5370	0.2635	0.0426	0.4543
0.8782	0.8502	1.5558	0.2353	1.9698	0.5535	0.3781	1.6151	1.8896	0.6320
0.3076	2.5999	2.2703	0.4799	0.8810	0.7712	2.5527	0.0961	0.1181	0.6099
0.5845	2.8184	0.9106	0.6406	0.0958	2.5912	0.2465	1.6289	0.4201	1.1714
0.4383	1.2734	0.2024	0.2033	0.2913	2.1051	1.1084	0.8697	0.0176	0.1087
0.3505	0.6409	2.3410	1.3040	0.5340	0.4504	0.9270	3.5887	0.0176	4.5769
1.2887	1.4947	1.1008	1.2816	0.2288	0.5403	0.9440	1.6718	1.2628	0.2549
2.2799	1.8089	0.7045	0.4570	1.3930	2.3095	1.3981	0.0348	1.9592	5.4993
0.8962	0.1314	0.3085	0.4515	0.2083	0.2435	0.2162	2.0085	1.2198	2.7758
1.5377	0.9400	1.7723	0.6106	0.0424	0.0028	0.2508	0.9324	0.5952	2.8823
0.7063	0.7870	0.5889	0.2872	0.5340	4.6029	3.3972	0.1139	0.5627	0.4867
1.2034	0.9080	0.2172	0.0169	0.0644	1.8354	0.7608	1.8465	1.2756	1.1670
0.4955	0.2511	0.0872	1.5460	1.8302	0.9005	0.1636	0.3825	3.2568	0.3192
0.4295	0.5593	0.4639	2.7039	0.4622	0.6043	2.0626	0.2370	0.8121	0.1569
0.2431	0.1724	0.2445	0.6547	0.1757	0.2650	0.7652	0.3943	0.3859	0.0425
1.0792	1.1186	0.4233	0.9735	0.2673	0.5750	1.5016	2.7067	0.8731	0.3399
0.4208	1.1568	0.9394	0.4712	0.3919	0.4354	0.2549	0.1986	2.6522	0.0408
1.9315	2.2283	0.4783	0.5306	1.5770	0.2343	0.8735	4.0513	0.2107	0.7635
3.6057	4.8781	0.8152	2.5874	0.2012	0.8300	0.2237	2.2275	1.2300	0.1360
8.4972	1.0523	1.6771	3.3247	0.0836	4.2645	0.9052	0.2409	1.8666	0.0302
0.1932	0.3425	0.4364	1.3182	1.8740	0.3963	0.0274	1.4935	0.9083	0.5664
1.0321	0.7576	0.1101	0.7726	1.9979	0.8429	2.7080	0.6552	0.9545	0.3399
0.1924	0.6298	0.2231	0.0313	0.0132	1.2678	0.6509	0.6169	2.6451	1.3429
0.4964	0.1706	0.8262	1.0182	1.1400	0.0837	0.2614	0.0949	0.7188	1.2772
0.5938	2.2960	0.0643	1.6260	1.1426	3.5810	1.2046	0.4621	1.6437	1.7404
0.7921	2.0433	0.8042	2.3699	0.2993	0.9512	0.2510	3.4331	0.0986	2.9672
0.0627	2.7120	0.1561	4.4399	0.2363	0.5129	0.4065	0.0241	0.5858	0.4217
0.1051	0.2881	0.3167	1.2225	0.1422	0.2583	0.0032	1.4266	0.9464	2.2253
1.0219	3.1888	1.3849	0.6060	0.0858	0.7613	1.4090	0.3017	0.5856	0.2055
2.0679	0.1217	0.1310	0.5555	1.4606	0.4404	0.8616	0.5933	0.4423	0.0088
0.8709	0.8439	0.0513	0.8451	0.5618	1.2807	3.4832	0.2536	0.2057	1.0591
0.6493	0.3464	0.4102	0.8092	0.0219	1.0808	0.2915	0.0460	2.8171	3.6311
2.3879	1.1853	1.2833	0.3341	0.8964	1.6769	0.9345	1.6264	0.2299	0.0826
0.0137	2.0478	1.1258	0.1467	0.5040	0.0605	0.0947	0.0074	0.2776	1.3126
1.2043	0.0229	0.8479	0.1928	0.6007	0.6812	0.2069	2.4509	0.0461	0.1587
0.8571	0.0065	0.9830	4.8011	1.3846	0.7025	0.3283	0.7490	0.6792	0.2111
0.1703	0.0516	0.5281	2.2538	1.5120	0.1247	1.1168	0.2596	0.9805	0.3836
1.1553	0.0295	0.5172	0.9408	1.1564	0.3959	3.6302	1.0429	0.0681	0.8513
0.5860	0.1763	0.0754	0.3237	0.6777	4.4005	0.0454	0.2653	1.3221	0.9655
0.4648	0.1211	0.1251	0.3464	2.1544	3.0242	1.2974	0.4552	0.0796	1.8947

0.3225	1.0650	0.0161	1.6670	0.4674	0.3242	0.0734	0.7709	0.2395	0.3337
1.1911	2.7358	0.6790	0.7600	0.5138	0.1291	0.4222	0.6654	0.6381	1.3100
0.3451	0.2419	3.6555	0.0186	0.0236	0.3375	1.8152	0.1544	1.4845	1.2158
3.5440	2.8569	0.5313	0.4481	0.2044	0.3136	1.2577	1.2306	1.9160	0.0678
0.9039	0.2563	2.6067	0.2537	0.0510	0.2955	0.1571	1.6562	0.2477	1.0020
1.6963	0.9860	2.5091	1.5547	0.9164	0.2915	2.7050	0.4800	0.8942	1.4413
0.2417	0.6101	0.7853	1.4039	0.8590	1.1795	1.2945	0.4256	0.5444	0.4838
0.5561	1.2005	0.0198	0.3045	0.4920	0.0642	0.2946	1.1166	2.6119	0.9079
3.0961	0.5537	0.7118	0.6817	1.6405	1.4653	3.7672	1.2467	1.4338	1.4290
1.1939	1.3661	2.8130	2.9333	0.3383	0.2054	0.4964	0.1413	2.1046	1.1361
0.1992	0.7719	0.2674	1.2932	0.4058	1.1339	1.3010	0.3364	0.9962	0.5214
0.3979	0.6721	0.1517	1.3605	0.0244	0.9889	0.9673	0.4509	0.0230	0.0679
0.4236	0.1123	0.0203	0.1660	1.4428	2.5517	1.8990	0.7391	0.3259	0.1074
0.0059	0.1269	0.0016	0.1855	0.2136	0.7185	0.3999	0.1773	1.8161	0.4270
0.8531	0.9818	0.3410	0.2880	0.3711	1.4076	2.5961	1.0038	0.1402	2.0252
0.2479	3.0019	0.2209	0.5164	1.5471	1.0552	0.2849	0.3364	1.1884	1.1964
1.6535	0.1193	1.3134	0.1291	0.3013	0.8232	2.3469	1.4267	0.8353	1.2455
1.3252	2.5751	2.3465	0.1882	1.8173	0.8323	0.1345	2.3552	0.5207	1.4660
1.9167	0.1135	0.1964	2.2752	0.2066	1.1800	1.0843	0.2859	0.9249	1.9323
1.8726	1.1492	0.4916	1.3554	0.0602	0.9880	1.5715	0.1458	1.1776	2.0868
0.1910	0.7606	0.8482	1.2514	4.2987	0.6483	2.3567	1.0607	0.2287	0.0882
0.0400	0.4862	0.2019	0.0050	1.9056	0.1178	0.2582	3.2765	1.5651	4.7954
0.1453	0.3025	2.4653	4.3034	1.8665	0.2716	1.0640	0.6671	0.5612	1.8890
0.3911	0.3436	1.1703	0.3679	0.1875	1.6154	1.7808	0.8341	1.7471	4.9056
0.0852	2.9668	0.4222	0.1814	0.5304	1.1749	1.2840	0.9586	4.5554	1.3274
1.0043	1.1371	0.8428	0.6316	0.5952	0.9318	0.4654	2.1684	0.3810	1.5733
0.3512	1.1913	0.2460	2.3531	1.8260	0.8944	0.3884	0.3803	0.6048	1.6582
0.0116	0.2001	1.2422	1.2906	1.0025	0.3055	1.9163	2.3040	0.2066	1.0513
0.6008	0.3449	0.3110	0.0055	0.6480	1.5093	0.7446	0.1825	0.1556	1.1540
0.7628	2.5791	0.0502	0.3400	1.4389	0.2751	0.0550	0.5778	4.1245	0.7293

SOLUTIONS AND COMMENTS
FOR SELECTED EXERCISES

Chapter 1

1.1 (a) As n increases, the ratio r/n is seen to stabilize. The 'limiting frequency' definition of probability is in terms of the limits of such ratios as $n \to \infty$.

(b) See, e.g., Feller (1957, p. 84). Note the occurrence of 'long leads', i.e. once $(2r - n)$ becomes positive (or negative) it frequently stays so for many consecutive trials.

1.2 Buffon's needle is discussed further in Exercises 7.1–7.4. Estimate Prob(crossing) by:

no. of crossings/no. of trials

and solve to estimate π ($\hat{\pi}$, say).

(i) $\dfrac{2}{\hat{\pi}} = \dfrac{254}{390}$; $\hat{\pi} = 3.071$

(ii) $\dfrac{2}{\hat{\pi}} = \dfrac{638}{960}$; $\hat{\pi} = 3.009$

1.3 Effects of increasing the traffic at a railway station, due to re-routing of trains.

Effects of instituting fast-service tills at a bank/supermarket.

Effects of promoting more lecturers to senior lecturers in universities (see Morgan and Hirsch, 1976).

The improved stability, in high winds, of high-sided vehicles with roofs with rounded edges.

Election voting patterns.

1.5 The story is taken up by McArthur *et al.* (1976) (see Fig. 8.3).

1.6 Appointment systems are now widely used in British surgeries. For a small-scale study resulting from data collected from a provincial French practice, see Example 8.1.

1.7   If a model did not result in a simplification then it would not be a model. Clearly models such as that of Exercise 1.6 ignore features which would make the models more realistic. Thus one might expect more women patients than men (adults) in the morning, as compared with the afternoon. Were this true, and if there was a sex difference regarding consultation times, or lateness factors, the model should be modified accordingly.

1.9   A system with small mean waiting time may frequently give rise to very small waiting times, but occasionally result in very large waiting times. An alternative system with slightly larger mean waiting time, but no very large waiting times, could be preferable. An example of this kind is discussed by Gross and Harris (1974, p. 430). In some cases a multivariate response may be of interest (see Schruben, 1981).

1.10   The following two examples are taken from Shannon and Weaver (1964, pp. 43–44).
     (a) Words chosen independently but with their appropriate frequencies: 'Representing and speedily is an good apt or come can different natural here he the a in came the to of to expert gray came to furnishes the line message had be these.'
     (b) If we simulate to match the first-order transition frequencies, i.e. matching the frequencies of what follows what, we get: 'The head and in frontal attack on an English writer that the character of this point is therefore another method for the letters that the time of who ever told the problem for an unexpected.'

Chapter 2

2.1 $X = -\log_e U; f_X(x) = f_Y(y)\left|\dfrac{dy}{dx}\right|$

in general, and so here,

$$f_X(x) = 1\left|\frac{du}{dx}\right| = \frac{1}{u^{-1}}$$

but $u = e^{-x}$, and so $f_X(x) = e^{-x}$, for $x \geq 0$.

2.2   Note that for the exponential, gamma and normal cases X remains, respectively, exponential, gamma and normal.

2.3 $f_X(x) = e^{-x}$ for $x \geq 0$

      $W = \gamma X^{1/\beta}$

$(W/\gamma)^\beta = X$

$f_W(w) = f_X(x)|dx/dw|$

$dw/dx = \dfrac{\gamma}{\beta} x^{1/\beta - 1}$

$f_W(w) = \dfrac{\beta e^{-x}}{\gamma} x^{1 - 1/\beta} = \dfrac{\beta}{\gamma} e^{-(w/\gamma)^\beta} \left(\dfrac{w}{\gamma}\right)^{(\beta - 1)}$

i.e. $f_W(w) = \dfrac{\beta}{\gamma^\beta} w^{\beta - 1} e^{-(w/\gamma)^\beta}$ for $w \geq 0$.

2.4 $Y = N^2$; $\Pr(0 \leq Y \leq y) = \Pr(-y^{1/2} \leq N \leq y^{1/2})$

$$= \Phi(y^{1/2}) - \Phi(-y^{1/2})$$

and so

$$f_Y(y) = \tfrac{1}{2} y^{-1/2} \phi(y^{1/2}) + \tfrac{1}{2} y^{-1/2} \phi(-y^{1/2}) = y^{-1/2} \phi(y^{1/2})$$

$$= y^{-1/2} e^{-y/2} / \sqrt{(2\pi)} \qquad \text{for } y \geq 0, \text{ i.e. } \chi_1^2.$$

Note that $Y = N^2$ is not 1–1 and so rote application of Equation (2.3) gives the wrong answer.

2.5 $M_Y(\theta) = \mathscr{E}[e^{Y\theta}] = \displaystyle\int_0^\infty \dfrac{y^{-1/2} e^{y(\theta - 1/2)}}{\sqrt{(2\pi)}} \, dy$

Suppose that $\theta < \tfrac{1}{2}$. Let $z = -2y(\theta - \tfrac{1}{2})$,
i.e. $z = y(1 - 2\theta)$; $dz = dy(1 - 2\theta)$,

$$M_Y(\theta) = \int_0^\infty \left(\dfrac{z}{1 - 2\theta}\right)^{-1/2} e^{-z/2} \dfrac{dz}{\sqrt{(2\pi)}(1 - 2\theta)}$$

$$= \dfrac{(1 - 2\theta)^{-1/2}}{\sqrt{(2\pi)}} \int_0^\infty z^{-1/2} e^{-z/2} \, dz,$$

i.e. $M_Y(\theta) = (1 - 2\theta)^{-1/2}$ (see also Table 2.1).

Hence, in the notation of this question,

$M_{\sum_{i=1}^n Y_i^2}(\theta) = (1 - 2\theta)^{-n/2}$

i.e. (from Table 2.1), the m.g.f. of a χ_n^2 random variable. (See also Exercise 2.21.)

2.7   Solution:

If $S = \dfrac{1}{n} \displaystyle\sum_{i=1}^n Y_i$ then $f_S(s) = \dfrac{1}{\pi(1 + s^2)}$ for $-\infty \leq s \leq \infty$, i.e.
S has the same Cauchy p.d.f. as the component $\{Y_i\}$ random variables.

2.8 (a) $N(\mu_1 + \mu_2, \sigma_1^2 + \sigma_2^2)$

(b) Poisson with parameter $(\lambda + \mu)$

(c) If $\lambda = \mu$ the solution is given by Example 2.6, with a generalization given in Exercise 2.6. If $\lambda \neq \mu$,

$$M_{X+Y}(\theta) = \left(\frac{\lambda}{\lambda - \theta}\right)\left(\frac{\mu}{\mu - \theta}\right) = \frac{1}{(\lambda - \theta)}\frac{\lambda\mu}{(\mu - \lambda)} + \frac{1}{(\mu - \theta)}\frac{\lambda\mu}{(\lambda - \mu)}$$

and so,

$$f_{X+Y}(z) = \left(\frac{\lambda\mu}{\mu - \lambda}\right)e^{-\lambda z} + \left(\frac{\lambda\mu}{\lambda - \mu}\right)e^{-\mu z} \qquad \text{for } z \geq 0$$

i.e. a mixture of exponential densities.

While the solutions for (a) and (b) also follow easily from using generating functions, the solution to any one of these parts of this question is also readily obtained from using the convolution integral. For (a), if the convolution integral is used, note that:

$$\int_{-\infty}^{\infty} \exp\{-\tfrac{1}{2}(ax^2 + 2bx + c)\}\,dx = \sqrt{\left(\frac{2\pi}{a}\right)}\exp\left(\frac{b^2 - ac}{2a}\right).$$

2.9 $\Pr(X = r \mid X + Y = n) = \dfrac{\Pr(X = r \text{ and } Y = n - r)}{\Pr(X + Y = n)}$

$$= \frac{e^{-\lambda}\lambda^r}{r!}\frac{e^{-\mu}\mu^{n-r}}{(n-r)!}\frac{n!}{e^{-(\lambda+\mu)}(\lambda + \mu)^n} = \binom{n}{r}\frac{\lambda^r \mu^{n-r}}{(\lambda + \mu)^n}$$

i.e. the conditional distribution of X is binomial:

$$B(n, \lambda/(\lambda + \mu)).$$

2.10 $\Pr(Z \leq z) = \Pr(\max(X, Y) \leq z) = \Pr(X \leq z \text{ and } Y \leq z)$

$$F_Z(z) = F_X(z)F_Y(z).$$

2.11 $\Pr(Y \leq y) = (1 - e^{-y})^n$

Therefore $f_Y(y) = n(1 - e^{-y})^{n-1}e^{-y}$

Now $M_Z(\theta) = \dfrac{1}{(1 - \theta)}\dfrac{2}{(2 - \theta)}\cdots\dfrac{n}{(n - \theta)} = \displaystyle\sum_{i=1}^{n}\left(\frac{i}{i - \theta}\right)\prod_{\substack{j \neq i \\ j = 1}}^{n}\left(\frac{j}{j - i}\right)$

Note that

$$\prod_{\substack{j \neq i \\ j = 1}}^{n}\left(\frac{j}{j - i}\right) = \frac{1 \cdot 2 \cdot \,\ldots\, (i-1)(i+1)(i+2)\,\ldots\, n}{(1 - i)(2 - i)\,\ldots\,(-1)1 \cdot 2 \cdot \,\ldots\,(n - i)} = (-1)^{i-1}\binom{n}{i}$$

therefore $M_Z(\theta) = \displaystyle\sum_{i=1}^{n}(-1)^{i-1}\binom{n}{i}\left(\frac{i}{i - \theta}\right)$

and so $f_Z(z) = \sum_{i=1}^{n} \binom{n}{i}(-1)^{i-1} i e^{-iy}$

and as $n\binom{n-1}{i-1} = i\binom{n}{i}$ then it is clear that Y and Z have the same distribution.

Note that

$$M_Y(\theta) = n \int_0^\infty (1 - e^{-y})^{n-1} e^{-y} e^{\theta y} \, dy$$

If we let $u = e^{-y}$, $du = -e^{-y} dy$, and so

$$M_Y(\theta) = n \int_0^1 (1-u)^{n-1} u^{-\theta} \, du,$$

i.e. is proportional to the beta integral, so that

$$M_Y(\theta) = \frac{n\Gamma(n)\Gamma(1-\theta)}{\Gamma(n-\theta+1)}$$

$$= \frac{n!}{(1-\theta)(2-\theta)\ldots(n-\theta)}$$

as above. To prove this result by induction, note that if

$$W = X/(n-1), \quad f_W(w) = (n+1)e^{-(n+1)w} \qquad \text{for } w \ge 0.$$

Let $Z_{n+1} = Z_n + W$, in an obvious notation, then using the convolution integral

$$f_{Z_{n+1}}(z) = n \int_0^z (1-e^{-y})^{n-1} e^{-y} (n+1) e^{-(n+1)(z-y)} \, dy$$

$$= n(n+1)e^{-(n+1)z} \int_0^z (1-e^{-y})^{n-1} e^{ny} \, dy$$

$$= n(n+1)e^{-(n+1)z} \int_0^z (e^y - 1)^{n-1} e^y \, dy$$

$$= (n+1)e^{-(n+1)z}(e^z - 1)^n = (n+1)(1-e^{-z})^n e^{-z}.$$

Thus if the result is true for n, it is true for $(n+1)$. But it is clearly true for $n = 0$, and so it is true for $n \ge 0$.

Both Y and Z may be interpreted as the time to extinction for a population of size n, living according to the rules of a linear birth-process (see Cox and Miller, 1965, p. 156, or Bailey, 1964, p. 88, and the solution to Exercise 8.4).

2.12 $\left. \begin{array}{l} X_1 = Y_1 - Y_2 \\ \\ X_2 = Y_1 + Y_2 \end{array} \right\}$ $\quad \dfrac{\partial x_1}{\partial y_1} = 1; \ \dfrac{\partial x_1}{\partial y_2} = -1$

$\qquad\qquad\qquad\qquad \dfrac{\partial x_2}{\partial y_1} = 1; \ \dfrac{\partial x_2}{\partial y_2} = 1$

$$f_{X_1 X_2}(x_1 x_2) = f_{Y_1 Y_2}(y_1 y_2) \begin{vmatrix} \dfrac{\partial y_1}{\partial x_1} & \dfrac{\partial y_1}{\partial x_2} \\ \dfrac{\partial y_2}{\partial x_1} & \dfrac{\partial y_2}{\partial x_2} \end{vmatrix}$$

$$= \frac{\lambda^2}{2} e^{-\lambda(y_1 + y_2)} = \frac{\lambda^2}{2} e^{-\lambda x_2}.$$

It is now that this problem becomes tricky—we must determine the possible region for X_1 and X_2. Clearly, X_1 and X_2 are not independent, and the ranges of each depend upon the value taken by the other.

If $x_1 \geq 0$, then $x_2 \geq x_1$, while if $x_1 \leq 0$, $x_2 \geq -x_1$. So, to obtain the marginal distributions:

$$f_{X_1}(x_1) = \lambda^2 \int_{x_1}^{\infty} \frac{e^{-\lambda x_2}}{2} dx_2 \qquad \text{for } x_1 \geq 0$$

$$= \lambda^2 \int_{-x_1}^{\infty} \frac{e^{-\lambda x_2}}{2} dx_2 \qquad \text{for } x_1 \leq 0$$

$$= \frac{\lambda}{2} e^{-\lambda x_1} \qquad \text{for } x_1 \geq 0$$

$$= \frac{\lambda}{2} e^{\lambda x_1} \qquad \text{for } x_1 \leq 0$$

and $\quad f_{X_2}(x_2) = \displaystyle\int_{-x_2}^{x_2} \frac{\lambda^2}{2} e^{-\lambda x_2} dx_1 = \lambda^2 x_2 e^{-\lambda x_2} \qquad \text{for } x_2 \geq 0$

i.e. $\Gamma(2, \lambda)$, as anticipated.

2.13 $\Pr(|X_1 - X_2| \leq x, \min(X_1, X_2) \leq y)$

$\quad = \Pr(|X_1 - X_2| \leq x, \min(X_1, X_2) \leq y, X_1 < X_2) + \cdots$

$\quad = 2 \Pr(X_1 - X_2 \leq x, X_2 \leq y, X_1 > X_2)$

$\quad = 2 \displaystyle\int_0^y f(u) \Pr(X_1 - X_2 \leq x, X_2 \leq y, X_1 > X_2 | X_2 = u) \, du$

$\quad = 2 \displaystyle\int_0^y f(u) \Pr(X_1 \leq x + u, X_1 > u) \, du = 2 \displaystyle\int_0^y f(u)[F(x + u) - F(u)] \, du$

Therefore the joint p.d.f. is

$$\frac{d}{dx} [2f(y)\{F(x + y) - F(y)\}] = 2f(y) f(x + y) = g(x, y), \text{ say.}$$

$$f(x) = \lambda e^{-\lambda x} \Rightarrow g(x, y) = 2\lambda e^{-\lambda y} \lambda e^{-\lambda(x + y)} = 2\lambda^2 e^{-2\lambda y} e^{-\lambda x}$$

$$\Rightarrow \text{independence}$$

Independence $\Rightarrow g(x, y) = 2f(y)f(x+y) = l(x)h(y)$
for some functions l and h.

Put $x = 0$, and let $K = l(0)$, then

$$2[f(y)]^2 = Kh(y) \Rightarrow h(y) = \frac{2}{K}[f(y)]^2$$

Also, $\quad h(y) = \int_0^\infty g(x, y)\,dx = 2f(y)[1 - F(y)]$

Thus, $\quad 2f(y)[1 - F(y)] = \frac{2}{K}[f(y)]^2$

$$1 - F(y) = f(y)/K$$

$$F(y) + \frac{dF}{dy}\bigg| K = 1$$

$F(y) = 1 + e^{-Ky}$, as required, Finally,

$$\Pr(X_1 + X_2 \le 3\min(X_1, X_2) \le 3b) = \Pr(U + 2V < 3V, V \le b)$$

where $U = |X_1 - X_2|$ and $V = \min(X_1, X_2)$.

U, V are independent, from above, and $X_1 + X_2 = U + 2V$. Hence required probability $= \Pr(U < V \le b)$

$$f_U(u) = \lambda e^{-\lambda u} \text{ (see Exercise 2.12) for some } \lambda > 0$$

(cf. Exercise 2.10), and $f_V(v) = 2\lambda e^{-2\lambda v}$, and using the independence property, we have required probability

$$\int_0^b \int_0^v \lambda e^{-\lambda u} 2\lambda e^{-2\lambda v}\, du\, dv = \frac{1}{3} - e^{-2\lambda b} + \frac{2}{3} e^{-3\lambda b}.$$

2.14 $\quad f_X(x) = \dfrac{e^{-x/2}\, x^{a-1}}{2^a\,\Gamma(a)} \qquad$ for $x \ge 0$

$f_Y(y) = \dfrac{e^{-y/2}\, y^{b-1}}{2^b\,\Gamma(b)} \qquad$ for $y \ge 0$

$$\begin{cases} S = X + Y \\[2mm] T = X/(X+Y) \end{cases} \qquad \begin{aligned} \frac{\partial s}{\partial x} &= 1 \\[2mm] \frac{\partial t}{\partial x} &= \frac{y}{(x+y)^2} \end{aligned} \qquad \begin{aligned} \frac{\partial s}{\partial y} &= 1 \\[2mm] \frac{\partial t}{\partial y} &= -\frac{x}{(x+y)^2} \end{aligned}$$

therefore $ST = X$, $Y = S(1 - T)$.

$$f_{ST}(s, t) = f_{XY}(x, y)(x + y) = \frac{e^{-((x+y)/2)}x^{a-1}y^{b-1}(x+y)}{2^{a+b}\Gamma(a)\Gamma(b)}$$

$$= \frac{se^{-s/2}(st)^{a-1}(s(1-t))^{b-1}}{2^{a+b}\Gamma(a)\Gamma(b)}$$

i.e. $$f_{ST}(s, t) = \frac{s^{a+b-1}e^{-s/2}t^{a-1}(1-t)^{b-1}\Gamma(a+b)}{2^{a+b}\Gamma(a+b)\Gamma(a)\Gamma(b)}.$$

We see that S and T are independent. T is $\chi^2_{2(a+b)}$ and S is $B_e(a, b)$.

2.15 (a) $$M_{N_1 N_2}(\theta) = \int_{-\infty}^{\infty} \int_{-\infty}^{\infty} \frac{\exp[\theta x_1 x_2]}{2\pi} \exp[-\tfrac{1}{2}(x_1^2 + x_2^2)] \, dx_1 \, dx_2$$

$$= \frac{1}{\sqrt{(2\pi)}} \int_{-\infty}^{\infty} \exp[-x_1^2/2]$$

$$\times \left(\frac{1}{\sqrt{(2\pi)}} \int_{-\infty}^{\infty} \exp[\theta x_1 x_2 - x_2^2/2] \, dx_2 \right) dx_1$$

$$= \frac{1}{\sqrt{(2\pi)}} \int_{-\infty}^{\infty} \exp[-x_1^2/2] \exp[+\tfrac{1}{2}\theta^2 x_1^2]$$

$$\times \left(\frac{1}{\sqrt{(2\pi)}} \int_{-\infty}^{\infty} \exp[-\tfrac{1}{2}(\theta x_1 - x_2)^2] \, dx_2 \right) dx_1$$

therefore $$= \frac{1}{\sqrt{(2\pi)}} \frac{1}{\sqrt{(1-\theta^2)}} \int_{-\infty}^{\infty} \exp[-y^2/2] \, dy$$

$$\left(\text{from setting } x_1 = \frac{y}{\sqrt{(1-\theta^2)}} \right)$$

$$= \frac{1}{\sqrt{(1-\theta^2)}}$$

therefore $$M_{N_1 N_2 + N_3 N_4}(\theta) = \frac{1}{(1-\theta^2)}$$

But this is the m.g.f. of a random variable with the Laplace distribution:

$$f_X(x) = \tfrac{1}{2}\exp[-|x|] \qquad \text{for } -\infty \le x \le \infty,$$

To see this:

$$M_X(\theta) = \frac{1}{2} \int_{-\infty}^{0} e^x e^{\theta x} \, dx + \frac{1}{2} \int_{0}^{\infty} e^{\theta x - x} \, dx$$

$$= \frac{1}{2(1+\theta)} - \frac{1}{2(\theta-1)}, \qquad \text{for } \theta < 1$$

$$= \frac{1}{(1-\theta^2)}$$

(This is the distribution of $Y_1 - Y_2$ in Exercise 2.12.)

Hence

$$f_{|N_1N_2+N_3N_4|}(x) = e^{-x} \qquad \text{for } x \geq 0$$

(b) $C = N_1/N_2$. Also, set $D = N_1$, say, to give a 1–1 transformation from (N_1, N_2) to (C, D)

$$\frac{\partial c}{\partial n_1} = \frac{1}{n_2} \qquad \frac{\partial c}{\partial n_2} = -\frac{n_1}{n_2^2} \qquad \frac{\partial d}{\partial n_1} = 1 \qquad \frac{\partial d}{\partial n_2} = 0$$

Hence, $\quad f_{C,D}(c, d) = \dfrac{\exp\left[-\tfrac{1}{2}(n_1^2 + n_2^2)\right]}{2\pi} \begin{vmatrix} n_2^2 \\ n_1 \end{vmatrix}$

$$(d \equiv y, x \equiv c) = \frac{\exp\left[-\tfrac{1}{2}y^2(1 + 1/x^2)\right]}{2\pi} \begin{vmatrix} y \\ x^2 \end{vmatrix} = f_{X,Y}(x, y),$$

and now we form the marginal distribution of X:

$$f_X(x) = \int_{-\infty}^{\infty} f_{X,Y}(x, y)\,dy = \int_{-\infty}^{0} f_{X,Y}(x, y)\,dy + \int_{0}^{\infty} f_{X,Y}(x, y)\,dy$$

$$= \int_{-\infty}^{0} \frac{1}{2\pi x^2} (-y) \exp\left[-(y^2/2)(1 + 1/x^2)\right] dy$$

$$+ \int_{0}^{\infty} \frac{1}{2\pi x^2} y \exp\left[(-y^2/2)(1 + 1/x^2)\right] dy$$

i.e. $f_X(x) = \dfrac{1}{2\pi x^2} \dfrac{1}{(1 + 1/x^2)} + \dfrac{1}{2\pi x^2} \dfrac{1}{(1 + 1/x^2)} = \dfrac{1}{\pi(1 + x^2)}$,

for $-\infty \leq x \leq \infty$,

i.e. X has the standard Cauchy distribution.

If X, Y are independent, identically distributed *exponential* random variables, with probability density function, e^{-x} for $x \geq 0$, then for $Z = X/Y, f_Z(z) = (1 + z)^{-2}$ for $z \geq 0$, and $\mathscr{E}[Z] = \infty$.

2.16 $\phi(\mathbf{x}) = (2\pi)^{-p/2} |\Sigma|^{-1/2} \exp\{-\tfrac{1}{2}\mathbf{x}'\Sigma^{-1}\mathbf{x}\}\, d\mathbf{x}$

Suppose, first of all, that $\mu = \mathbf{0}$:

$$\mathbf{x} = \mathbf{A}^{-1}\mathbf{z} \qquad \text{and the Jacobian is: } \|\mathbf{A}^{-1}\|$$

and so, $\phi(\mathbf{z}) = (2\pi)^{-p/2} |\Sigma|^{-1/2} \exp\{-\tfrac{1}{2}(\mathbf{A}^{-1}\mathbf{z})'\Sigma^{-1}\mathbf{A}^{-1}\mathbf{z}\} \|\mathbf{A}^{-1}\|$

i.e. $\phi(\mathbf{z}) = (2\pi)^{-p/2} |\Sigma|^{-1/2} \|\mathbf{A}\|^{-1} \exp\{-\tfrac{1}{2}(\mathbf{A}^{-1}\mathbf{z})'\Sigma^{-1}\mathbf{A}^{-1}\mathbf{z}\}$

$$= (2\pi)^{-p/2} |\mathbf{A}\Sigma\mathbf{A}'|^{-1/2} \exp\{-\tfrac{1}{2}\mathbf{z}'(\mathbf{A}^{-1})'\Sigma^{-1}\mathbf{A}^{-1}\mathbf{z}\}$$

$$= (2\pi)^{-p/2} |\mathbf{A}\Sigma\mathbf{A}'|^{-1/2} \exp\{-\tfrac{1}{2}\mathbf{z}'(\mathbf{A}\Sigma\mathbf{A}')^{-1}\mathbf{z}\}$$

i.e. $\mathbf{Z} = \mathbf{A}\mathbf{X}$ has an $N(\mathbf{0}, \mathbf{A}\Sigma\mathbf{A}')$ distribution.

If $\Sigma = I$, as in the question, Z has an $N(0, AA')$ distribution. The translation: $\tilde{Z} = Z + \mu$ is readily shown to result in $N(\mu, AA')$, as required.

2.17
$$\Pr(k \text{ events}) = \int_0^\infty \lambda_1 e^{-\lambda_1 t} e^{-\lambda_2 t} \frac{(\lambda_2 t)^k}{k!} \, dt \qquad \text{for } k \geq 0$$

$$= \frac{\lambda_2^k \lambda_1}{k!} \int_0^\infty e^{-(\lambda_1 + \lambda_2) t} t^k \, dt = \frac{\lambda_2^k \lambda_1}{k!(\lambda_1 + \lambda_2)^{k+1}} \int_0^\infty e^{-\theta} \theta^k d\theta$$

$$= \frac{\lambda_1 \lambda_2^k}{(\lambda_1 + \lambda_2)^{k+1}} \qquad \text{for } k \geq 0$$

i.e. a geometric distribution.

2.18
$$\Pr(Y \geq n) = \sum_{k=n}^{n+m} \binom{n+m}{k} (1-\theta)^k \theta^{n+m-k}$$

$$= \sum_{k=0}^{m} \binom{n+m}{n+k} (1-\theta)^{n+k} \theta^{m-k}$$

while
$$\Pr(X \leq n) = \sum_{k=0}^{m} \binom{n+k-1}{k} \theta^k (1-\theta)^n$$

Therefore we require

$$\sum_{k=0}^{m} \binom{n+k-1}{k} \theta^k = \sum_{k=0}^{m} \binom{n+m}{n+k} (1-\theta)^k \theta^{m-k}$$

i.e. we require

$$\binom{n+i-1}{i} = \sum_{k=m-i}^{m} \binom{n+m}{n+k} \binom{k}{m-i} (-1)^{k-m+i} \qquad \text{for } 0 \leq i \leq m,$$

and this follows simply from considering the coefficient of z^i on both sides of the identity

$$(1+z)^{n+i-1} = (1+z)^{n+m}/(1+z)^{m+1-i}$$

2.19 $e^{-n} \sum_{r=0}^{n} \frac{n^r}{r!} = \Pr(X_n \leq n)$, where X_n has a Poisson distribution of parameter n. Now (see Exercise 2.8(b)), if X_n is of this form, we can write

$$X_n = \sum_{i=1}^{n} Z_i$$

where the Z_i are independent, identically distributed Poisson random variables of parameter 1. Hence by a simple central limit theorem,

$$\Pr(X_n \leq n) \rightarrow \Phi(0) = \tfrac{1}{2} \qquad \text{as } \mathscr{E}[X_n] = n$$

Therefore $\lim_{n \to \infty} \left(e^{-n} \sum_{r=0}^{n} \frac{n^r}{r!} \right) = \frac{1}{2}$

2.22   See *ABC*, p. 389.

2.25 $y = g(x)$ does not, in this example, have a continuous derivative. Thus Equation (2.3) is not appropriate. However, we always have:

$$\Pr(Y \leq y) = \Pr(X \leq x)$$

and for $y \leq 1$,

$$\Pr(Y \leq y) = 1 - e^{-y}$$

while for $y \geq 1$,

$$\Pr(Y \leq y) = 1 - e^{-(1+y)/2}$$

2.26   See Exercise 7.25.

2.27   For $\lambda < \mu$ the queue settles down to an 'equilibrium' state. For $\lambda \geq \mu$ no such state exists, and the queue size increases ultimately without limit. See Exercise 7.24.

Chapter 3

3.1 $\Pr(\text{respond 'Yes'}) = \Pr(\text{respond 'Yes' to (i)})\Pr((\text{i}) \text{ is the question})$

$$+ \Pr(\text{respond 'Yes' to (ii)})\Pr((\text{ii}) \text{ is the question}).$$

If one responds 'Yes' to (i) then one responds 'No' to (ii)
   $\Pr(\text{respond 'Yes' to (ii)}) = 1 - \Pr(\text{respond 'Yes' to (i)}) = 1 - \eta$, say
   $\Pr(\text{respond 'Yes'}) = \eta \Pr((\text{i})) + (1 - \eta) \Pr(\text{ii})$
$\Pr(\text{i})$ and $\Pr(\text{ii})$ are determined by a randomization device, and $\Pr(\text{respond 'Yes'})$ is estimated from the responses. For example, from a class survey of first-year mathematics students, in which group X approved of sit-ins as a form of protest (a question topical at the time), $\Pr(\text{i}) = 3/10; \Pr(\text{ii}) = 7/10$, and 34 responded 'Yes' out of 56, resulting in $\hat{\eta} = 0.38$. The students also wrote their opinion anonymously on slips which were collected, and which gave rise to $\hat{\eta} = 0.33$. See Warner (1965) for further discussion on confidence intervals, etc.

3.2   Conduct two surveys with two different probabilities of answering the innocent question, resulting in two equations in two unknowns. For further discussion, see Moors (1971). Innocent questions with known frequency of response might include month of birth, or whether an identity-number of some kind is even or odd (Campbell and Joiner, 1973).

3.3   Let $\theta = \Pr(\text{respondent has had an abortion})$
Let proportions of balls be: p_r, p_w, p_b.
Then $\Pr(\text{respond 'Yes'}) = \theta p_r + p_w$.
See Greenberg *et al.* (1971).

3.4 With the question as stated in Example 3.1, 'Yes' is potentially incriminating, whereas 'No' is not. See Abdul-Ela *et al.* (1967) and Warner (1971).

3.5 In experiments of this kind, individuals typically overestimate μ, introducing a judgement bias. Here we find:

judgement: $\bar{x} = 45.56; s = 13.69$
random   : $\bar{x} = 37.21; s = 10.28$.

3.6 Let X denote the recorded value.

$$\Pr(X = 0) = \sum_{i=4}^{6} \Pr(\text{1st die} = i) \ \Pr(\text{2nd die} = 10 - i)$$

$$= 3 \cdot \tfrac{1}{6} \cdot \tfrac{1}{6} = \tfrac{1}{12} \qquad \text{(assuming independence)}$$

$$\Pr(X = 1) = \sum_{i=5}^{6} \Pr(\text{1st die} = i) \ \Pr(\text{2nd die} = 11 - i) = \tfrac{1}{18}.$$

For $2 \le i \le 9$,

$$\Pr(X = i) = \sum_{j=\max(1, i-6)}^{\min(6, i-1)} \Pr(\text{1st die} = j) \ \Pr(\text{2nd die} = i - j).$$

Thus $\Pr(X = 2) = \tfrac{1}{36}$, $\Pr(X = 3) = \tfrac{1}{18}$, $\Pr(X = 4) = \tfrac{1}{12}$, etc. For equiprobable random digits, the method suggested is clearly unsuitable.

3.7 $\Pr(\text{2nd coin is } H \mid \text{two tosses differ}) = \dfrac{pq}{pq + qp} = \tfrac{1}{2}$

$(p = \Pr(H) = 1 - q)$.

3.8 There are 6 possibilities for both A and B, and so there are 36 possibilities in all. If we reject (i, i) results, for $1 \le i \le 6$, we get 30 possibilities, which may be used to generate uniform random digits from 0–9 as follows:

Possible outcomes from dice			Digit selected
(1, 2)	(1, 3)	(1, 4)	0
(1, 5)	(1, 6)	(2, 1)	1
(2, 3)	(2, 4)	(2, 5)	2
(2, 6)	(3, 1)	(3, 2)	3
(3, 4)	(3, 5)	(3, 6)	4
(4, 1)	(4, 2)	(4, 3)	5
(4, 5)	(4, 6)	(5, 1)	6
(5, 2)	(5, 3)	(5, 4)	7
(5, 6)	(6, 1)	(6, 2)	8
(6, 3)	(6, 4)	(6, 5)	9

Thus, for example, we choose digit 5 if we get one of (4, 1), (4, 2), (4, 3), and the conditional probability of this is:

$$(3 \times (\tfrac{1}{6}) \times (\tfrac{1}{6}))/(30/36) = \tfrac{1}{10} \qquad \text{as required.}$$

The pairs given thus result in the sequence:

$$0, 3, 1, 5, -, 9, 6, 6, 4, 0.$$

If we take these digits in fours we obtain the required numbers; e.g. 0315, 9664.

3.9 Regard HCY 7F as HCY 007F, etc. One set of 600 digits obtained in this manner is given below:

157	741	602	823	438	455	816	493	681	241
260	765	308	684	564	918	422	772	471	072
217	192	159	274	946	068	017	230	889	812
235	801	517	582	277	573	808	623	641	770
601	319	100	153	976	015	506	460	342	357
485	803	335	844	370	556	724	900	935	195
800	360	263	427	280	419	515	991	296	712
297	122	007	388	186	876	581	793	352	053
285	307	996	988	973	794	981	677	212	464
246	893	373	113	723	725	778	645	028	611
395	288	291	370	744	142	486	374	548	580
591	454	733	986	484	423	594	938	670	323
792	355	642	059	803	356	278	500	840	383
416	453	461	412	851	560	978	483	772	615
885	520	441	909	435	802	055	933	659	554
801	726	501	651	828	941	570	164	104	380
253	882	072	848	909	249	147	309	522	503
015	813	421	805	702	342	920	170	226	312
832	562	730	801	704	965	728	387	761	360
028	331	334	202	479	916	953	930	462	369

3.11 It is easy to demonstrate degeneration of this method:

(i) 55 02 00

(ii) 66 35 22 48 30 90 10 10 10

3.12 If this sequence is operated to small decimal-place accuracy then it can degenerate, as the following example shows:

0.1   0.925   0.309   0.180   0.326   0.349   0.198   0.401

0.964   0.485   0.327   0.073   0.265   0.776   0.776

3.13 We require sample mean and variance for the sample: $\{i/m, 0 \le i \le m-1\}$

$$\bar{x} = \frac{1}{m^2}\,\frac{m(m-1)}{2} = \frac{1}{2}\left(1 - \frac{1}{m}\right)$$

$$(m-1)s^2 = \left(\frac{1}{m^2}\sum_{i=1}^{m-1} i^2 - m\bar{x}^2\right) = \left(\frac{1}{6m^2}(m-1)m(2m-1) - m\bar{x}^2\right)$$

whence $s^2 = \dfrac{1}{12}\left(1 + \dfrac{1}{m}\right).$

3.14 If $A \times U0 + B = (k \times 1000) + r$

where $0 \le r < 1000$, $U1 = k + r/1000 + \varepsilon$

where $\varepsilon \ll 0.001$ is the round-off error, and $INT(U1) = r/1000 + \varepsilon$,

$$Y = (U1 - INT(U1)) \times 1000 = r + 1000\varepsilon$$

We require r, therefore set $U1 = INT(r + 1000\varepsilon + \theta)$
where θ is such that $\theta + 1000\varepsilon > 0$, but $\theta + 1000\varepsilon < 1$.
$\theta = 0.5$ will do.
An example is:

$$1 \quad 168 \quad 595 \quad 82 \quad 429 \quad 435.964 \quad 874.781 \ldots$$

With the additional line we get:

$$1 \quad 168 \quad 595 \quad 82 \quad 429 \quad 436 \quad 903 \ldots$$

3.15 $x_{i+1} = ax_i + b \quad (\bmod m)$

i.e. $x_{i+1} = ax_i + b - \kappa m \qquad$ for some κ and $0 \le ax_i + b < m$

and so
$$\left(\frac{x_{i+1}}{m}\right) = a\left(\frac{x_i}{m}\right) + \frac{b}{m} - \kappa$$

i.e. $u_{i+1} = a u_i + \dfrac{b}{m} \quad (\bmod 1) \qquad$ and $\quad 0 \le a u_i + \dfrac{b}{m} < 1$.

3.16 For any $n \ge 0$, $x_{n+1} = ax_n + b \qquad (\bmod m)$

Therefore $ax_n + b = \gamma m + x_{n+1} \qquad$ for some integral $\gamma \ge 0$ and
$0 \le x_{n+1} < m$

and so $x_{n+2} = ax_{n+1} + b \qquad (\bmod m)$

$$= a^2 x_n + ab - \gamma am + b \qquad (\bmod m)$$

$$= a^2 x_n + (a+1)b \qquad (\bmod m)$$

and this is the approach which may be used to prove this result by
induction on k for any n:

$$x_{n+k+1} = \left(a^{k+1} x_n + \left(\frac{a^k - 1}{a - 1}\right) ba + b\right) \qquad \bmod(m)$$

$$= a^{k+1} x_n + (a^{k+1} - 1)b \, (a - 1)^{-1} \qquad \bmod(m)$$

Thus every kth term of the original series is another congruential series,
with multiplier a^k and increment $(a^k - 1)(a - 1)^{-1}b$, or, equivalently,
$a^k \ (\bmod m)$ and $(a^k - 1)(a - 1)^{-1}b \ (\bmod m)$, respectively.

3.17 The illustration below is taken from Wichmann and Hill (1982a):

Value of U_2 to 1 d.p.	Value of U_1 to 1 d.p.									
	0.0	0.1	0.2	0.3	0.4	0.5	0.6	0.7	0.8	0.9
0.0	0.0	0.1	0.2	0.3	0.4	0.5	0.6	0.7	0.8	0.9
0.1	0.1	0.2	0.3	0.4	0.5	0.6	0.7	0.8	0.9	0.0
0.2	0.2	0.3	0.4	0.5	0.6	0.7	0.8	0.9	0.0	0.1
0.3	0.3	0.4	0.5	0.6	0.7	0.8	0.9	0.0	0.1	0.2
0.4	0.4	0.5	0.6	0.7	0.8	0.9	0.0	0.1	0.2	0.3
0.5	0.5	0.6	0.7	0.8	0.9	0.0	0.1	0.2	0.3	0.4
0.6	0.6	0.7	0.8	0.9	0.0	0.1	0.2	0.3	0.4	0.5
0.7	0.7	0.8	0.9	0.0	0.1	0.2	0.3	0.4	0.5	0.6
0.8	0.8	0.9	0.0	0.1	0.2	0.3	0.4	0.5	0.6	0.7
0.9	0.9	0.0	0.1	0.2	0.3	0.4	0.5	0.6	0.7	0.8

The values above give the fractional part of $(U_1 + U_2)$. If U_1 and U_2 are independent, then whatever the value of U_2, if U_1 is uniform then so is the fractional part of $(U_1 + U_2)$, and U_2 need not be uniform.

3.20 Note the generalization of this result.

3.21 (a) Numbers in the two half-periods differ by 16:

    0 13   2 31   4 17   6   3   8 21 10   7 12 25 14 11
  16 29 18 15 20   1 22 19 24   5 26 23 28   9 30 27

We have:

u_r	$9u_r + 13$	u_{r+1}	Binary form of u_{r+1}	Decimal form of $u_{r+1}/32$
0	13	13	01101	0.40625
13	130	2	00010	0.06250
2	31	31	11111	0.96875
31	292	4	00100	0.12500
4	49	17	10001	0.53125
17	166	6	00110	0.18750
6	67	3	00011	0.09375
3	40	8	01000	0.25000
8	85	21	10101	0.65625
21	202	10	01010	0.31250
10	103	7	00111	0.21875
7	76	12	01100	0.37500
12	121	25	11001	0.78125
25	238	14	01110	0.43750
14	139	11	01011	0.34375
11	112	16	10000	0.50000
16	157	29	11101	0.90625
29	274	18	10010	0.56250
⋮	⋮	⋮	⋮	⋮

revealing further clear patterns.

3.22 (a) Procedures are equivalent if the indicator digit is from a generator of period g.

(b) The sequence is: 1, 8, 11, 10, 5, 12, 15, 14, 9, 0, 3, 2, 13, 4, 7, 6, 1.
Suppose $g = 4$, for illustration. We start with (1, 8, 11, 10).
Let X denote the next number in the sequence, and suppose that if

$$(*)\begin{cases} 0 \le X \le 3 & \text{we replace the 1st stored term} \\ 4 \le X \le 7 & \text{we replace the 2nd stored term} \\ 8 \le X \le 11 & \text{we replace the 3rd stored term} \\ 12 \le X \le 15 & \text{we replace the 4th stored term} \end{cases}$$

This gives:

$$\left.\begin{array}{ll} (1, 8, 11, 10) & \\ (1, 5, 11, 10) & \text{use } 8 \\ (1, 5, 11, 12) & \text{use } 10 \\ (1, 5, 11, 15) & \text{use } 15 \end{array}\right\}$$

etc.

Once the store contains (1, 5, 9, 14) in this example then it enters a (full) cycle. Note that before a cycle can commence, the numbers in the store must correspond, in order, to the four different ranges in $(*)$, and the sequence of numbers for any part of the store must correspond to that in the original sequence for the numbers in that range.

3.23 Omit trailing decimal places, when the x_i are divided by m, to give: $u_i = x_i/m$.

3.25 For suitable α, β, θ,

$$\begin{aligned} x_{i+1} &= (2^{16} + 3)x_i + \alpha 2^{31} \\ &= 6x_i + x_i(2^{16} - 3) + \alpha 2^{31} \\ &= 6x_i + (2^{16} + 3)(2^{16} - 3)x_{i-1} + (2^{16} - 3)\beta 2^{31} + \alpha 2^{31} \\ &= 6x_i + (2^{32} - 9)x_{i-1} + \theta 2^{31} \\ &= 6x_i - 9x_{i-1} + \theta 2^{31} \\ &= 6x_i - 9x_{i-1} \pmod{2^{31}}. \end{aligned}$$

3.26 (b) One-sixth of the time.
$0 \le x_i < m$
therefore $x_n + x_{n-1} = x_{n+1} + \kappa m$ where $\kappa = 0$ or $\kappa = 1$.
Now suppose $x_{n-1} < x_{n+1} < x_n$ $(*)$
then $x_{n-1} + \kappa m < x_n + x_{n-1} < x_n + \kappa m$
If $\kappa = 1$, this implies $x_n > m$ $\Big\}$
If $\kappa = 0$, this implies $x_{n-1} < 0$ $\Big\}$ neither can occur, so $(*)$ is false.

3.27 (b) No zero values with a multiplicative congruential generator.

3.28 FORTRAN programs are provided by Law and Kelton (1982, p. 227); see also Nance and Overstreet (1975). Some versions of FORTRAN allow one to declare extra-large integers.

3.29 Set $y_n = \theta^n$ and solve the resulting quadratic equation in θ (roots θ_1 and θ_2). The general solution is then $y_n = A\theta_1^n + B\theta_2^n$, where A and B are determined by the values of y_0 and y_1.

3.32 From considering the outcome of the first two tosses,

$$p_n = \tfrac{1}{2}p_{n-1} + \tfrac{1}{4}p_{n-2} \qquad \text{for } n \geq 2.$$

$$np_n = \frac{n}{2}p_{n-1} + \frac{n}{4}p_{n-2}$$

$$= \frac{(n-1)}{2}p_{n-1} + \tfrac{1}{2}p_{n-1} + \frac{(n-2)}{4}p_{n-2} + \tfrac{1}{2}p_{n-2} \qquad \text{for } n \geq 2$$

Summing over n: if $\mu = \sum_{n=1}^{\infty} np_n$,

$$\mu - 2p_2 - p_1 = \frac{\mu}{2} + \frac{\mu}{4} + \tfrac{1}{2}(1-p_1) + \tfrac{1}{2}; \ p_1 = 0, p_2 = \tfrac{1}{4}, \mu = 6.$$

Chapter 4

4.2

```
10   LET P = .5
20   LET Q = 1-P
30   LET S = 0
40   RANDOMIZE
50   LET U = RND
60   FOR I = 1 TO 3
70   IF U < P THEN 100
80   LET U = U/Q
90   GOTO 120
100    LET U = U/P
110    LET S = S+1
120  NEXT I
130  PRINT S
140  END
```

4.6 $N_1 = (-2\log_e U_1)^{1/2} \cos 2\pi U_2$

$N_2 = (-2\log_e U_1)^{1/2} \sin 2\pi U_2$

$\dfrac{n_2}{n_1} = \tan(2\pi u_2); \ u_1 = \exp\left[-\tfrac{1}{2}(n_1^2 + n_2^2)\right]$

$$\begin{vmatrix} \dfrac{\partial u_1}{\partial n_1} & \dfrac{\partial u_1}{\partial n_2} \\[2mm] \dfrac{\partial u_2}{\partial n_1} & \dfrac{\partial u_2}{\partial n_2} \end{vmatrix} = \begin{vmatrix} -n_1 u_1 & -n_2 u_1 \\[2mm] \dfrac{-n_2}{n_1^2 2\pi \sec^2(2\pi u_2)} & \dfrac{1}{2\pi n_1 \sec^2(2\pi u_2)} \end{vmatrix}$$

$$= \frac{u_1(1 + n_2^2/n_1^2)}{2\pi \sec^2(2\pi u_2)} = \frac{u_1}{2\pi} = \frac{\exp\left[-\tfrac{1}{2}(n_1^2 + n_2^2)\right]}{2\pi}, \text{ as required.}$$

4.8 The simplest case is when $n = 2$, resulting in the triangular distribution:

$$f(x) = x \qquad \text{for} \qquad 0 \leq x \leq 1$$
$$f(x) = 2 - x \qquad \text{for} \qquad 1 \leq x \leq 2$$

For illustrations and applications, see Section 5.5.

4.9 Define $Y_0 = 0$.

$K = i$ if and only if $Y_i \leq 1 < Y_{i+1}$ for $i \geq 0$.

Thus $K \leq k$ if and only if $Y_{k+1} > 1$. Therefore

$$\Pr(K \leq k) = \Pr(Y_{k+1} > 1) = \int_1^\infty \frac{\lambda(\lambda y)^k e^{-\lambda y} dy}{k!}$$

and $\Pr(K = k) = \Pr(K \leq k) - \Pr(K \leq k-1)$ for $k \geq 0$

$[\Pr(K \leq -1) \equiv \Pr(Y_0 > 1) = 0]$

$$= \Pr(Y_{k+1} > 1) - \Pr(Y_k > 1)$$

$$= \int_1^\infty \left\{ \frac{\lambda^{k+1} y^k e^{-\lambda y}}{k!} - \frac{\lambda^k y^{k-1} e^{-\lambda y}}{(k-1)!} \right\} dy$$

But

$$\int_1^\infty \frac{\lambda^{k+1} y^k e^{-\lambda y}}{k!} dy = -\frac{\lambda^k}{k!} \int_1^\infty y^k d(e^{-\lambda y})$$

$$= -\left[\frac{\lambda^k}{k!} y^k e^{-\lambda y} \right]_1^\infty + \frac{\lambda^k}{(k-1)!} \int_1^\infty e^{-\lambda y} y^{k-1} dy$$

and so (integrals are finite)

$$\Pr(K = k) = \frac{e^{-\lambda} \lambda^k}{k!} \qquad \text{for } k \geq 0.$$

4.10 Lenden–Hitchcock (1980) considered the following four-dimensional method.

Let N_i, $1 \leq i \leq 4$, be independent random variables, each with the half-normal density:

$$f_{N_i}(x) = \sqrt{\left(\frac{2}{\pi}\right)} e^{-x^2/2} \qquad \text{for } x \geq 0.$$

Changing from Cartesian to polar co-ordinates we get:

$$(*) \begin{cases} N_1 = R \cos \Theta_1 \cos \Theta_2 \cos \Theta_3 \\ N_2 = R \sin \Theta_1 \cos \Theta_2 \cos \Theta_3 \qquad 0 < \Theta_i < \dfrac{\pi}{2}, \text{ for } 1 \leq i \leq 4 \\ N_3 = R \sin \Theta_2 \cos \Theta_3 \qquad\qquad 0 < R < \infty. \\ N_4 = R \sin \Theta_3 \end{cases}$$

The Jacobian of the transformation is $R^3 \cos \Theta_2 \cos^2 \Theta_3$, and so

$$f_{R,\Theta_1,\Theta_2,\Theta_3}(r, \theta_1, \theta_2, \theta_3) = \frac{4}{\pi^2} r^3 e^{-r^2/2} \cos \theta_1 \cos^2 \theta_2.$$

Thus R, Θ_1, Θ_2 and Θ_3 are independent, and to obtain the $\{N_i\}$ we simply simulate R, Θ_1, Θ_2 and Θ_3 using their marginal distributions and transform back to the $\{N_i\}$ by (∗). R^2 has a χ_4^2 distribution, and so is readily simulated, as described in Section 4.3.

(Had a three-dimensional generalization been used, R^2 would have had a χ_3^2 distribution, which is less readily simulated.) Lenden–Hitchcock suggested that the four-dimensional method is more efficient than the standard two-dimensional one.

4.12 Let $\Psi = \text{Pr(Accept } (V_2, V_3) \text{ and Accept } (V_1, V_2))$

$$= \text{Pr}(V_2^2 + V_3^2 \leq 1 \text{ and } V_2^2 + V_1^2 \leq 1)$$

$$f_{V^2}(x) = \frac{1}{2x^{1/2}} \quad \text{for } 0 \leq x \leq 1; \qquad F_{V^2}(x) = x^{1/2}$$

Therefore $\Psi = \displaystyle\int_0^1 \text{Pr}(V_3^2 \leq 1 - x) \, \text{Pr}(V_1^2 \leq 1 - x) \frac{1}{2x^{1/2}} \, dx$

$$= \frac{1}{2} \int_0^1 (x^{-1/2} - x^{1/2}) \, dx = 1 - \frac{1}{3} = \frac{2}{3} \neq \frac{\pi^2}{16}.$$

4.13 Logistic.

4.14 It is more usual to obtain the desired factorization using Choleski's method (see Conte and de Boor, 1972, p. 142), which states that we can obtain the desired factorization with a lower triangular matrix **A**. The elements of **A** can be computed recursively as follows.

Let $\Sigma = \{\sigma_{ij}\}$, $\mathbf{A} = \{a_{ij}\}$. We have $a_{ij} = 0$ for $j > i$, and so $\sigma_{ij} = \displaystyle\sum_{k=1}^{\min(i,j)} a_{ik} a_{jk}; \sigma_{11} = a_{11}^2$, so that $a_{11} = \sigma_{11}^{1/2}$, and then $a_{i1} = (\sigma_{i1}/\sigma_{11}^{1/2})$ for $i > 1$. This gives the first column of **A**, which may be used to give the second column, and so on. Given the first $(j-1)$ columns of **A**,

$$a_{jj} = \left(\sigma_{jj} - \sum_{k=1}^{j-1} a_{jk}^2 \right)^{1/2}$$

and $\qquad a_{ij} = \left(\sigma_{ij} - \displaystyle\sum_{k=1}^{j-1} a_{ik} a_{jk} / a_{jj} \right) \qquad$ for $i > j$.

4.16 The result is easily shown by the transformation-of-variable theory, with the Jacobian of the transformation $= (1 - \rho^2)^{1/2}$. Alternatively, the joint p.d.f. of Y_1 and X_1 can be written as the product of the conditional p.d.f. of $Y_1 | X_1$ and the marginal distribution of X_1, to yield, directly:

$$f_{Y_1 X_1}(y, x) = \frac{\exp[-\tfrac{1}{2}(y - \rho x)^2/(1 - \rho^2)]}{(1 - \rho^2)^{1/2} \sqrt{(2\pi)}} \frac{\exp[-\tfrac{1}{2}x^2]}{\sqrt{(2\pi)}}$$

(See also the solution to Exercise 6.5.)

4.17 Median $= m$ if and only if one value $= m$, $(n - 1)$ of the other values are less than m, and $(n - 1)$ are greater than m. The value to be m can be chosen $(2n - 1)$ ways; the values to be $< m$ can be chosen $\binom{2n - 2}{n - 1}$ ways, hence,

$$f_M(m)\,dm = (2n - 1)\binom{2n - 2}{n - 1} m^{n-1}(1 - m)^{n-1}\,dm$$

i.e. M has a $B_e(n, n)$ distribution.

4.19
$$\Pr(X = k) = \frac{1}{k!} \int_0^\infty \frac{e^{-\lambda} \lambda^k e^{-\theta \lambda} \theta^n \lambda^{n-1}}{\Gamma(n)}\,d\lambda$$

$$= \frac{\theta^n}{k!\Gamma(n)} \int_0^\infty \lambda^{n+k-1} e^{-\lambda(\theta+1)}\,d\lambda$$

$$= \frac{\theta^n}{k!\Gamma(n)} \frac{\Gamma(n+k)}{(\theta+1)^{n+k}} \qquad \text{as required.}$$

4.21 $y = e^x$; $\dfrac{dy}{dx} = e^x = y$; $x = \log_e(y)$

$$f_Y(y) = \frac{1}{y\sigma\sqrt{(2\pi)}} \exp\left\{-\frac{1}{2}\left(\frac{\log_e y - \mu}{\sigma}\right)^2\right\} \qquad \text{for } y \geq 0.$$

4.22 $M(\theta) = \displaystyle\sum_{k=1}^\infty e^{\theta k} p_k = -\frac{1}{\log(1 - \alpha)} \sum_{k=1}^\infty \frac{(\alpha e^\theta)^k}{k}$

Note that $\qquad dM(\theta)/d\theta = -\dfrac{1}{\log(1 - \alpha)} \displaystyle\sum_{j=0}^\infty (\alpha e^\theta)^j \alpha e^\theta$

i.e. $\qquad \dfrac{dM(\theta)}{d\theta} = \dfrac{\alpha e^\theta}{\log(1 - \alpha)(\alpha e^\theta - 1)} \qquad \text{for } \alpha e^\theta < 1$

Therefore $\qquad M(\theta) = \kappa + \dfrac{1}{\log(1 - \alpha)} \log(1 - \alpha e^\theta)$

$M(0) = 1$ so $\kappa = 0$.

$$\Pr(X = k) = -\frac{1}{\log(1 - \alpha)} \int_0^\alpha y^{k-1}\,dy = -\frac{\alpha^k}{k \log(1 - \alpha)} \qquad \text{for } k \geq 1.$$

This question is continued in Exercise 5.14.

Chapter 5

5.1 Let $W = 1 - U$; $\quad f_W(w) = f_U(u)\left|\dfrac{du}{dw}\right|$

$$\left|\frac{du}{dw}\right| = 1 \qquad \text{and the result is proved.}$$

5.2 $\quad F_{\tilde{X}}(w) = \Pr(\tilde{X} \le w) = \Pr(\tilde{X} \le w \,|\, \tilde{X} = X)\,\Pr(\tilde{X} = X)$

$$+ \Pr(\tilde{X} \le w \,|\, \tilde{X} = -X)\,\Pr(\tilde{X} = -X)$$

$$= \tfrac{1}{2}\Pr(X \le w) + \tfrac{1}{2}\Pr(-X \le w)$$

$$= \tfrac{1}{2}\Pr(X \le w) + \tfrac{1}{2}\Pr(X \ge -w)$$

Therefore if $w \ge 0$,

$$F_{\tilde{X}}(w) = \frac{1}{2}\int_0^w f_X(x)\,dx + \frac{1}{2} = \Phi(w)$$

and if $w \le 0$,

$$F_{\tilde{X}}(w) = 0 + \tfrac{1}{2}\Pr(X \ge -w) = \Phi(-w).$$

5.3 Poisson random variables with large mean values, μ, say, can be simulated as the sum of independent Poisson variables with means which sum to μ.

5.4 We can take a Poisson random variable X, with mean 3 as an illustration:

i	0	1	2	3	4	5	6
$\Pr(X = i)$	0.0498	0.1494	0.2240	0.2240	0.1680	0.1008	0.0504

i	7	8	9	10	≥ 11
$\Pr(X = i)$	0.0216	0.008	0.003	0.0008	0.0002

(b) Here we can set $\theta = 2$, say; $\Pr(X \le 2) = 0.4232$.
 As a first stage, check to see if $U > 0.4232$. If so, then it is not necessary to check U against $\Sigma_{i=0}^{j}\Pr(X = i)$, for $j \le \theta = 2$.

(c) Here we could take $\theta_1 = 1$ and $\theta_2 = 4$
 $\Pr(X \le 1) = 0.1992$; $\Pr(X \le 4) = 0.8152$, and as a first stage we check to see where U lies:

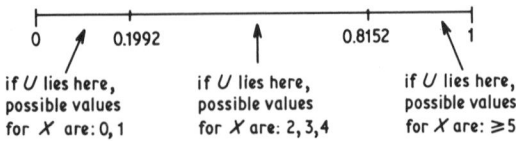

if U lies here, possible values for X are: 0, 1

if U lies here, possible values for X are: 2, 3, 4

if U lies here, possible values for X are: ≥ 5

(d) When the probabilities are ordered we have:

i	2	3	4	1	5	6
$\Pr(X = i)$	0.224	0.224	0.168	0.1494	0.1008	0.0504

i	0	7	8	9	10	≥ 11
$\Pr(X = i)$	0.0498	0.0216	0.008	0.003	0.0008	0.0002

Ordering is a time-consuming operation but once it is done we obtain the obvious benefit of checking the most likely intervals first. As a first step one might check to see if $U \leq 0.616$. If so, X takes one of the values 2, 3 or 4.

5.5   (a) Illustrate $F(x)$ by means of a diagram. Note the symmetry.
If $U \leq 0.5$, set $X = \sqrt{(2U)}$; if $U \geq 0.5$, set $X = 2 - \sqrt{(2 - 2U)}$.
(b) We also obtain such an X by setting $X = U_1 + U_2$, where U_1 and U_2 are independent, identically distributed $U(0, 1)$ random variables (see Exercise 4.8).

5.7 $F_X(x) = 6 \int_0^x y(1 - y)\,dy = 6\left[\dfrac{y^2}{2} - \dfrac{y^3}{3}\right]_0^x = 3x^2 - 2x^3$

Therefore set $U = 3X^2 - 2X^3$ and solve this cubic equation in X.

5.8   (a) $F(x) = \dfrac{1}{\pi} \int_{-\infty}^x \dfrac{dy}{(1 + y^2)}$, $-\infty \leq x \leq \infty$.

Set $y = \tan \theta$; $F(x) = \dfrac{1}{\pi} \int_{-\pi/2}^{\tan^{-1} x} d\theta = \dfrac{1}{\pi}\left(\tan^{-1} x + \dfrac{\pi}{2}\right)$

for $-\dfrac{\pi}{2} \leq \theta \leq \dfrac{\pi}{2}$

Therefore set $U\pi = \tan^{-1} x + \dfrac{\pi}{2}$,

i.e. $\tan(U\pi + \pi/2) = X$
or, equivalently,
$X = \tan(\pi U)$

(b) We saw, from Section 4.2.1, that we can write
$N_1 = (-2\log_e U_1)^{1/2} \sin 2\pi U_2$
$N_2 = (-2\log_e U_1)^{1/2} \cos 2\pi U_2$
where U_1 and U_2 are independent $U(0, 1)$ random variables. Thus $N_1/N_2 = \tan(2\pi U_2)$, which clearly has the same distribution as $X = \tan(\pi U)$ above, i.e. the standard Cauchy.

5.9   Let $u^{-1} = 1 + \exp(-2a_1 \tilde{x}(1 + a_2 \tilde{x}^2))$
Therefore we need to solve

$$\log_e(u^{-1} - 1) = -2a_1 \tilde{x}(1 + a_2 \tilde{x}^2)$$

Use the result: If $ax^3 + x - b = 0$, $x = c - 1/(3ac)$
where

$$2ac^3 = b + \left(b^2 + \frac{4}{27a}\right)^{1/2} \qquad \text{(see Page, 1977).}$$

5.10   See Kemp and Loukas (1978a, b). They suggest truncating the distribution—e.g. to give values of $\{p_{ij}\}$ summing, over i and j, to give unity to, say, 4 decimal places. Then use the approach of Exercise 5.4(d). They found this approach faster than first simulating a marginal variable and then simulating the bivariate distribution via the conditional distribution.

5.11   One approach is via the conditional and marginal distributions, as above, in the comments on Exercise 5.10.

5.12 $\Pr(Y = i - 1) = e^{-\lambda(i-1)} - e^{-\lambda i} = e^{-\lambda i}(e^\lambda - 1) \qquad \text{for } i \geq 1$

$$= e^{-\lambda(i-1)}(1 - e^{-\lambda}) \qquad \text{for } i \geq 1.$$

Hence, from Section 2.6, $(Y+1)$ has a geometric distribution with $q = e^{-\lambda}$. This relationship is not surprising since both geometric and exponential random variables measure waiting-times, as stated in Section 2.10.

5.13   (a) $F(x) = 1/(1 + e^{-x})$
Therefore set $U = 1/(1 + e^{-X})$

$$U^{-1} = 1 + e^{-X}$$

$$X = -\log_e(U^{-1} - 1).$$

(b) $F(x) = \dfrac{\beta}{\gamma^\beta} \displaystyle\int_0^x w^{\beta-1} \exp[-(w/\gamma)^\beta]\, dw$

$$= \left[\exp[-(w/\gamma)^\beta]\right]_x^0 = 1 - \exp[-(x/\gamma)^\beta]$$

Therefore set $U = 1 - \exp[-(X/\gamma)^\beta]$

i.e. $\exp[-(X/\gamma)^\beta] = 1 - U;$ $-\left(\dfrac{X}{\gamma}\right)^\beta = \log_e(1 - U),$

$$X = \gamma(-\log_e(1 - U))^{1/\beta}$$

or equivalently, and more simply (see Exercise 5.1)

$$X = \gamma(-\log_e U)^{1/\beta}$$

(c) Set $U = 1 - (k/X)^a$; $1 - U = (k/X)^a$
$$k(1 - U)^{-1/a} = X$$

or equivalently, as in (b)
$$X = kU^{-1/a}$$

(d) Set $U = \exp(-\exp((\xi - X)/\theta)$
$-\log_e U = \exp((\xi - X)/\theta)$
$\theta \log_e(-\log_e U) = \xi - X$,
$X = \xi - \theta \log_e(-\log_e U)$

5.14 To complete the simulation of a random variable with the logarithmic distribution of Exercise 4.22, we need to simulate the random variable Y, for which
$$F_Y(y) = \log(1 - y)/\log(1 - \alpha) \qquad \text{for } 0 \le y \le \alpha.$$

To do this by the inversion method, set $U = \log(1 - Y)/\log(1 - \alpha)$, i.e.
$\log(1 - Y) = U \log(1 - \alpha)$; $1 - Y = (1 - \alpha)^U$;
$$Y = 1 - (1 - \alpha)^U$$

5.15 (a) First we find $f(x, y)$:
$$f(x, y) = f(x)f(y)(1 - \alpha(1 - F(x))(1 - F(y))) + f(x)F(y)\alpha f(y)$$
$$\times (1 - F(x))$$
$$= \alpha F(x)f(y)f(x)(1 - F(y)) - \alpha f(x)f(y)F(x)F(y)$$

Now make the transformation of variable: $U = F(X)$, $V = F(Y)$:
$$f(u, v) = 1 - \alpha(1 - u)(1 - v) + \alpha v(1 - u) + \alpha u(1 - v) - \alpha uv$$
$$= 1 - \alpha(1 - 2u)(1 - 2v)$$

(b) $f(u, v) = (1 - \alpha \log u)(1 - \alpha \log v) - \alpha) vu^{-\alpha \log v}$

(c) $f(u, v) = \dfrac{\pi}{2} \dfrac{\{1 + \tan^{-2}(\pi u)\}\{1 + \tan^{-2}(\pi v)\}}{\{1 + \tan^{-2}(\pi u) + \tan^{-2}(\pi v)\}^{3/2}}$

5.16 Rejection, with a high probability of rejection, can be applied simply by generating a uniform distribution of points over an enveloping rectangle, and accepting the abscissae of the points that lie below the curve. High rejection here is not too important as this density is only sampled with probability 0.0228.

5.17 $X = \cos \pi U$; $f_X(x) = \left| \dfrac{du}{dx} \right|$

$\dfrac{dx}{du} = -\pi \sin(\pi u)$; $f_X(x) = \dfrac{1}{\pi \sin \pi u} = \dfrac{1}{\pi \sqrt{(1 - x^2)}}$

$$\text{for } -1 \le x \le 1.$$

To simulate X without using the 'cos' function we can use the same approach as in the Polar Marsaglia method:

$$\cos \pi U = 2 \cos^2 \left(\frac{\pi U}{2} \right) - 1$$

Thus, if U_1, U_2 are independent $U(0, 1)$ random variables

if $U_1^2 + U_2^2 < 1$, set $X = \frac{2U_1^2}{U_1^2 + U_2^2} - 1 = \left(\frac{U_1^2 - U_2^2}{U_1^2 + U_2^2} \right).$

5.18 Using a rectangular envelope gives an acceptance probability of 2/3. Using a symmetric triangular envelope gives an acceptance probability of 2/3 also, but the acceptance probability of 8/9 results from using a symmetric trapezium. Simulation from a trapezium density is easily done using inversion.

5.19 For the exponential envelope we simulate from the half-logistic density:

$$f(x) = 2e^{-x}/(1 + e^{-x})^2 \qquad \text{for } x \geq 0$$

We need to choose $k > 1$ so that $ke^{-x} > 2e^{-x}/(1 + e^{-x})^2$ for all $x \geq 0$, i.e. choose k so that $k > 2(1 + e^{-x})^{-2}$ for all $x \geq 0$, and this is done for smallest k by setting $k = 2$ ($x = \infty$).

If U_1, U_2 are independent $U(0, 1)$ random variables, set $X = -\log_e U_1$ if $U_2 < (1 + U_1)^{-2}$. Finally transform to $(-\infty, \infty)$ range.

5.20
$$
\left.
\begin{array}{l}
f(x) = \dfrac{1}{\sqrt{(2\pi)}}\, e^{-x^2/2} \\[2mm]
h(x) = e^{-x}(1 + e^{-x})^{-2}
\end{array}
\right\} \qquad -\infty \leq x \leq \infty
$$

Set $q(x) = \dfrac{1}{\sqrt{(2\pi)}}\, e^{-x^2/2}\, (1 + e^{-x})^2\, e^x.$

If $l(x) = \log_e q(x)$

$$l(x) = -\frac{1}{2} \log_e (2\pi) + x - \frac{x^2}{2} + 2 \log_e (1 + e^{-x})$$

$$\frac{dl(x)}{dx} = 1 - x - \frac{2}{(1 + e^x)} = 0 \qquad \text{when } x = 0$$

$$\frac{d^2 l(x)}{dx^2} = -1 + \frac{2e^x}{(1 + e^x)^2} < 0 \qquad \text{when } x = 0$$

Therefore $x = 0$ maximizes $q(x)$, and $k = \max_x (q(x)) = \dfrac{4}{\sqrt{(2\pi)}}$

$= 1.596.$

To operate the rejection method here we need to simulate from $h(x)$, and this is readily done by the inversion method of Exercise 5.13(a); probability of rejection $= 1 - 1/k = 0.37$. This may seem surprisingly high, but recall that the two distributions have different variances (1 and $\pi^2/3$) and a better approach would be to use a logistic distribution of unit variance (see Fig. 5.14).

5.21 $k = \max\limits_{x} \left\{ \sqrt{\left(\dfrac{2}{\pi}\right)} \exp\left[-x^2/2\right] / (\lambda \exp\left[-\lambda x\right]) \right\}$

i.e. $k = \max\limits_{x} \left\{ \left(\dfrac{1}{\lambda}\sqrt{\left(\dfrac{2}{\pi}\right)}\right) \exp(\lambda x - x^2/2) \right\}$

Let $y = \lambda x - x^2/2$; $\dfrac{dy}{dx} = \lambda - x = 0$ when $x = \lambda$, $\dfrac{d^2 y}{dx^2} = -1$.

Hence k is obtained for $x = \lambda$, to give:

$$k = \frac{1}{\lambda}\sqrt{\left(\frac{2}{\pi}\right)} \exp(\lambda^2/2)$$

The method becomes: accept $X = -\dfrac{1}{\lambda} \log_e U_1$ if

$U_1 U_2 \le \exp\left[-(\lambda^2 + X^2)/2\right]$,   for independent $U(0, 1)$ random variables U_1 and U_2.

Probability of rejection $= 1 - \lambda \sqrt{\left(\dfrac{\pi}{2}\right)} \exp(-\lambda^2/2)$

Let $y = \lambda \exp(-\lambda^2/2)$

let $z = \log y = \log \lambda - \lambda^2/2$; $\dfrac{dz}{d\lambda} = \dfrac{1}{\lambda} - \lambda = 0$ when $\lambda = 1$.

$\dfrac{d^2 z}{d\lambda^2} = -\dfrac{1}{\lambda^2} - 1, < 0$ when $\lambda = 1$

Thus taking $\lambda = 1$ minimizes the probability of rejection.

5.22   (a) Use the inversion method:

$$F(x) = \lambda\mu \int_0^x y^{\lambda-1} (\mu + y^\lambda)^{-2} \, dy = \left[\mu(\mu + y^\lambda)^{-1}\right]_x^0$$

$$= 1 - \mu(\mu + x^\lambda)^{-1} \qquad \text{for } x \ge 0.$$

Set $U = 1 - \mu(\mu + X^\lambda)^{-1}$; $1 - U = \mu(\mu + X^\lambda)^{-1}$

or equivalently

$$\frac{\mu}{U} = \mu + X^\lambda; \quad X = \left\{ \mu\left(\frac{1}{U} - 1\right) \right\}^{1/\lambda}$$

(b) $k = 4\alpha^{\alpha} e^{-\alpha} (\Gamma(\alpha) \sqrt{(2\alpha - 1)})^{-1}$

resulting in $1 - k^{-1}$ as the probability of rejection.

5.23 The ratio of the k values from Example 5.6, and Exercise 5.22 is:
$4e^{-1} (\sqrt{(2\alpha - 1)})^{-1}$, which is > 1 when $4 > e \sqrt{(2\alpha - 1)}$,
i.e. $\alpha < 8e^{-2} + 0.5 = 1.5827$.

 Thus for values of $1 < \alpha < 1.5827$, Cheng's algorithm has the smaller probability of rejection. A full comparison also would include the relative speeds of the algorithms.
 Cheng's algorithm becomes:

 (i) Set $V = (2\alpha - 1)^{-1/2} \log_e [U_1/(1 - U_1)]$, and $X = \alpha e^V$
 (ii) Accept X if $\alpha - \log_e 4 + (\alpha + (2\alpha - 1)^{1/2}) V - X \geq \log_e (U_1^2 U_2)$

 Cheng noted that $\log_e x$ is a concave function of $x > 0$, and so for any given $\theta > 0$, $\theta x - \log \theta - 1 \geq \log x$.
 Thus at stage (ii) the left-hand-side can first of all be checked against $\theta U_1^2 U_2 - \log \theta - 1$, for some fixed θ (he suggested $\theta = 4.5$, irrespective of α). If the inequality is satisfied for $\theta U_1^2 U_2 - \log \theta - 1$, then *a fortiori* it is satisfied for $\log(U_1^2 U_2)$ and a time-consuming logarithm need not be evaluated. Cheng's algorithm can also be adapted for the case $\alpha < 1$. Cheng and Atkinson and Pearce provide timings of these algorithms run on two different computers (see also Atkinson, 1977a).

5.24 Set $X = 9 - 2 \log_e U$.
 If this is not obvious, show that it results from applying the inversion method to the conditional density:

$$f(x) = \tfrac{1}{2} \exp((9 - x)/2) \qquad \text{for } x \geq 9$$

5.25 The objective is, as in the solution to Exercise 5.23, to improve efficiency, by evaluating simpler functions much of the time.

5.26 $f_X(x) = \left| \dfrac{du}{dx} \right|; \quad x^2 = a^2 - 2 \log u; \quad u = \exp[\tfrac{1}{2}(a^2 - x^2)]$

$2x \dfrac{dx}{du} = -\dfrac{2}{u}; \quad f_X(x) = x \exp[\tfrac{1}{2}(a^2 - x^2)]$

$\Pr(U_2 X < a) = a \displaystyle\int_a^{\infty} \dfrac{1}{x} x \exp[\tfrac{1}{2}(a^2 - x^2)] \, dx$

$\qquad\qquad = a \exp[a^2/2] \displaystyle\int_a^{\infty} \exp[-x^2/2] \, dx$

$\qquad\qquad = \sqrt{(2\pi)} a \exp[a^2/2] (1 - \Phi(a))$

Therefore with the conditioning event: $\{U_2 X < a\}$,

$$f_X(x) = \int_0^{a/x} \frac{x \exp[\frac{1}{2}(a^2 - x^2)] \, du}{\sqrt{(2\pi)} a \exp[a^2/2](1 - \Phi(a))} = \frac{\phi(x)}{(1 - \Phi(a))} \qquad \text{for } x \geq a.$$

5.27 Solution follows from the Polar Marsaglia algorithm, symmetry and Exercise 5.8(b).

5.28 $f_{X_1 X_2}(x_1 x_2) = e^{-x_1 - x_2}$ for $x_1 \geq 0$; $x_2 \geq 0$.

Probability of conditioning event $= \displaystyle\int_0^\infty e^{-x_1} \int_{\frac{1}{2}(x_1 - 1)^2}^\infty e^{-x_2} dx_2 dx_1,$

$$= \int_0^\infty e^{-x_1} e^{-\frac{1}{2}(x_1 - 1)^2} dx_1 = e^{-1/2} \int_0^\infty e^{-\frac{1}{2}x_1^2} dx_1 = \frac{\sqrt{(2\pi)} e^{-1/2}}{2}$$

Therefore under the conditioning

$$f_{X_1}(x_1) = \frac{e^{-x_1} \displaystyle\int_{\frac{1}{2}(x_1 - 1)^2}^\infty e^{-x_2} dx_2}{\sqrt{(\pi/(2e))}}$$

$$= \sqrt{\left(\frac{2}{\pi}\right)} e^{-x_1^2/2} \qquad \text{for } x_1 \geq 0, \text{ as required.}$$

5.29 This is simply an alternative way of writing the rejection method: in the notation of Section 5.3,

$$k = \max_x \left\{\frac{f(x)}{h(x)}\right\}$$

Therefore $f(x) \leq kh(x)$

$$f(x) = kh(x) \left\{\frac{f(x)}{kh(x)}\right\} = kh(x)s(x) \qquad \text{where } s(x) \leq 1.$$

In this case we have $h(x)$ as a p.d.f. We simulate X from $h(x)$, and then accept it if $Ukh(x) < f(X)$, i.e. $U < f(x)/kh(x)$. We obtain the situation of the question by setting:

$$c = k/m,$$

$$g(x) = h(x)$$

$$\gamma(x) = (mf(x))/(kh(x))$$

Probability of rejection $= 1 - \dfrac{1}{k} = 1 - \dfrac{1}{(mc)}$.

5.31 See the discussion of Example 5.5.

5.32 (a) One possibility is to set $Y = V_1 + V_2 - 2/\alpha$

where V_1 and V_2 are independent $U(0, 2/\alpha)$ random variables. Set $X = |Y|$.

(b) Set $q(x) = \dfrac{f(x)}{f_1(x)} = \dfrac{\lambda e^{-\lambda x}}{\left(\alpha - \dfrac{\alpha^2 x}{2}\right)}$

Let $w(x) = \log q(x) = \kappa - \lambda x - \log(1 - \alpha x/2)$, for constant κ.

$$\frac{dw(x)}{dx} = -\lambda + \frac{\alpha}{2 - \alpha x} = 0 \text{ when } \lambda = \alpha/(2 - \alpha x),$$

i.e. when $x = (2/\alpha - 1/\lambda)$. At this point,

$$\frac{d^2 w(x)}{dx^2} > 0$$

and so we have a minimum. Note that as $2\lambda > \alpha, \dfrac{2}{\alpha} > \dfrac{1}{\lambda}$.

The shrinking factor is $q\left(\dfrac{2}{\alpha} - \dfrac{1}{\lambda}\right) = \dfrac{2\lambda^2}{\alpha^2} e^{1 - 2\lambda/\alpha}$.

Let $u(\alpha) = \log\left(q\left(\dfrac{2}{\alpha} - \dfrac{1}{\lambda}\right)\right) = \kappa - \dfrac{2\lambda}{\alpha} - 2\log\alpha$, for constant κ.

$$\frac{du(\alpha)}{d\alpha} = +\frac{2\lambda}{\alpha^2} - \frac{2}{\alpha} = 0 \text{ when } \lambda = \alpha.$$

This value maximizes $u(\alpha)$, and so the probability of simulating from $f_1(x)$ is less than or equal to $2/e$.

5.33 (b) $f(y) = \dfrac{e^y}{(e-1)} = (e-1)^{-1} \displaystyle\sum_{i=0}^{\infty} \dfrac{y^i}{i!} = \displaystyle\sum_{i=0}^{\infty} \dfrac{(i+1)y^i}{(i+1)!(e-1)}$

5.34 Sample from the density:

$$f_m(x) = \frac{e^{m-x}}{(e-1)} \qquad \text{for } (m-1) < x \le m,$$

and m is an integer ≥ 1, with probability $(e-1)e^{-m}$ for $m \ge 1$.

5.35 For (ii), $\Pr(W \le w) = \displaystyle\prod_{i=1}^{I+1} \Pr(U_i \le w) = w^{I+1}$

Therefore $f_W(w) = (I+1)w^I \qquad \text{for } 1 \ge w \ge 0$

Thus we see, from Exercise 5.33(b) that steps (i) and (ii) return a random variable W with p.d.f.

$$f_W(y) = \frac{e^y}{(e-1)} \qquad \text{for } 0 \le y < 1$$

But we see from Exercise 5.34 that to simulate from e^{-x} we simulate from $e^{m-x}/(e-1)$ with probability $(e-1)e^{-m}$. Step (iii) selects m and then, from Exercise 5.33(a), step (iv) converts from W to X.

This is a way of simulating exponential random variables without evaluating a logarithm. The two discrete distributions involved have the forms:

i	$(e-1)e^{-i}$	$\dfrac{1}{i!(e-1)}$
1	0.63	0.58
2	0.23	0.29
3	0.09	0.10
4	0.03	0.02
⋮	⋮	⋮

On average only 1.582 $U(0, 1)$ random variables need to be considered at the maximization stage.

5.37 This provides a fast algorithm but many constants have to be stored.

5.38 This algorithm utilizes a number of interesting features for improving efficiency. Step 3 uses the algorithm of Marsaglia (1964) as modified by Ahrens and Dieter (1972) for sampling X values in the tail, $|X| > \zeta$. cf. Exercise 5.26. Steps 5, 7 and 8 correspond to the three triangle rejection procedures for the three main triangles illustrated in Fig. 5.16 (see Exercise 5.25). 'Generate' means select an independent $U(0, 1)$ random variable.

5.42 This proof is rather difficult—see the source paper for details.

Chapter 6

6.2 In order, values of $(n-1)s^2/\bar{x}$ are: 514.97, 464.07, 452.09, 472.05, 551.89, 470.56. Using the approximation to χ^2 of Exercise 6.24 gives approximate standard normal values in the range $(-1.51, 1.65)$.

6.3 If dice are fair, and thrown independently,

Pr(0 sixes) = 25/36; Pr(1 six) = 5/18; Pr(2 sixes) = 1/36, resulting in:

No. of sixes	0	1	2	Total
Expected frequency	150	60	6	216

$\chi_2^2 = 9.6$, significant at the 1% level using a 1-tail test.
If p is unknown, the likelihood of the data L is given by:

$$L \propto (1-p)^{n_1 + 2n_0} p^{n_1 + 2n_2}$$

and the maximum-likelihood estimator of p is then $\hat{p} = (n_1 + 2n_2)/\{2(n_0 + n_1 + n_2)\}$ and here $\hat{p} = 0.2222$, resulting in the expected values:

No. of sixes	0	1	2	Total
Expected frequency	130.7	74.7	10.7	216.1

Now $\chi_1^2 = 0.072$, not significant at the 5% level, using a 1-tail test.

6.4 $\Pr(R \leq r) = \displaystyle\int_{-\infty}^{\infty} \Pr(n-1$ values lie in $(x, x+r) |$ smallest of n values $= x) m(x) dx$ where $m(x)$ is the p.d.f. of the smallest of n values.

Thus

$$\Pr(R \leq r) = \int_{-\infty}^{\infty} \left(\frac{\Phi(x+r) - \Phi(x)}{1 - \Phi(x)} \right)^{n-1} n(1 - \Phi(x))^{n-1} \phi(x) dx$$

(cf. Exercise 6.13)

$$= n \int_{-\infty}^{\infty} (\Phi(x+r) - \Phi(x))^{n-1} \phi(x) dx$$

6.5   For a p-dimensional proof, see Kendall and Stuart (1969, p. 355). Equation (6.3) states that

$$D^2 = \frac{1}{(1-\rho^2)} (N_1^2 - 2\rho N_1 N_2 + N_2^2)$$

is χ_2^2, where N_1, N_2 are bivariate normal, with zero means, unit variances, and correlation ρ. By Exercise 4.16, we can therefore write

$$\left. \begin{array}{l} N_1 = X_1 \\ N_2 = \rho X_1 + (1-\rho^2)^{1/2} X_2 \end{array} \right\} \begin{array}{l} \text{where } X_1 \text{ and } X_2 \text{ are independent and} \\ N(0, 1) \end{array}$$

and so

$$D^2 = \frac{1}{(1-\rho^2)} \{X_1 + X_2(X_2 - 2\rho X_1)\} = \frac{1}{(1-\rho^2)} \{X_1^2 + (\rho X_1$$
$$+ (1-\rho^2)^{1/2} X_2) \times ((1-\rho^2)^{1/2} X_2 - \rho X_1)\}$$

$$= \frac{1}{(1-\rho^2)} \{(1-\rho^2)(X_1^2 + X_2^2)\},$$

i.e. D^2 has a χ_2^2 distribution.

6.6 (i) Time to first new digit is simply 1. Time to next new digit is X_2, etc.

$$G_{X_i}(z) = (11 - i)z(10 - (i-1)z)^{-1} \qquad \text{for } 2 \le i \le 10,$$

and so $G_S(z) = z \prod_{i=1}^{9} \dfrac{iz}{(10 - (10-i)z)}$

$$G_S(z) = \sum_{j=10}^{\infty} z^j \frac{9}{10^8} \sum_{i=1}^{9} \binom{8}{i-1} \frac{i^8}{(10-i)} (-1)^{9-i}$$

$$\times \left(1 - \frac{i}{10}\right)\left(\frac{i}{10}\right)^{j-10}$$

$$= \sum_{j=10}^{\infty} z^j \frac{9}{10^9} \sum_{i=1}^{9} \binom{8}{i-1} (-1)^{9-i} \frac{i^{j-2}}{10^{j-10}}$$

$$= \sum_{j=10}^{\infty} z^j 9 \cdot 10^{1-j} \sum_{i=1}^{9} \binom{8}{i-1} (-1)^{9-i} i^{j-2}$$

$$= \sum_{j=10}^{\infty} z^j 10^{1-j} \sum_{v=1}^{9} (-1)^{v+1} \binom{9}{v-1} (10-v)^{j-1}.$$

(ii) Pr(at least one cell is empty) $= S_1 - S_2 + S_3 - S_4 \dots \pm S_n$,

where $S_1 = \sum_i p_i$; $S_2 = \sum \sum_{i<j} p_{ij}$, etc., and

$$p_i = \left(1 - \frac{1}{n}\right)^r$$

$$p_{ij} = \left(1 - \frac{2}{n}\right)^r$$

$$p_{ijk} = \left(1 - \frac{3}{n}\right)^r, \text{ etc.} \qquad \text{(see Feller, 1957, p. 89).}$$

Therefore for every $v \le n$, $S_v = \binom{n}{v}\left(1 - \frac{v}{n}\right)^r$

and so $u(r, n) = 1 - S_1 + S_2 \dots$

$$= \sum_{v=0}^{n} (-1)^v \binom{n}{v}\left(1 - \frac{v}{n}\right)^r$$

and so Pr$(j) = u(j, 10) - u(j-1, 10)$

$$\text{Pr}(j) = 10^{-j} \sum_{v=0}^{10} v(-1)^{v+1} \binom{10}{v} (10-v)^{j-1}$$

$$= 10^{1-j} \sum_{v=1}^{9} (-1)^{v+1} \binom{9}{v-1} (10-v)^{j-1}$$

for $j \ge 10$.

6.7 Failure of the frequency test does not necessarily imply failure of other tests.

6.12 For example, take an acceptable sequence of digits and replace i by ii, for all digits i.

6.13 $\Pr(M \le x) = x^n = F_M(x); \quad f_M(x) = nx^{n-1} \quad$ for $0 \le x \le 1$ (cf. Exercises 2.10 and 2.11).

Let $Y = M^n; \dfrac{dy}{dm} = nm^{n-1}, \quad$ and Y is $U(0, 1)$, facilitating the application of this test.

6.18 Use $-2\sin^2(2\pi U_2) \log_e U_1$ and $-2(1 - \sin^2 2\pi U_2) \log_e U_1$, thus avoiding the use of cos and square root.

6.22 Variables can be summed in groups and tests applied to the sums, using the results of Exercise 4.17. We saw in Fig. 4.5 that putting exponential variates end-to-end is equivalent to simulating a Poisson process. Conditional upon k events in a Poisson process (in time, say) occurring within some fixed interval, then the times of occurrence are uniformly distributed over that interval, and for a standard-length interval, when an odd number of events occurs the median time may be referred to the appropriate beta p.d.f.

6.24 Due to Fisher, this result is discussed by Kendall and Stuart (1969, p. 371). A more accurate approximation, due to Wilson and Hilferty, is that if X has a χ_v^2 p.d.f.,

$$(X/v)^{1/2} \approx N(1 - 2/9v, 2/9v)$$

6.28 As the entire cycle is used, the frequency test would reject the numbers as too uniform. In practice one only uses a fraction of a cycle (cf. comments in Section 3.5).

6.29 See Knuth (1981, p. 517) for an efficient procedure.

Chapter 7

7.1 $2l/\pi d$

7.2 When $l \le d$,

$$\Pr \text{(needle intersects line)} = 2 \int_{-\pi/2}^{\pi/2} \frac{l \cos\theta}{2\pi d} d\theta = \frac{2l}{\pi d}$$

When $l \ge d$,

$$\Pr \text{(needle intersects line)} = \int_{-\pi/2}^{\pi/2} \frac{\min(l \cos\theta, d)}{2\pi d} d\theta$$

$$= \frac{2}{\pi d} \left\{ d \cos^{-1}\left(\frac{d}{l}\right) + l(1 - \sqrt{(1 - d^2/l^2)}) \right\}$$

7.4

	Estimates of π		
Experiment 1:	2.72	3.013	3.125
Experiment 2:	3.212	3.212	3.207

7.5 If Pr (coin lands totally within square) = 0.5, then $(a-b)^2/a^2 = \frac{1}{2}$

$$2(a-b)^2 = a^2; \quad 2b^2 - 4b + 1 = 0$$

if we set $a = 1$ and $b = 1 - 1/\sqrt{2}$.

7.6 See Fletcher (1971).

7.7 The variability of the integrand is greater when $x > 0$ and the range of integration is $(-\infty, x)$. For $x > 0$, it is better to set

$$\Phi(x) = \frac{1}{2} + \int_0^x \phi(x)\,dx$$

and then evaluate the integral over the $(0, x)$ range.

7.8 Write

$$\int_0^1 e^{-x^2}\,dx = \int_0^1 (1 + (1 - e^{-1})x)\,dx + \int_0^1 (e^{-x^2} - 1 - (1 - e^{-1})x)\,dx.$$

The first integral on the right-hand side is known $= (3 - e^{-1})/2$, while the variability of $e^{-x^2} - 1 - (1 - e^{-1})x$ is less than e^{-x^2}.

7.10 With just two strata we might have:

$$\tilde{\theta} = \sum_{i=1}^{n_1} \frac{\alpha}{n_1} f(\alpha U_i) + \sum_{j=n_1+1}^{n} \frac{(1-\alpha)}{(n-n_1)} f(\alpha + (1-\alpha)U_j)$$

If we choose $n_1 = n - n_1 = m$ then we can incorporate antithetic variates from

$$\tilde{\theta} = \frac{1}{m} \sum_{i=1}^{m} \{\alpha f(\alpha U_i) + (1-\alpha)f[\alpha + (1-\alpha)(1-U_i)]\}$$

7.11 Let $I = \int_0^1 [x(x-2)(x^2-1)]^{1/2}\,dx$

Set $x = z + \frac{1}{2}$

$$I = \int_{-1/2}^{1/2} [(z^2 - \frac{9}{4})(z^2 - \frac{1}{4})]^{1/2}\,dz$$

Set $2z = \sin\theta$

$$I = \frac{3}{8} \int_{-\pi/2}^{\pi/2} \left(1 - \frac{\sin^2\theta}{9}\right)^{1/2} \cos^2\theta\,d\theta$$

i.e., $I = \dfrac{3}{8} \displaystyle\int_{-\pi/2}^{\pi/2} \cos^2\theta \left\{ 1 - \dfrac{\sin^2\theta}{18} - \sum_{k=2}^{\infty} \dfrac{\sin^{2k}\theta \, (2k-3)!}{k! \, 3^{2k} \, 2^{2k-2} \, (k-2)!} \right\} d\theta$

Now use:

$$\int_0^{\pi/2} \cos^2\theta \, d\theta = \frac{\pi}{4}$$

$$\int_0^{\pi/2} \sin^{2k}\theta \, d\theta = \pi \binom{2k-1}{k} 2^{-2k}$$

$$\int_0^{\pi} \cos^2\theta \sin^{2k}\theta \, d\theta = \frac{\pi 2^{-2k-1} (2k-1)!}{(k+1)! \, (k-1)!}$$

and sum the infinite series using an appropriate truncation rule.

7.12   For related discussion, see Simulation I (1976, pp. 44–46).

7.13 $\hat{\pi} = \dfrac{4}{n} \displaystyle\sum_{i=1}^{n} \sqrt{(1 - U_i^2)}$

$\mathscr{E}[\hat{\pi}] = \pi$

$\mathrm{Var}(\hat{\pi}) = \dfrac{16}{n} \mathrm{Var}(\sqrt{1 - U^2}) = \dfrac{16}{n}\left(\dfrac{2}{3} - \dfrac{\pi^2}{16}\right) \approx \dfrac{16(0.0498)}{n}$

By a central limit theorem, $\hat{\pi}$ is approximately distributed as $N(\pi, 16(0.0498)/n)$, and so an approximate 95 % confidence interval for $\hat{\pi}$ has width $\approx 3.4997/\sqrt{n} = 0.01$, for example, when $n \approx 122\,480$.

7.16   Variance reduction y, is given by

$$y = -c^2 \mathrm{Var}(Z) + 2c \mathrm{Cov}(X, Z)$$

$$\frac{dy}{dc} = 2(-c \mathrm{Var}(Z) + \mathrm{Cov}(X, Z))$$

$= 0$ when $c = \mathrm{Cov}(X, Z)/\mathrm{Var}(Z)$ and $\dfrac{d^2 y}{dc^2} < 0$, so we have a maximum.

7.17   Proof from Kleijnen (1974, p. 254):

$$x_i = ax_{i-1} \pmod{m} \qquad i \geq 1$$

and so antithetic variates are: $r_i = 1 - x_i/m$.

Consider the sequence: $y_i = ay_{i-1} \pmod{m}$, where $y_0 = m - x_0$, and let $\tilde{r}_i = y_i/m$. First show, by induction, that $x_i + y_i = m$, for $i \geq 0$. Then deduce that $r_i - \tilde{r}_i$ is an integer, but $0 < r_i, \tilde{r}_i < 1$, and so we must have $r_i = \tilde{r}_i$.

7.21 Cf. Example 7.3. See Cox and Smith (1961, p. 135) and Rubinstein (1981, p. 150) for discussion.

Chapter 8

8.3

Type of event	Time from start of simulation	Queue size	Total waiting-times
A1	0 ⎤	1	
D1	0.30 ⎦	0	0.30
A2	1.34 ⎤	1	
A3	⌐1.56	2	
A4	⌐2.98	3	
D2	3.04 ⎦	2	1.70
A5	3.58 ⌐	3	
D3	⌐3.68 ⎤	2	2.12
A6	4.92 ⌐	3	
A7	5.31 ⌐⎤⎤	4	
D4	5.56 ↓↓↓	3	2.58

8.4 Let the n individual lifetimes be: $X_1, X_2, \ldots, X_n$. From the specification of the process, these are independent random variables with an exponential, e^{-x} p.d.f. for $x \geq 0$. The time to extinction is therefore $Y = \max(X_1, X_2, \ldots, X_n)$. Also, time to first death has the exponential p.d.f., ne^{-nx} for $x \geq 0$, and the time between the $(i-1)$th and ith deaths has the exponential p.d.f., $(n-i+1)e^{-(n-i+1)x}$, for $1 \leq i \leq n$, which is the p.d.f. of $X/(n-i+1)$, where X has the e^{-x} p.d.f. Thus we see that we can also write the time to extinction as,

$$Z = X_1 + \frac{X_2}{2} + \ldots + \frac{X_n}{n}$$

8.8
$$\text{Var}(\bar{X}) = \text{Var}\left(\sum_{i=1}^{n} X_i/n\right)$$

$$= \frac{1}{n^2}\left\{\sum_{i=1}^{n}\text{Var}(X_i) + 2\sum\sum_{i<j}\text{Cov}(X_i, X_j)\right\}$$

$$= \frac{\sigma^2}{n} + \frac{2}{n^2}\sum_{i=1}^{n-1}\sum_{s=1}^{n-i}\text{Cov}(X_i, X_{i+s})$$

$$= \frac{\sigma^2}{n} + \frac{2}{n^2}\sum_{s=1}^{n-1}\sum_{i=1}^{n-s}\text{Cov}(X_i, X_{i+s})$$

$$= \frac{\sigma^2}{n} + \frac{2\sigma^2}{n^2}\sum_{s=1}^{n-1}\rho_s(n-s)$$

$$= \frac{\sigma^2}{n}\left(1 + 2\sum_{s=1}^{n-1}\left(1-\frac{s}{n}\right)\rho_s\right)$$

8.10   We can envelop e^{-x} by a suitable multiple, k, of the 'half-logistic' p.d.f.

$$f(x) = \frac{2e^{-x}}{(1 + e^{-x})^2} \qquad \text{for } x \geq 0$$

of Exercise 5.19. Thus here we are reversing the rôles played by these p.d.f.'s in that exercise. Clearly we take $k = 2$, and the probability of rejection is 0.5. Using this method we would expect a greatly reduced variance reduction, as the likelihood of rejection interferes with the antithetic variates.

8.11   The following flow diagram is taken from Gross and Harris (1974, p. 384). We see that traffic selects first of all a size of queue. The test $U \leq 0.4$, checks which drivers have the correct change for the automatic booth. Automatic booths are assumed to have constant service-times,

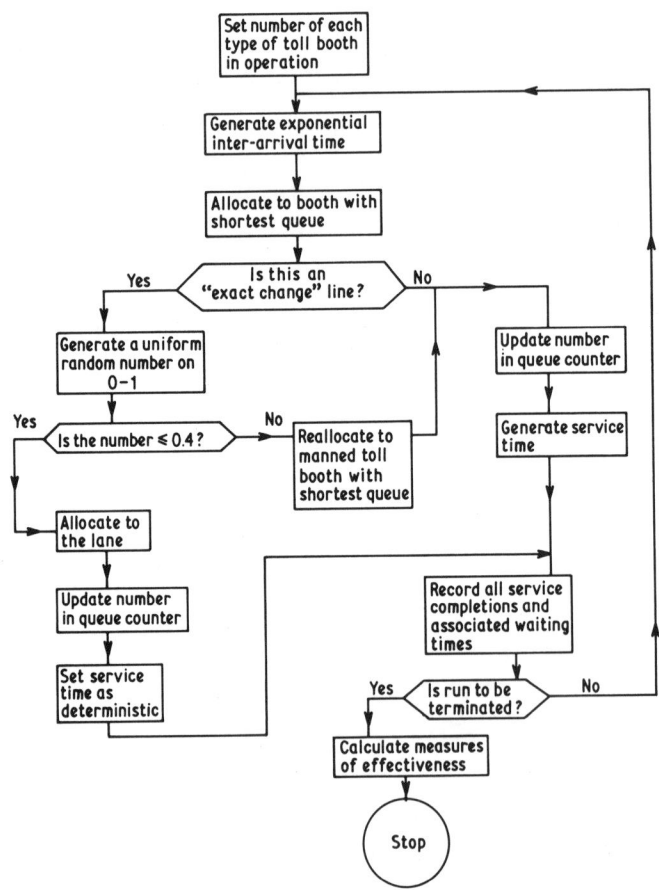

while manual booths have variable service-times from some distribution. Changing booths is a complex procedure, and an alternative possibility would be to select first of all on the availability of the correct change.

8.12 The numbers of individuals turned away per month are all rather large, whatever the distribution of beds.

8.14 Using the formula derived in solution to Exercise 8.8,

$$\text{Var}(\bar{X}) = \frac{\sigma^2}{n}\left(1 + 2\sum_{s=1}^{n-1}\left(1 - \frac{s}{n}\right)\rho_s\right)$$

and the required expression follows from setting $\rho_s = \rho^s$.

8.16 See Donnelly and Shannon (1981) and Schruben (1981) for recent discussion.

8.17 A good discussion is provided by Fishman (1973, p. 233). Empirical distributions are useful in a validation context, but more general results are likely when simulating from a fitted distribution.

Chapter 9

9.1 Ten $N(0, 1)$ variates are put in column vector C1. C2 contains the squares of these, which are summed and put in the constant K1. K1 thus is a realization of a χ^2_{10} random variable, by the result of Exercise 2.5. 100 such values are put into C3 and finally a histogram of these values is produced.

For logistic random variables:

```
NOPRINT
URANDOM 100 OBSERVATIONS, PUT IN C1
LET C2=-LOGE(1/C1-1)
HISTO C2
```

Here the second line simulates 100 $U(0, 1)$ variates and puts them in C1. The third line implements the inversion method, described in the solution to Exercise 5.13, on each element of C1, putting the result in the corresponding element of C2.

For beta random variables: use the MINITAB program of the exercise and the result of Exercise 2.14.

9.3 As stated in Section 1.6, an analytic solution is more elegant and powerful than a solution obtained using simulation. However, as this exercise shows, an analytic solution can be quite complicated, and needs a computer program for its complete enumeration.

9.4   If the second-stage sampling occurs with replacement (i.e. animals are released after capture) then if $p = m/n$, if one samples X animals until $\tilde{m}$ marked animals are recaptured, then

$$\Pr(X = \tilde{m} + k) = \binom{\tilde{m} + k - 1}{\tilde{m} - 1} p^{\tilde{m}}(1 - p)^k, \qquad \text{for} \qquad k \geq 0$$

i.e. X has the negative-binomial distribution of Section 2.6, and

$$\mathscr{E}[X] = \tilde{m}/p = \tilde{m}n/m$$

Thus we can estimate n by:

$$\hat{n} = Xm/\tilde{m}$$

an unbiased estimator of n. For further discussion, see Chapman (1952).

9.5   (i) The percentages of variance explained by the first three principal components are given below:

Principal component	Whole class (%)	13 men (%)	15 women (%)
1	67.6	59.6	53.2
2	11.8	18.0	19.5
3	9.9	9.4	13.8
	89.3	87.0	86.5

The coefficients in the components are given by:

Men

	Principal component		
Measurement	1	2	3
Chest	0.90	0.09	−0.01
Waist	0.94	−0.08	0.07
Wrist	0.60	0.59	−0.34
Hand	0.46	0.70	0.52
Head	0.83	−0.10	−0.41
Height	0.87	−0.23	0.03
Forearm	0.67	−0.59	0.31

Women

Measurement	Principal component		
	1	2	3
Chest	0.86	−0.30	−0.12
Waist	0.91	0.03	0.03
Wrist	0.51	0.73	0.11
Hand	0.63	0.66	0.03
Head	0.53	−0.22	0.80
Height	0.77	−0.51	−0.00
Forearm	0.79	−0.06	−0.54

(ii) The linear discriminant function for all the measurements has the coefficients:

0.78	chest
−1.02	waist
0.25	wrist
−1.47	hand
0.19	head
−0.53	height
0.01	forearm

If we use only chest and waist measurements, the linear discriminant function has the coefficients:

−0.81	chest
2.20	waist

If both of these discriminant functions are applied to the data that gave rise to them, then in each case only one individual is misclassified.

9.6 One would expect a high positive correlation between $\hat{\delta}$ and $\Phi(\hat{\delta}/2)$, and the distribution of $\hat{\delta}^2$ is known, as it is proportional to Hotelling's T^2 statistic—see Mardia *et al.* (1979, p. 77). This therefore suggests using $\hat{\delta}$ as a control variate for $\Phi(\hat{\delta}/2)$, as described in Section 9.4.2.

9.8 For the best subsets, the minimum value R_m of the multiple correlation coefficient (see Mardia *et al.*, 1979, pp. 167–169) between the variables of the set and any of the other variables is maximized, at a value of R_m^*, say. For the 'good' subsets, $0.7 R_m^* \le R_m < R_m^*$, as shown by Jolliffe (1972a).

9.9 The cluster centres for the samples were obtained from sample averages. The two A samples agree well with expectation; however, the C statistic for the second sample from population A does not perform as well as in the other structured examples, possibly due to overlap of the sub-samples. The C statistic provides no clear indication for the unstructured data of population B.

9.10   To be conservative, we can take the first row of Table 9.3 ($n = 8$). The two-dimensional stress value of 6.44% is ≈ 3 standard deviations smaller than the average shown in Table 9.3, adding weight to the 'good' interpretation of this stress value.

9.12   Again we see that the method has worked well in recapturing much of the known structure in the simulated population. Here we would also want to investigate structure using a measure as in Exercise 9.9, and consider 2-, 3-, etc., cluster solutions.

9.13

Winter temperature

October mean pressure (terciles)	Quintiles	1–2	Quintiles	3–5	Total
Upper	(6.4)	11	(9.6)	5	16
Middle	(5.6)	3	(8.4)	11	14
Lower	(6.0)	4	(9)	11	15
Total		18		27	45

Expected values under the hypothesis of no association are given in parentheses. Ignoring the ordering of both the row and column categories, the standard chi-square test of association results in a statistic 8.63, significant at the 2.5% level when referred to χ_2^2 tables. The known (asymptotic) chi-square distribution of the pooled test-statistic may be used to improve precision if that statistic is used as a control variate, as in Section 9.4.2.

9.14   See Tocher (1975, p. 92) for an example with sample median and sample mean, for samples of size 3 from a normal distribution. Here the sample mean is the control variate.

9.16   A BASIC program is given below (cf. Patefield, 1981).

```
10   REM PROGRAM TO PERFORM THE MONTE-CARLO TEST OF
20   REM EXAMPLE 9.3
30   RANDOMIZE
40   DIM A(3,5),B(499)
50   LET N = 499
60   REM 499 RANDOM TABLES WILL BE SIMULATED
70   FOR L = 1 TO N
80     FOR I = 1 TO 3
```

```
 90     FOR J = 1 TO 5
100       LET A(I,J) = -3
110     NEXT J
120     NEXT I
130     REM FOR EACH COLUMN WE DISTRIBUTE THE COLUMN TOTAL
140     REM OF 9, UNIFORMLY OVER THE 3 ROWS
150     FOR J = 1 TO 5
160       FOR K = 1 TO 9
170       LET U = RND
180       IF U < .3333333 THEN 220
190       IF U > .666666667 THEN 240
200       LET A(2,J) = A(2,J)+1
210       GOTO 250
220       LET A(1,J) = A(1,J)+1
230       GOTO 250
240       LET A(3,J) = A(3,J)+1
250       NEXT K
260     NEXT J
270     REM S CONTAINS THE RELEVANT PART OF THE CHI-
280     REM SQUARE STATISTIC
290     LET S = 0
300     FOR I = 1 TO 3
310       FOR J = 1 TO 5
320       LET S = S+A(I,J)*A(I,J)/3
330       NEXT J
340     NEXT I
350     LET B(L) = S
360     NEXT L
370     REM THE N VALUES OF THE S STATISTIC ARE NOW
380     REM ORDERED USING A SHELL-SORT ALGORITHM,
390     REM TAKEN FROM COOKE,CRAVEN AND CLARKE, 1982, P18.
400     LET G = N
410     LET G = INT(G/2)
420     IF G = 0 THEN 550
430     FOR I = 1 TO (N-G)
440       FOR J = I TO 1 STEP -G
450       LET J1 = J+G
460       IF B(J) > B(J1) THEN 490
470       LET J = 0
480       GOTO 520
490       LET W = B(J)
500       LET B(J) = B(J1)
510       LET B(J1) = W
520       NEXT J
530     NEXT I
540     GOTO 410
550     FOR I = 1 TO N
560       PRINT B(I)
570     NEXT I
580     END
```

9.17 For small values of the probability, the binomial distribution may be approximated by the Poisson, and $Var(\hat{p}) \approx p/n$.

9.18 The modern substitute for pieces of cardboard:

```
NOPRINT
STORE
NRANDOM 4 OBSERVATIONS, WITH MU = 0.0, SIGMA=1.0, PUT IN C1
AVERAGE C1, PUT IN K1
STANDARD DEV. C1, PUT IN K2
LET K3=2*K1/K2
JOIN K3 TO C4, PUT IN C4
END
EXECUTE 750 TIMES
HISTO C4
```

9.19
```
NOPRINT
STORE
URANDOM 3 OBSERVATIONS, PUT IN C1
SUM C1, PUT IN K1
JOIN K1 TO C2, PUT IN C2
END
EXECUTE 100 TIMES
HISTO C2
```

This program draws a histogram of 100 sums of k $U(0, 1)$ variates, where $k = 3$. The program may be repeated for different values of k.

BIBLIOGRAPHY

This bibliography contains all the books, papers and reports referred to in the text. In addition, it contains references to other work relevant to simulation, some of which provide further examples of applications. Bibliographies which concentrate on particular areas are to be found in Nance and Overstreet (1972) and Sowey (1972, 1978).

Abdul-Ela, A-L, A., Greenberg, B. G., and Horvitz, D. G. (1967) A multi-proportions randomized response model. *J. Amer. Statist. Assoc.*, **62**, 990–1008.

Abramowitz, M., and Stegun, I. A. (Eds.) (1965) *Handbook of Mathematical Functions*. Dover Publications, New York.

Adam, N. and Dogramaci, A. (Eds.) (1979) *Current Issues in Computer Simulation*. Academic Press, New York.

Ahrens, J. H., and Dieter, U. (1972) Computer methods for sampling from the exponential and normal distributions. *Commun. Ass. Comput. Mach.*, **15**, 873–882.

Ahrens, J. H., and Dieter, U. (1973) Extensions of Forsythe's method for random sampling from the normal distribution. *Math. Comput.*, **27**, 927–937.

Ahrens, J. H., and Dieter, U. (1974a) *Non-uniform Random Numbers*. Technische Hoshschule in Graz, Graz, Austria.

Ahrens, J. H., and Dieter, U. (1974b) Computer methods for sampling from gamma, beta, Poisson and binomial distributions. *Computing (Vienna)*, **12**, 223–246.

Ahrens, J. H., and Dieter, U. (1980) Sampling from binomial and Poisson distributions: a method with bounded computation times. *Computing*, **25**, 193–208.

Ahrens, J. H., and Dieter, U. (1982) Generation of Poisson deviates from modified normal distributions. *Assoc. Comput. Mach. Trans. Math. Soft.*, **8**, 163–179.

Aitchison, J., and Brown, J. A. C. (1966) *The Lognormal Distribution, With Special Reference to its Uses in Economics.* C.U.P.

Akima, H. (1970) A new method of interpolation and smooth curve fitting based on local procedures. *J. Assoc. Comput. Mach.*, **17** (4), 589–602.

Anderson, O. D. (1976) *Time Series Analysis and Forecasting.* Butterworths, London and Boston.

Andrews, D. F. (1976) Discussion of: 'The computer generation of beta, gamma and normal random variables' by A. C. Atkinson and M. C. Pearce. *J. Roy. Statist. Soc.*, **139**, 4, 431–461.

Andrews, D. F., Bickel, P. J., Hampel, F. R., Huber, P. J., Rogers, W. H., and Tukey, J. W. (1972) *Robust Estimates of Location.* Princeton University Press, Princeton.

Apostol, T. M. (1963) *Mathematical Analysis.* Addison-Wesley, Tokyo.

Appleton, D. R. (1976) Discussion of paper by Atkinson and Pearce. *J. Roy. Statist. Soc.*, A, **139**, 4, 449–451.

Arabie, P. (1973) Concerning Monte-Carlo evaluations of nonmetric multi-dimensional scaling algorithms. *Psychometrika*, **38**, 4, 607–608.

Arvidson, N. I., and Johnsson, T. (1982) Variance reduction through negative correlation; a simulation study. *J. Stat. Computation and Simulation*, **15**, 119–128.

Ashcroft, H. (1950) The productivity of several machines under the care of one operator. *J. Roy. Statist. Soc.*, B, **12**, 145–151.

Ashton, W. D. (1971) Gap acceptance problems at a traffic intersection. *Appl. Stats.*, **20** (2), 130–138.

Atkinson, A. C. (1977a) An easily programmed algorithm for generating gamma random variables. *J. Roy. Statist. Soc.*, A, **140**, 232–234.

Atkinson, A. C. (1977b) Discussion of: 'Modelling spatial patterns' by B. D. Ripley. *J. Roy. Statist. Soc.*, B, **39** (2), 172–212.

Atkinson, A. C. (1979a) The computer generation of Poisson random variables. *Appl. Stats.*, **28** (1), 29–35.

Atkinson, A. C. (1979b) A family of switching algorithms for the computer generation of beta random variables. *Biometrika*, **66** (1), 141–146.

Atkinson, A. C. (1979c) Recent developments in the computer generation of Poisson random variables. *Appl. Stats.*, **28** (3), 260–263.

Atkinson, A. C. (1980) Tests of pseudo-random numbers. *Appl. Stats.*, **29** (2), 164–171.

Atkinson, A. C., and Pearce, M. C. (1976) The computer generation of beta, gamma and normal random variables. *J. Roy. Statist. Soc.*, A, **139** (44), 431–461.

Atkinson, A. C., and Whittaker, J. (1976) A switching algorithm for the generation of beta random variables with at least one parameter less than 1. *J. Roy. Statist. Soc.*, A, **139** (4), 462–467.

Bailey, B. J. R. (1981) Alternatives to Hastings' approximation to the inverse of the normal cumulative distribution function. *Appl. Stats.*, **30**, 275–276.

Bailey, N. T. J. (1952) A study of queues and appointment systems in hospital out-patient departments with special reference to waiting times. *J. Roy. Statist. Soc.*, B, **14**, 185–199.

Bailey, N. T. J. (1964) *The Elements of Stochastic Processes with Applications to the Natural Sciences*. Wiley, New York.

Bailey, N. T. J. (1967) The simulation of stochastic epidemics in two dimensions. *Proc. 5th Berkeley Symp. Math. Statist. Prob.*, **4**, 237–257.

Bailey, N. T. J. (1975) *The Mathematical Theory of Infectious Diseases and its Applications* (2nd edn). Griffin, London.

Barnard, G. A. (1963) Discussion of Professor Bartlett's paper. *J. Roy. Statist. Soc.*, B, **25**, 294.

Barnett, V. D. (1962a) The Monte Carlo solution of a competing species problem. *Biometrics*, **18**, 76–103.

Barnett, V. D. (1962b) The behaviour of pseudo-random sequences generated on computers by the multiplicative congruential method. *Math. Comp.*, **16**, 63–69.

Barnett, V. D. (1965) *Random Negative Exponential Deviates*. Tracts for Computers XXVII. C.U.P.

Barnett, V. (1974) *Elements of Sampling Theory*. The English Universities Press Ltd, London.

Barnett, V. (1976) Discussion of: 'The computer generation of beta, gamma and normal random variables' by A. C. Atkinson and M. C. Pearce. *J. Roy. Statist. Soc.*, A, **139** (4), 431–461.

Barnett, V. (1980) Some bivariate uniform distributions. *Commun. Statist. – Theory Meth.*, A, **9** (4), 453–461.

Barnett, V., and Lewis, T. (1978) *Outliers in Statistical Data*. Wiley, Chichester.

Barrodale, I., Roberts, F. D. K., and Ehle, B. L. (1971) *Elementary Computer Applications in Science, Engineering and Business*. Wiley, New York.

Bartholomew, D. J., and Forbes, A. F. (1979) *Statistical Techniques for Manpower Planning*. Wiley, Chichester.

Bartlett, M. S. (1953) Stochastic processes or the statistics of change. *Appl. Stats.*, **2**, 44–64.

Bartlett, M. S., Gower, J. C., and Leslie, P. H. (1960) A comparison of theoretical and empirical results for some stochastic population models. *Biometrika*, **47**, 1–11.

Bays, C., and Durham, S. D. (1976) Improving a poor random number generator. *Assoc. Comput. Math. Trans. Math. Soft.*, **2**, 59–64.

Beale, E. M. L. (1969) Euclidean cluster analysis. *Bull. I.S.I.*, **43**, Book 2, 92–94.

Beasley, J. D., and Springer, S. G. (1977) Algorithm AS111. The percentage points of the normal distribution. *Appl. Stats.*, **26**, 118–121.

Bebbington, A. C. (1975) A simple method of drawing a sample without replacement. *Appl. Stats.*, **24** (1), 136.

Bebbington, A. C. (1978) A method of bivariate trimming for robust estimation of the correlation coefficient. *Appl. Stats.*, **27** (3), 221–226.

Besag, J., and Diggle, P. J. (1977) Simple Monte Carlo tests for spatial pattern. *Appl. Stats.*, **26** (3), 327–333.

Best, D. J. (1979) Some easily programmed pseudo-random normal generators. *Aust. Comp. J.*, **11**, 60–62.

Best, D. J., and Fisher, N. I. (1979) Efficient simulation of the von Mises distribution. *Appl. Stats.*, **28** (2), 152–157.

Best, D. J., and Winstanley, J. (1978) *Random Number Generators.* CSIRO DMS Newsletter 42.

Bevan, J. M., and Draper, G. J. (1967) *Appointment Systems in General Practice.* Oxford Univ. Press for Nuffield Provincial Hospitals Trust, London.

Bishop, J. A., and Bradley, J. S. (1972) Taxi-cabs as subjects for a population study. *J. Biol. Educ.*, **6**, 227–231.

Bishop, J. A., and Sheppard, P. M. (1973) An evaluation of two capture–recapture models using the technique of computer simulations. In, M. S. Bartlett and R. W. Hiorns (eds.). *The Mathematical Theory of the Dynamics of Biological Populations.* Academic Press, London, pp. 235–252.

Blackman, R. G., and Tukey, J. W. (1958) *The Measurement of Power Spectra.* Dover, New York.

Blake, I. F. (1979) *An Introduction to Applied Probability.* Wiley, New York.

Blanco White, M. J., and Pike, M. C. (1964) Appointment systems in out-patients' clinics and the effect of patients' unpunctuality. *Medical Care*, **2**, 133–142.

Bofinger, E., and Bofinger, V. J. (1961) A note on the paper by W. E. Thomson on 'Ernie – a mathematical and statistical analysis'. *J. Roy. Statist. Soc.*, A, **124** (2), 240–243.

Bowden, D. C., and Dick, N. P. (1973) Maximum likelihood estimation for mixtures of 2 normal distributions. *Biometrics*, **29**, 781–790.

Box, G. E. P., Hunter, W. G., and Hunter, J. S. (1978) *Statistics for Experimenters: An introduction to design, data analysis and model building.* Wiley, New York.

Box, G. E. P., and Müller, M. E. (1958) A note on the generation of random normal deviates. *Ann. Math. Stat.*, **29**, 610–611.

Boyett, J. M. (1979) Random $r \times c$ tables with given row and column totals. Algorithm AS144. *Appl. Stats.*, **28**, 329–332.

Bratley, P., Fox, B. L., and Schrage, L. E. (1983) *A Guide to Simulation.* Springer–Verlag, Berlin.

Bremner, J. M. (1981) Calculator warning. *R.S.S. News & Notes,* **8** (1), 10–11.

Brent, R. P. (1974) A Gaussian pseudo random number generator. *Comm. Assoc. Comput. Mach.,* **17**, 704–706.

Brown, D., and Rothery, P. (1978) Randomness and local regularity of points in a plane. *Biometrika,* **65**, 115–122.

Brown, S., and Holgate, P. (1974) The thinned plantation. *Biometrika,* **61**, 2, 253–261.

Buckland, S. T. (1982) A mark–recapture survival analysis. *J. Animal Ecol.,* **51**, 833–847.

Bunday, B. D., and Mack, C. (1973) Efficiency of bi-directionally traversed machines. *Appl. Stats.,* **22**, 74–81.

Burnham, K. P., Anderson, D. R., and Laake, J. L. (1980) Estimation of density from line transect sampling of biological populations. *Wildlife Monographs,* No. 72. Supplement to *The Journal of Wildlife Management,* **44**, No. 2.

Burnham, K. P., and Overton, W. S. (1978) Estimation of the size of a closed population when capture probabilities vary among animals. *Biometrika,* **65** (3), 625–634.

Burr, I. W. (1967) A useful approximation to the normal distribution function, with application to simulation. *Technometrics,* **9**, 647.

Butcher, J. C. (1960) Random sampling from the normal distribution. *Comp. J.,* **3**, 251–253.

Calinski, T., and Harabasz, J. (1974) A dendrite method of cluster analysis. *Comm. in Statis.,* **3**, 1–27.

Cammock, R. M. (1973) *Health Centres, Reception, Waiting and Patient Call.* HMSO, London.

Campbell, C., and Joiner, B. L. (1973) How to get the answer without being sure you've asked the question. *The Amer. Statist.,* **27** (5), 229–230.

Cane, V. R., and Goldblatt, P. O. (1978) The perception of probability. *Manchester/Sheffield res. report. 69/VRC/6.*

Carlisle, P. (1978) Investigation of a doctor's waiting-room using simulation. Unpublished undergraduate dissertation, Univ. of Kent.

Carter, C. O. (1969) *Human Heredity.* Penguin Books Ltd, Harmondsworth.

Chambers, J. M. (1970) Computers in statistical research: Simulation and computer-aided mathematics. *Technometrics,* **12** (1), 1–15.

Chambers, J. M. (1977) *Computational Methods for Data Analysis.* Wiley, New York.

Chambers, J. M. (1980) Statistical computing: History and trends. *The Amer. Statist.,* **34** (4), 238–243.

Chambers, J. M., Mallows, C. L., and Sturk, B. W. (1976) A method for simulating stable random variables. *J. Amer. Statist. Assoc.,* **71**, 340–344.

Chapman, D. G. (1952) Inverse multiple and sequential sample censuses. *Biometrics*, **8**, 286–306.

Chapman, D. G. (1956) Estimating the parameters of a truncated gamma distribution. *Ann. Math. Stat.*, **27**, 498–506.

Chay, S. C., Fardo, R. D., and Mazumdar, M. (1975) On using the Box–Müller transform with congruential pseudo-random generators. *Appl. Stats.*, **24** (1), 132–134.

Chen, H., and Asau, Y. (1974) Generating random variates from an empirical distribution. *Amer. Inst. Indust. Engineers Trans.*, **6**, 163–166.

Cheng, R. C. H. (1977) The generation of gamma variables with non-integral shape parameters. *Appl. Stats.*, **26** (1), 71–74.

Cheng, R. C. H. (1978) Generating beta variates with non-integral shape parameters. *Communications of the ACM*, **21** (4), 317–322.

Cheng, R. C. H. (1982) The use of antithetic variates in computer simulations. *J. Opl. Res. Soc.*, **33**, 229–237.

Cheng, R. C. H., and Feast, G. M. (1979) Some simple gamma variate generators. *Appl. Stats.*, **28** (3), 290–295.

Cheng, R. C. H., and Feast, G. M. (1980) Control variables with known mean and variance. *J. Opl. Res. Soc.*, **31**, 51–56.

Chernoff, H. (1969) Optimal design applied to simulation. *Bull. Int. Statist. Inst.*, **43**, 2264.

Chernoff, H., and Lieberman, G. J. (1956) The use of generalized probability paper for continuous distributions. *Ann. Math. Stat.*, **27**, 806–818.

Clark, C. E. (1964) Sampling efficiency in Monte Carlo analyses. *Ann. Inst. Stat. Math.*, **15**, 197.

Clarke, G. M., and Cooke, D. (1983) *A Basic Course in Statistics*. (2nd Edn) Edward Arnold, London.

Clements, G. (1978) The study of the estimation of animal population sizes. Unpublished undergraduate dissertation, Univ. of Kent, Canterbury.

Cochran, W. G. (1952) The χ^2 test of goodness of fit. *Ann. Math. Stat.*, **23**, 315–346.

Cochran, W. G. (1952) The χ^2 test of goodness of fit. *Ann. Math. Stat.*, **23**, 315–346.

Conte, S. D., and de Boor, C. (1972) *Elementary Numerical Analysis*. McGraw-Hill, Kogakusha, Tokyo.

Conway, R. W. (1963) Some tactical problems in digital simulation. *Management Science*, **10**, 47–61.

Cooke, D., Craven, A. H., and Clarke, G. M. (1982) *Basic Statistical Computing*. Edward Arnold, London.

Cooper, B. E. (1976) Discussion of 'Computer generation of beta, gamma and normal random variables' by A. C. Atkinson and M. C. Pearce. *J. Roy. Statist. Soc.*, A, **139** (4), 431–461.

Cormack, R. M. (1966) A test for equal catchability. *Biometrics*, **22**, 330–342.

Cormack, R. M. (1968) The statistics of capture–recapture methods. *Oceanogr. Mar. Bio. Ann. Rev.*, **6**, 455–506.

Cormack, R. M. (1971) A review of classification. *J. Roy. Statist. Soc.*, A, **134**, 321–367.

Cormack, R. M. (1973) Commonsense estimates from capture-recapture studies. pp. 225–234, In '*The Mathematical Theory of the Dynamics of Biological Populations*' Eds. M. S. Bartlett and R. W. Hiorns. Academic Press, London.

Coveyou, R. R., and MacPherson, R. D. (1967) Fourier analysis of uniform random number generators. *J. Assoc. Comput. Mach.*, **14**, 100–119.

Cox, D. R. (1955) Some statistical methods connected with series of events. *J. Roy. Statist. Soc.*, B, **17**, 129–157.

Cox, D. R. (1958) *Planning of Experiments*. Wiley, New York.

Cox, D. R. (1966) Notes on the analysis of mixed frequency distributions. *Brit. J. Math. & Stat. Psychol.*, **19** (1), 39–47.

Cox, D. R., and Lewis, P. A. W. (1966) *The Statistical Analysis of Series of Events*. Chapman and Hall, London.

Cox, D. R., and Miller, H. D. (1965) *The Theory of Stochastic Processes*. Chapman and Hall, London.

Cox, D. R., and Smith, W. L. (1961) *Queues*. Chapman and Hall, London.

Cox, M. A. A., and Plackett, R. L. (1980) Small samples in contingency tables. *Biometrika*, **67**, 1, 1–14.

Craddock, J. M., and Farmer, S. A. (1971) Two robust methods of random number generation. *The Statistician*, **20** (3), 55–66.

Craddock, J. M., and Flood, C. R. (1970) The distribution of the χ^2 statistic in small contingency tables. *Appl. Stats.*, **19**, 173–181.

Cramér, H. (1954) *Mathematical Methods of Statistics*. Univ. Press, Princeton.

Cugini, J. V., Bowden, J. S., and Skall, M. W. (1980) NBS Minimal BASIC Test programs – version 2, Users Manual. Vol. 2, Source Listings and Sample Output. National Bureau of Standards, Special Publication 500/70/2.

Cunningham, S. W. (1969) From normal integral to deviate. Algorithm AS24. *Appl. Stats.*, **18**, 290–293.

Daley, D. J. (1968) The serial correlation coefficients of waiting times in a stationary single server queue. *J. Austr. Math. Soc.*, **8**, 683–699.

Daley, D. J. (1974) Computation of bi- and tri-variate normal integrals. *Appl. Stats.*, **23** (3), 435–438.

Daniels, H. E. (1982) A tribute to L. H. C. Tippett. *J. Roy. Statist. Soc.*, A, **145** (2), 261–263.

Darby, K. V. (1984) Statistical analysis of British bird observatory data. Unpublished Ph.D. thesis, University of Kent, Canterbury.

Darroch, J. N. (1958) The multiple-recapture census I: Estimation of a closed population. *Biometrika*, **45**, 343–359.

David, F. N. (1973) p. 146 in *Chamber's Encyclopedia*. International Learning Systems Ltd., London.

Davis, P. J., and Rabinowitz, P. (1967) *Numerical Integration*. Blaisdell Pub. Co., Waltham, Mass.

Davis, P. J., and Rabinowitz, P. (1975) *Methods of Numerical Integration*. Academic Press, London.

Day, N. E. (1969) Estimating the components of a mixture of normal distributions. *Biometrika*, **56**, 463–474.

Déak, I. (1981) An economical method for random number generation and a normal generator. *Computing*, **27**, 113–121.

Devroye, L. (1981) The computer generation of Poisson random variables. *Computing*, **26**, 197–207.

Dewey, G. (1923) *Relative Frequency of English Speech Sounds*. Harvard University Press.

de Wit, C. T., and Goudriaan, J. (1974) Simulation of ecological processes. Simulation Monographs, Wageningen; Centre for Agricultural Pub. and Doc.

Diaconis, P., and Efron, B. (1983) Computer-intensive methods in statistics. *Scientific American*, **248** (5), 96–109.

Dieter, U., and Ahrens, J. H. (1974) *Uniform random numbers*. Produced by: Institut für Math. Statistik, Technische Hochschule in Graz. A 8010 Graz, Hamerlinggasse 6, Austria.

Diggle, P. J. (1975) Robust density estimation using distance methods. *Biometrika*, **62**, 39–48.

Diggle, P. J., Besag, J., and Gleaves, J. T. (1976) Statistical analysis of spatial point patterns by means of distance methods. *Biometrics*, **32**, 3, 659–668.

Dixon, W. J. (1971) Notes on available materials to support computer-based statistical teaching. *Rev. Int. Statist. Inst.*, **39**, 257–286.

Donnelly, J. H., and Shannon, R. E. (1981) Minimum mean-squared-error estimators for simulation experiments. *Comm. Assoc. Comp. Mach.*, **24** (4), 253–259.

Donnelly, K. (1978) Simulations to determine the variance and edge effect of total nearest neighbour distance. In: *Simulation Methods in Archaeology*. I. Hodder (Ed.), Cambridge Univ. Press, London.

Douglas, J. B. (1980) *Analysis with Standard Contagious Distributions*. International Co-operative Publishing House, Fairland, Maryland.

Downham, D. Y. (1969) The runs up and down test. *Comp. J.*, **12**, 373–376.

Downham, D. Y. (1970) Algorithm AS29. The runs up and down test. *Appl. Stats.*, **19**, 190–192.

Downham, D. Y., and Roberts, F. D. K. (1967) Multiplicative congruential pseudo-random number generators. *Comp. J.*, **10**, 74–77.

Dudewicz, E. J., and Ralley, T. G. (1981) *The Handbook of Random Number Generation and Testing with TESTRAND Computer Code.* American Sciences Press, Columbus, Ohio.

Duncan, I. B., and Curnow, R. N. (1978) Operational research in the health and social services. *J. Roy. Statist. Soc.*, A, **141**, 153–194.

Dunn, O. J., and Varady, P. D. (1966) Probabilities of correct classification in discriminant analysis. *Biometrics*, **22**, 908–924.

Dwass, M. (1972) Unbiased coin tossing with discrete random variables. *Ann. Math. Stats.*, **43** (3), 860–864.

Efron, B. (1979a) Bootstrap methods: Another look at the jackknife. *Ann. Stat.* **7**, 1–26.

Efron, B. (1979b) Computers and the theory of statistics: Thinking the unthinkable. *SIAM Review*, **21**, 460–480.

Efron, B. (1981) Nonparametric estimates of standard error: The jackknife, the bootstrap and other methods. *Biometrika*, **68** (3), 589–600.

Egger, M. J. (1979) Power transformations to achieve symmetry in quantal bioassay. *Technical Report No. 47*, Division of Biostatistics, Stanford, California.

Ehrenfeld, S., and Ben–Turia, S. (1962) The efficiency of statistical simulation procedures. *Technometrics*, **4**, 257–275.

Ekblom, H. (1972) A Monte Carlo investigation of mode estimators in small samples. *Appl. Stats.*, **21** (2), 177–184.

Ernvall, J., and Nevalainen, O. (1982) An algorithm for unbiased random sampling. *Comp. J.*, **25**, 45–47.

Estes, W. K. (1975) Some targets for mathematical psychology. *J. Math. Psychol.*, **12**, 263–282.

Evans, D. H., Herman, R., and Weiss, G. H. (1964) The highway merging and queueing problem. *Oper. Res.*, **12**, 832–857.

Evans, D. H., Herman, R., and Weiss, G. H. (1965) Queueing at a stop sign. *Proc. 2nd. Internat. Symp. on the Theory of Traffic Flow, London, 1963.* OECD, Paris.

Everitt, B. S. (1977) *The analysis of contingency tables.* Chapman and Hall, London.

Everitt, B. S. (1979) Unresolved problems in cluster analysis. *Biometrics*, **35** (1), 169–182.

Everitt, B. S. (1980) *Cluster Analysis* (2nd edn). Heinemann Educational, London.

Farlie, D. J. and Keen, J. (1967) Quick ways to the top: a team game illustrating steepest ascent techniques. *Appl. Stats.*, **16**, 75–80.

Feller, W. (1957) *An Introduction to Probability Theory and its Applications*, Vol. 1 (2nd edn). Wiley, New York.

Feller, W. (1971) *An Introduction to Probability Theory and its Applications*, Vol. 2 (2nd edn). Wiley, New York.

Felton, G. E. (1957) Electronic computers and mathematicians, pp. 12–21, in: *Oxford Mathematical Conference for Schoolteachers and Industrialists.* Oxford university delegacy for extra-mural studies. Times. Pub. Co. Ltd., London.

Fieller, E. C., and Hartley, H. O. (1954) Sampling with control variables. *Biometrika*, **41**, 494–501.

Fieller, E. C., Lewis, T., and Pearson, E. S. (1955) Correlated random normal deviates. *Tracts for Computers*, No. 26, Cambridge University Press.

Fienberg, S. E. (1980) *The Analysis of Cross-classified Categorical Data.* M.I.T. Press.

Fine, P. E. M. (1977) A commentary on the mechanical analogue to the Reed–Frost epidemic model. *American J. Epidemiology*, **106** (2), 87–100.

Fisher, L., and Kuiper, F. K. (1975) A Monte Carlo comparison of six clustering procedures. *Biometrics*, **31**, 777–784.

Fisher, R. A., and Yates, F. (1948) *Statistical Tables.* Oliver & Boyd, pp. 104–109.

Fishman, G. S. (1973) *Concepts and Methods in Discrete Event Digital Simulation.* Wiley-Interscience, New York.

Fishman, G. S. (1974) Correlated simulation experiments. *Simulation*, **23**, 177–180.

Fishman, G. S. (1976) Sampling from the Poisson distribution on a computer. *Computing*, **17**, 147–156.

Fishman, G. S. (1978) *Principles of Discrete Event Simulation.* Wiley, New York.

Fishman, G. S. (1979) Sampling from the binomial distribution on a computer. *J. Amer. Statist. Assoc.*, **74**, 418–423.

Fletcher, R. (1971) A general quadratic programming algorithm. *J. Inst. Math. Applic.*, **7**, 76–91.

Folks, J. L. (1981) *Ideas of Statistics.* Wiley, New York.

Forsythe, G. E. (1972) Von Neumann's comparison method for random sampling from the normal and other distributions. *Math. Comput.*, **26**, 817–826.

Fox, B. L. (1963) Generation of random samples from the beta F distributions. *Technometrics*, **5**, 269–270.

Freeman, P. R. (1979) Exact distribution of the largest multinomial frequency. Algorithm AS145. *Appl. Stats.*, **28**, 333–336.

Fuller, A. T. (1976) The period of pseudo-random numbers generated by Lehmer's congruential method. *Comp. J.*, **19**, 173–177.

Fuller, M. F., and Lury, D. A. (1977) *Statistics Workbook for Social Science Students.* Philip Allan Pub. Ltd., Oxford.

GAMS (1981) *Guide to Available Mathematical Software.* Centre for applied mathematics. National Bureau of Standards, Washington, DC. 20234.

Gani, J. (1980) The problem of Buffon's needle. *Mathematical Spectrum*, **13**, 14–18.

Gardner, G. (1978) A discussion of the techniques of cluster analysis with reference to an application in the field of social services. Unpublished M.Sc. dissertation, Univ. of Kent at Canterbury.

Gardner, M. (1966) *New Mathematical Diversions from Scientific American.* Simon & Schuster.

Gates, C. E. (1969) Simulation study of estimators for the line transect sampling method. *Biometrics*, **25**, 317–328.

Gauss, C. F. (1809) *Theoria Motus Corporum Coelestium.* Perthes & Besser, Hamburg.

Gaver, D. P. and Thompson, G. L. (1973) *Programming and Probability Models in Operations Research.* Brooks/Cole, Monterey, California.

Gebhardt, F. (1967) Generating pseudo-random numbers by shuffling a Fibonacci sequence. *Math. Comp.* **21**, 708–709.

Gerontidis, I., and Smith, R. L. (1982) Monte Carlo generation of order statistics from general distributions. *Appl. Stats.*, **31** (3), 238–243.

Gerson, M. (1975) The technique and uses of probability plotting. *The Statistician*, **24** (4), 235–257.

Gibbs, K. (1979) Appointment systems in general practice. Unpublished undergraduate dissertation, Univ. of Kent at Canterbury.

Gibbs, K. (1980) Mathematical models for population growth with special reference to the African mountain gorilla. Unpublished M.Sc. dissertation, Univ. of Kent at Canterbury.

Gilchrist, R. (1976) The use of verbal call systems to call patients from waiting areas to consulting rooms. *Appl. Stats.*, **25**, 217–227.

Gipps, P. G. (1977) A queueing model for traffic flow. *J. Roy. Statist. Soc.*, B, **39** (2), 276–282.

Gleser, L. J. (1976) A canonical representation for the noncentral Wishart distribution useful for simulation. *J. Amer. Statist. Assoc.*, **71**, 690–695.

Gnanadesikan, R. (1977) *Methods for Statistical Data Analysis of Multivariate Observations.* Wiley, New York.

Gnedenko, B. V. (1976) *The Theory of Probability.* MIR Publishers, Moscow (English translation).

Golder, E. R. (1976a) Algorithm AS98. The spectral test for the evaluation of congruential pseudo-random number generators. *Appl. Stats.*, **25**, 173–180.

Golder, E. R. (1976b) Remark ASR18. The spectral test for the evaluation of congruential pseudo-random generators. *Appl. Stats.*, **25**, 324.

Golder, E. R., and Settle, J. G. (1976) The Box–Müller method for generating pseudo-random normal deviates. *Appl. Stats.*, **25** (1), 12–20.

Good, I. J. (1953) The serial test for sampling numbers and other tests for randomness. *Proc. Camb. Phil. Soc.*, **49**, 276–284.

Good, I. J. (1957) On the serial test for random sequences. *Ann. Math. Statist.*, **28**, 262–264.

Gordon, A. D. (1981) *Classification*. Chapman & Hall, London.

Gordon, R. (1970) On Monte Carlo algebra. *J. Appl. Prob.*, **7**, 373–387.

Gosper, W. G. (1975) Numerical experiments with the spectral test. *Stan-CS-75–490*. Computer Science Department, Stanford University.

Gower, J. C. (1968) Simulating multidimensional arrays in 1D. Algorithm AS1. *Appl. Stats.*, **17**, 180 (see also **18** (1), 116).

Gower, J. C., and Banfield, C. F. (1975) Goodness-of-fit criteria for hierarchical classification and their empirical distributions. In: *Proceedings of the 8th International Biometric Conference.* (Eds L.C.A. Corsten and T. Postelnicu), 347–361.

Grafton, R. G. T. (1981) Algorithm AS157: The runs-up and runs-down tests. *Appl. Stats.*, **30**, 81–85.

Green, P. J. (1978) Small distances and Monte Carlo testing of spatial pattern. *Advances in Applied Prob.*, **10**, 493.

Greenberg, B. G., Kuebler, R. T., Jr., Abernathy, J. R., and Horvitz, D. G. (1971) Application of randomized response technique in obtaining quantitative data. *J. Amer. Statist. Assoc.*, **66**, 243–250.

Greenberger, M. (1961) An a priori determination of serial correlation in computer generated random numbers. *Math. Comp.*, **15**, 383–389.

Greenwood, A. J. (1974) A fast generator for gamma distributed random variables. pp. 19–27 in *Compstat 1974* (G. Bruckman *et al.*, eds.) Physica–Verlag: Vienna.

Greenwood, J. A., and Durand, D. (1960) Aids for fitting the gamma distribution by maximum likelihood. *Technometrics*, **2**, 55–65.

Greenwood, R. E. (1955) Coupon collector's test for random digits. *Mathematical Tables and Other Aids to Computation*, **9**, 1–5.

Griffiths, J. D., and Cresswell, C. (1976) A mathematical model of a Pelican Crossing. *J. Inst. Maths. & Applics.*, **18** (3), 381–394.

Grimmett, G. R., and Stirzaker, D. R. (1982) *Probability and Random Processes*. Oxford University Press.

Grogono, P. (1980) *Programming in PASCAL*. Addison-Wesley Pub. Co. Inc., Reading, Massachusetts.

Gross, A. M. (1973) A Monte Carlo swindle for estimators of location. *Appl. Stats.*, **22** (3), 347–353.

Gross, D., and Harris, C. M. (1974) *Fundamentals of Queueing Theory*. Wiley, Toronto.

Gruenberger, F., and Mark, A. M. (1951) The d^2 test of random digits. *Math. Tables Other Aids Comp.*, **5**, 109–110.

Guerra, V. M., Tapia, R. A., and Thompson, J. R. (1972) A random number generator for continuous random variables. *ICSA Technical Report*, Rice University, Houston, Texas.

Haight, F. A. (1967) *Handbook of the Poisson distribution.* Wiley, New York.

Halton, J. H. (1970) A retrospective and prospective survey of the Monte Carlo method. *SIAM Rev.,* **12** (1), 1–63.

Hamaker, H. C. (1978) Approximating the cumulative normal distribution and its inverse. *Appl. Stats.,* **27** (1), 76–77.

Hammersley, J. M., and Handscomb, D. C. (1964) *Monte Carlo Methods.* Methuen, London.

Hammersley, J. M., and Morton, K. W. (1956) A new Monte Carlo technique: antithetic variates. *Proc. Camb. Phil. Soc.,* **52**, 449–475.

Hannan, E. J. (1957) The variance of the mean of a stationary process. *J. Roy. Statist. Soc.,* B, **19**, 282–285.

Harcourt, A. H., Stewart, K. J., and Fossey, D. (1976) Male emigration and female transfer in wild mountain gorilla. *Nature,* **263**, 226–227.

Harris, R., Norris, M. E., and Quaterman, B. R. (1974) Care of the elderly: structure of the group flow model. *Report 825.* Institute for Operational Research.

Harter, H. L. (1964) *New tables of the incomplete gamma-function ratio and of percentage points of the chi-square and beta distributions.* U.S. Government printing office, Washington, D.C.

Hastings, C., Jr. (1955) *Approximations for Digital Computers.* Princeton University Press, Princeton, N. J.

Hastings, W. K. (1970) Monte Carlo sampling methods using Markov chains and their applications. *Biometrika,* **57**, 97–109.

Hastings, W. K. (1974) Variance reduction and non-normality. *Biometrika,* **61**, 143–149.

Hauptman, H., Vegh, E., and Fisher, J. (1970) Table of all primitive roots for primes less than 5000. *Naval Research Laboratory Report No. 7070.* Washington, D.C.

Hawkes, A. G. (1965) Queueing for gaps in traffic. *Biometrika,* **52**, 1–6.

Hawkes, A. G. (1966) Delay at traffic intersections. *J. Roy. Statist. Soc.,* B, **28** (1), 202–212.

Hay, R. F. M. (1967) The association between autumn and winter circulations near Britain. *Met. Mag.,* **96**, 167–177.

Healy, M. J. R. (1968a) Multivariate normal plotting. *Appl. Stats.,* **17**, 157–161.

Healy, M. J. R. (1968b) Triangular decomposition of a symmetric matrix. Algorithm AS6. *Appl. Stats.,* **17**, 195–197.

Heathcote, C. R., and Winer, P. (1969) An approximation for the moments of waiting times. *Oper. Res.,* **17**, 175–186.

Heidelberger, P., and Welch, P. D. (1981) A spectral method for confidence interval generation and mean length control in simulations. *Comm. Assoc. Comp. Mach.,* **24** (4), 233–245.

Heiberger, R. M. (1978) Generation of random orthogonal matrices. Algorithm. AS127. *Appl. Stats.*, **27**, 199–206.

Hews, R. J. (1981) Stopping rules and goodness of fit criteria for hierarchical clustering methods. Unpublished M.Sc. dissertation. University of Kent.

Hoaglin, D. C., and Andrews, D. F. (1975) The reporting of computation-based results in statistics. *The American Statistician*, **29**, 122–126.

Hoaglin, D. C., and King, M. L. (1978) Remark ASR24. A remark on Algorithm AS98. The spectral test for the evaluation of congruential pseudo-random generators. *Appl. Stats.*, **27**, 375.

Hoeffding, W., and Simons, G. (1970) Unbiased coin tossing with a biased coin. *Ann. Math. Stat.*, **41**, 341–352.

Hoel, P. G. (1954) *Introduction to Mathematical Statistics* (3rd edn). Wiley, New York.

Holgate, P. (1981) Studies in the history of probability and statistics. XXXIX. Buffon's cycloid. *Biometrika*, **68** (3), 712–716.

Hollier, R. H. (1968) A simulation study of sequencing in batch production. *Op. Res. Quart.*, **19**, 389–407.

Hollingdale, S. H. (Ed.) (1967) *Digital Simulation in Operational Research*. English Universities Press, London.

Hope, A. C. A. (1968) A simplified Monte Carlo significance test procedure. *J. Roy. Statist. Soc.*, B, **30**, 582–598.

Hopkins, T. R. (1980) PBASIC—A verifier for BASIC. *Software Practice and Experience*, **10**, 175–181.

Hopkins, T. R. (1983a) A revised algorithm for the spectral test. In Algorithm AS 193, *Applied Statistics*, **32** (3), 328–335.

Hopkins, T. R. (1983b) The collision test—an empirical test for random number generators. Unpublished Computing Lab. Report, University of Kent.

Hordijk, A., Iglehart, D. L., and Schassberger, R. (1976) Discrete time methods for simulating continuous time Markov chains. *Adv. Appl. Prob.*, **8**, 772–788.

Hsu, D. A., and Hunter, J. S. (1977) Analysis of simulation-generated responses using autoregressive models. *Manag. Sci.*, **24**, 181–190.

Hsuan, F. (1979) Generating uniform polygonal random pairs. *Appl. Stats.*, **28**, 170–172.

Hull, T. E., and Dobell, A. R. (1962) Random number generators. *SIAM Rev.*, **4**, 230–254.

Hunter, J. S., and Naylor, T. H. (1970) Experimental designs for computer simulation experiments. *Manag. Sci.*, **16**, 422–434.

Hutchinson, T. P. (1979) The validity of the chi-square test when expected frequences are small: a list of recent research references. *Communications in Statistics*, Part A: *Theory and Methods*, **A8**, 327–335.

Hwang, F. K., and Lin, S. (1971) On generating a random sequence. *J. Appl. Prob.*, **8**, 366–373.

Inoue, H., Kumahora, H., Yoshizawa, Y., Ichimura, M. and Miyatake, O.

(1983) Random numbers generated by a physical device. *Applied Statistics*, **32** (2), 115–120.

Isida, M. (1982) Statistics and micro-computer. *Compstat 1982*, Part II, 141–142. Physica–Verlag, Vienna.

Jackson, R. R. P., Welch, J. D., and Fry, J. (1964) Appointment systems in hospitals and general practice. *Oper. Res. Quart.*, **15**, 219–237.

Jeffers, J. N. R. (1967) Two case studies in the application of principal component analysis. *Appl. Stats.*, **16**, 225–236.

Jeffers, J. N. R. (Ed.) (1972) *Mathematical Models in Ecology*. Blackwell Scientific Publications, Oxford.

Jenkinson, G. (1973) Comparing single-link dendrograms. Unpublished M.Sc. thesis, University of Kent, Canterbury.

Jöhnk, M. D. (1964) Erzeugung von Betarerteilten und Gammaverteilten Zuffallszahlen. *Metrika*, **8**, 5–15.

Johnson, N. L., and Kotz, S. (1969) *Distributions in Statistics: Discrete Distributions*. Houghton Mifflin, Boston.

Johnson, N. L., and Kotz, S. (1970a) *Continuous Univariate Distributions* 1. Houghton Mifflin, Boston.

Johnson, N. L., and Kotz, S. (1970b) *Continuous Univariate Distributions* 2. Houghton Mifflin, Boston.

Johnson, N. L., and Kotz, S. (1972) *Distributions in Statistics: Continuous Multivariate Distributions*. Wiley, New York.

Johnson, S. C. (1967) Hierarchical clustering schemes. *Psychometrika*, **32**, 241–254.

Johnston, W. (1971) The case history of a simulation study. *Appl. Stats.*, **20** (3), 308–312.

Jolliffe, I. T. (1972a) Discarding variables in a principal component analysis, I. Artificial data. *Appl. Stats.*, **21** (2), 160–173.

Jolliffe, I. T. (1972b) Discarding variables in a principal component analysis, II: Real data. *Appl. Stats.*, **22** (1), 21–31.

Jolliffe, I. T., Jones, B., and Morgan, B. J. T. (1982) Utilising clusters: a case-study involving the elderly. *J. Roy. Statist. Soc.*, A, **145** (2), 224–236.

Jolly, G. M. (1965) Explicit estimates from capture–recapture data with both death and immigration—stochastic model. *Biometrika*, **52**, 225–247.

Jones, G. T. (1972) *Simulation and Business Decisions*. Penguin Books, Harmondsworth.

Joseph, A. W. (1968) A criticism of the Monte Carlo method as applied to mathematical computations. *J. Roy. Statist. Soc.*, A, **131**, 226–228.

Jowett, G. H. (1955) The comparison of means of sets of observations from sections of independent stochastic series. *J. Roy. Statist. Soc.*, B, **17**, 208–227.

Kahan, B. C. (1961) A practical demonstration of a needle experiment designed to give a number of concurrent estimates of π. *J. Roy. Statist. Soc.*, A, **124**, 227–239.

Kahn, H. (1956) *Application of Monte Carlo*. Rand Corp., Santa Monica, California.

Kelker, D. (1973) A random walk epidemic simulation. *J. Amer. Statist. Assoc.*, **68**, 821–823.

Kelly, F. P. (1979) *Reversibility and Stochastic Networks*. Wiley, Chichester.

Kemp, A. W. (1981) Efficient generation of logarithmically distributed pseudo-random variables. *Appl. Stats.*, **30** (3), 249–253.

Kemp, C. D. (1982) Low-storage Poisson generators for microcomputers. *Compstat Proceedings*, Part II, 145–146. Physica–Verlag, Vienna.

Kemp, C. D., and Loukas, S. (1978a) The computer generation of bivariate discrete random variables. *J. Roy. Statist. Soc.*, A, **141** (4), 513–517.

Kemp, C. D., and Loukas, S. (1978b) Computer generation of bivariate discrete random variables using ordered probabilities. *Proc. Stat. Comp. Sec. Am. Stat. Assoc.*, San Diego Meeting, 115–116.

Kempton, R. A. (1975) A generalized form of Fisher's logarithmic series. *Biometrika*, **62** (1), 29–38.

Kendall, D. G. (1949) Stochastic processes and population growth. *J. Roy. Statist. Soc.*, **11** (2), 230–264.

Kendall, D. G. (1950) An artificial realisation of a simple birth and death process. *J. Roy. Statist. Soc.*, B, **12**, 116–119.

Kendall, D. G. (1965) Mathematical models of the spread of infection. In: *Mathematics and Computer Science in Biology and Medicine*. HMSO, Leeds.

Kendall, D. G. (1974) Pole-seeking Brownian motion and bird navigation. *J. Roy. Statist. Soc.*, B, **36** (3), 365–417.

Kendall, M. G., and Babbington–Smith, B. (1938) Randomness and random sampling numbers. *J. Roy. Statist. Soc.*, **101**, 147–166.

Kendall, M. G., and Babbington–Smith, B. (1939a) Tables of random sampling numbers. *Tracts for Computers*, XXIV. C.U.P.

Kendall, M. G., and Babbington-Smith, B. (1939b) Second paper on random sampling numbers. *J. Roy. Statist. Soc., Suppl.*, **6**, 51–61.

Kendall, M. G., and Moran, P. A. P. (1963) *Geometrical Probability*. Griffin & Co. Ltd., London.

Kendall, M. G., and Stuart, A. (1961) *The Advanced Theory of Statistics*, Vol. 2. Hafner Pub. Co., New York.

Kendall, M. G. and Stuart, A. (1969) *The Advanced Theory of Statistics*, Volume 1. Hafner Pub. Co., New York.

Kennedy, W. J. and Gentle, J. E. (1980) *Statistical Computing*. Marcel Dekker, New York.

Kenward, A. J. (1982) Computer simulation of two queueing systems. Unpublished undergraduate project, University of Kent at Canterbury.

Kermack, W. O., and McKendrick, A. G. (1937) Tests for randomness in a series of numerical observations. *Proc. Roy. Soc. of Edinburgh*.

Kerrich, J. E. (1946) *An Experimental Introduction to the Theory of Probability.* Einar Munksgaard, Copenhagen.

Kesting, K. W., and Mann, N. R. (1977) A simple scheme for generating multivariate gamma distributions with non-negative covariance matrix. *Technometrics*, **19**, 179–184.

Kimball, B. F. (1960) On the choice of plotting positions on probability paper. *J. Amer. Statist. Assoc.*, **55**, 546–560.

Kinderman, A. J., and Monahan, J. F. (1977) Computer generation of random variables using the ratio of normal deviates. *Assoc. Comput. Mach. Trans. Math. Soft.*, **3**, 257–260.

Kinderman, A. J., and Monahan, J. F. (1980) New methods for generating Student's *t* and gamma variables. *Computing*, **25**, 369–377.

Kinderman, A. J., Monahan, J. F., and Ramage, J. G. (1977) Computer methods for sampling from Student's *t* distribution. *Mathematics of Computation*, **31**, 1009–1017.

Kinderman, A. J., and Ramage, J. G. (1976) Computer generation of normal random variables. *J. Amer. Statist. Assoc.*, **71**, 356, 893–896.

Klahr, D. (1969) A Monte Carlo investigation of the statistical significance of Kruskal's nonmetric scaling procedure. *Psychometrika*, **34** (3), 319–330.

Kleijnen, J. P. C. (1973) Monte Carlo simulation and its statistical design and analysis. *Proceedings of the 29th I.S.I. Conference, Vienna*, pp. 268–279.

Kleijnen, J. P. C. (1974/5) *Statistical Techniques in Simulation* (Parts 1 and 2). Marcel Dekker, Inc., New York.

Kleijnen, J. P. C. (1977) Design and analysis of simulations: Practical Statistical Techniques. *Simulation*, **28**, 81–90.

Knuth, D. E. (1968) *The Art of Computer Programming*, Vol. 1: *Fundamental Algorithms.* Addison-Wesley, Reading, Massachusetts.

Knuth, D. E. (1981) *The Art of Computer Programming*, Vol. 2: *Seminumerical Algorithms.* Addison-Wesley, Reading, Massachusetts.

Kozlov, G. A. (1972) Estimation of the error of the method of statistical tests (Monte-Carlo) due to imperfections in the distribution of random numbers. *Theory of Probability and its Applications*, **17**, 493–509.

Kral, J. (1972) A new additive pseudorandom number generator for extremely short word-length. *Information Processing Letters* **1**, 164–167.

Kronmal, R. (1964) The evaluation of a pseudorandom normal number generator. *J. Assoc. Comp. Mach.*, **11**, 357–363.

Kronmal, R. A., and Peterson, A. V. Jr. (1979) On the alias method for generating random variables from a discrete distribution. *Amer. Statist.*, **33** (4), 214–218.

Kronmal, R. A., and Peterson, A. V. Jr. (1981) A variant of the acceptance–rejection method for computer generation of random variables. *J. Amer. Statist. Assoc.*, **76**, 446–451.

Kruskal, J. B. (1964) Nonmetric multidimensional scaling: a numerical method. *Psychometrika*, **29**, 115–129.

Kruskal, J. B., and Wish, M. (1978) Multidimensional scaling. Vol. 11 of: *Quantitative Applications in the Social Sciences*. Sage, Beverley Hills, California.

Krzanowski, W. J. (1978) Between-group comparison of principal components – some sampling results. *J. Stat. Comput. Simul.*, **15**, 141–154.

Kuehn, H. G. (1961) A 48-bit pseudo-random number generator. *Comm. Ass. Comp. Mach.*, **4**, 350–352.

Kuiker, F. K., and Fisher, L. (1975) A Monte-Carlo comparison of 6 clustering procedures. *Biometrics*, **31**, 777–784.

Lachenbruch, P. A. (1975) *Discriminant Analysis*. Hafner, New York.

Lachenbruch, P. A., and Goldstein, M. (1979) Discriminant analysis. *Biometrics*, **35** (1), 69–86.

Lack, D. (1965) *The Life of the Robin* (4th edn). M.F. & G. Witherby.

Laplace, P. S. (1774) Determiner le milieu que l'on doit prendre entre trois observations données d'un même phénomené. *Mémoires de Mathématique et Physique presentées a l'Académie Royale des Sciences par divers Savans*, **6**, 621–625.

Larntz, K. (1978) Small sample comparisons of exact levels for chi-square goodness-of-fit statistics. *J. Amer. Statist. Assoc.*, **73**, 253–263.

Lavenberg, S. S., and Welch, P. D. (1979) Using conditional expectation to reduce variance in discrete event simulation. *Proc. 1979 Winter Simulation Conference, San Diego, California*, 291–294.

Law, A. M., and Kelton, W. D. (1982) *Simulation Modelling and Analysis*. McGraw-Hill series in industrial engineering and management science, New York.

Learmonth, G. P., and Lewis, P. A. W. (1973) Naval Postgraduate School Random Number Generator Package LLRANDOM, NPS55LW73061A. Naval Postgraduate School, Monterey, California.

Lehman, R. S. (1977) *Computer Simulation and Modelling*. Wiley, Chichester.

Lehmer, D. H. (1951) Mathematical methods in large-scale computing units. *Ann. Comp. Lab. Harvard University*, **26**, 141–146.

Lenden–Hitchcock, K. J. (1980) Aspects of random number generation with particular interest in the normal distribution. Unpublished M.Sc. dissertation, University of Kent.

Leslie, P. H. (1958) A stochastic model for studying the properties of certain biological systems by numerical methods. *Biometrika*, **45**, 16–31.

Leslie, P. H., and Chitty, D. (1951) The estimation of population parameters from data obtained by means of the capture–recapture method, I. The maximum likelihood equations for estimating the death rate. *Biometrika*, **38**, 269–292.

Levene, H., and Wolfowitz, J. (1944) The covariance matrix of runs up and down. *Ann. Math. Statist.*, **15**, 58–69.

Lew, R. A. (1981) An approximation to the cumulative normal distribution with simple coefficients. *Appl. Stats.*, **30** (3), 299–300.

Lewis, J. G., and Payne, W. H. (1973) Generalized feedback shift register pseudorandom number algorithm. *J. Assoc. Comp. Mach.*, **20**, 456–468.

Lewis, P. A. W., Goodman, A. S., and Miller, J. M. (1969) A pseudo-random number generator for the System/360. *IBM Systems J.*, **8**, 136–145.

Lewis, P. A. W., and Shedler, G. S. (1976) Simulation of nonhomogeneous Poisson process with log linear rate function. *Biometrika*, **63**, 501–506.

Lewis, T. (1975) A model for the parasitic disease bilharziasis. *Advances in Applied Probability*, **7**, 673–704.

Lewis, T. G. (1975) *Distribution Sampling for Computer Simulation.* Lexington Books (P. C. Heath).

Lewis, T. G., and Payne, W. H. (1973) Generalized feedback shift register pseudo random number algorithm. *J. Assoc. Comp. Mach.*, **20** (3), 456–468.

Lewis, T. G., and Smith, B. J. (1979) *Computer Principles of Modelling and Simulation.* Houghton Mifflin, Boston.

Lieberman, G. J., and Owen, D. B. (1961) *Tables of the hypergeometric probability distribution.* Stanford University Press, Stanford, California.

Little, R. J. A. (1979) Maximum likelihood inference for multiple regression with missing values—a simulation study. *J. Roy. Statist. Soc.*, B, **41** (1), 76–87.

Lotka, A. J. (1931) The extinction of families. *J. Wash. Acad. Sci.*, **21**, 377–380 and 453–459.

Loukas, S. and Kemp, C. D. (1983) On computer sampling from trivariate and multivariate discrete distributions. *J. Stat. Comput. Simul.*, **17**, 113–123.

Luck, G. M., Luckman, J., Smith, B. W., and Stringer, J. (1971) *Patients, Hospitals and Operational Research.* Tavistock, London.

McArthur, N., Saunders, I. W., and Tweedie, R. L. (1976) Small population isolates: a micro-simulation study. *J. Polynesian Soc.*, **85**, 307–326.

Macdonell, W. R. (1902) On criminal anthropometry and the identification of criminals. *Biometrika*, **1**, 177–227.

MacLaren, M. D., and Marsaglia, G. (1965) Uniform random number generators. *J. Assoc. Comp. Mach.*, **12**, 83–89.

McLeod, A. I. and Bellhouse, D. R. (1983) A convenient algorithm for drawing a simple random sample. *Applied Statistics*, **32** (2), 182–184.

Maher, M. J., and Akçelik, R. (1977) Route control—simulation experiments. *Transportation Research*, **11**, 25–31.

Manly, B. F. J. (1971) A simulation study of Jolly's method for analysing capture–recapture data. *Biometrics*, **27**, 415–424.

Mann, H. B., and Wald, A. (1942) On the choice of the number of class intervals in the application of the chi-squared test. *Ann. Math. Stats.*, **13**, 306–317.

Mantel, N. (1953) An extension of the Buffon needle problem. *Ann. Math. Stats.*, **24**, 674–677.

Mardia, K. V. (1970) *Families of Bivariate Distributions.* Griffin, London.

Mardia, K. V., Kent, J. T., and Bibby, J. M. (1979) *Multivariate Analysis.* Academic Press, London.

Mardia, K. V. (1980) Tests of univariate and multivariate normality. pp. 279–320. In: P. R. Krishnaiah (Ed.) *Handbook of Statistics*, Vol. 1. North Holland, Amsterdam.

Mardia, K. V., and Zemroch, P. J. (1978) *Tables of the F- and Related Distributions with Algorithms.* Academic Press, London.

Marks, S., and Dunn, O. J. (1974) Discriminant functions when covariance matrices are unequal. *J. Amer. Statist. Assoc.*, **69**, 555–559.

Marriott, F. H. C. (1972) Buffon's problems for non-random distributions. *Biometrics*, **28**, 621–624.

Marriott, F. H. C. (1979) Barnard's Monte Carlo tests: How many simulations? *Appl. Stats.*, **28** (1), 75–77.

Marsaglia, G. (1961a) Generating exponential random variables. *Ann. Math. Stats.*, **32**, 899–900.

Marsaglia, G. (1961b) Expressing a random variable in terms of uniform random variables. *Ann. Math. Stats.*, **32**, 894–900.

Marsaglia, G. (1964) Generating a variable from the tail of the normal distribution. *Technometrics*, **6**, 101–102.

Marsaglia, G. (1968) Random numbers fall mainly in the planes. *Proc. Nat. Acad. Sci., USA*, **61**, 25–28.

Marsaglia, G. (1972a) Choosing a point from the surface of a sphere. *Ann. Math. Stats.*, **3** (2), 645–646.

Marsaglia, G. (1972b) The structure of linear congruential sequences. In: *Applications of Number Theory to Numerical Analysis* (Ed. S. K. Zaremba), pp. 249–285. Academic Press, London.

Marsaglia, G., and Bray, T. A. (1964) A convenient method for generating normal variables. *SIAM Rev.*, **6**, 260–264.

Marsaglia, G., MacLaren, M. D., and Bray, T. A. (1964) A fast procedure for generating normal random variables. *Communications of the Ass. Comp. Mach.*, **7** (1), 4–10.

Mawson, J. C. (1968) A Monte Carlo study of distance measures in sampling for spatial distribution in forest stands. *Forestry Science*, **14**, 127–139.

Mead, R., and Freeman, K. H. (1973) An experiment game. *Appl. Stats.*, **22**, 1–6.

Mead, R., and Stern, R. D. (1973) The use of a computer in the teaching of statistics. *J. Roy. Statist. Soc.*, A, **136** (2), 191–225.

Metropolis, N. C., Reitwiesner, G., and von Neumann, J. (1950) Statistical treatment of values of first 2000 decimal digits of e and π calculated on the ENIAC. *Math. Tables & Other Aids to Comp.*, **4**, 109–111.

Meynell, G. G. (1959) Use of superinfecting phage for estimating the division rate of lysogenic bacteria in infected animals. *J. Gen. Microbiol.*, **21**, 421–437.

Michael, J. R., Schucany, W. R., and Haas, R. W. (1976) Generating random variates using transformations with multiple roots. *Amer. Statist.*, **30** (2), 88–90.

Mikes, G. (1946) *How to be an Alien*. Andre Deutch, Tonbridge.

Miller, A. J. (1977a) Random number generation on the SR52. CSIRO, *DMS Newsletter*, No. 28, p. 4.

Miller, A. J. (1977b) On some unrandom numbers. CSIRO, *DMS Newsletter*, No. 29, p. 2.

Miller, A. J. (1980a) Another random number generator. CSIRO, *DMS Newsletter*, No. 65, p. 5.

Miller, A. J. (1980b) On random numbers. CSIRO, *DMS Newsletter*, No. 68, p. 7.

Miller, J. C. P., and Prentice, M. J. (1968) Additive congruential pseudo-random number generators. *Comp. J.*, **11** (3), 341–346.

Mitchell, B. (1971) Variance reduction by antithetic variates in GI/G/1 queueing simulations. *Oper. Res.*, **21**, 988–997.

Mitchell, G. H. (1969) Simulation. *Bull. Inst. Math. Applics.*, **5** (3), 59–62.

Mitchell, G. H. (1972) *Operational Research: Techniques and Examples*. The English Universities Press Ltd., London.

Mitchell, K. J. (1975) Dynamics and simulated yield of Douglas fir. *Forest. Science*, Monograph, 17.

Moder, J. J., and Elmaghraby, S. E. (1978) *Handbook of Operations Research: Foundations and Fundamentals*. Van Nostrand Reinhold, New York.

Mojena, R. (1977) Hierarchical grouping methods and stopping rules: an evaluation. *Computer J.*, **20**, 359–363.

Moore, P. G. (1953) A sequential test for randomness. *Biometrika*, **40**, 111–115.

Moors, J. A. A. (1971) Optimization of the unrelated question randomized response model. *J. Amer. Statist. Assoc.*, **66**, 627–629.

Moran, P. A. P. (1975) The estimation of standard errors in Monte Carlo simulation experiments. *Biometrika*, **62**, 1–4.

Moran, P. A. P., and Fazekas de St Groth, S. (1962) Random circles on a sphere. *Biometrika*, **49**, 384–396.

Morgan, B. J. T. (1974) On the distribution of inanimate marks over a linear birth-and-death process. *J. Appl. Prob.*, **11**, 423–436.

Morgan, B. J. T. (1976) Markov properties of sequences of behaviours. *Appl. Stats.*, **25** (1), 31–36.

Morgan, B. J. T. (1978) Some recent applications of the linear birth-and-death process in biology. *Math. Scientist*, **3**, 103–116.

Morgan, B. J. T. (1979a) A simulation model of the social life of the African Mountain Gorilla. Unpublished manuscript.

Morgan, B. J. T. (1979b) Four approaches to solving the linear birth-and-death (and similar) processes. *Int. J. Math. Educ. Sci. Technol.*, **10** (1), 51–64.

Morgan, B. J. T. (1981) Three applications of methods of cluster-analysis. *The Statistician*, **30** (3), 205–223.

Morgan, B. J. T. (1983) Illustration of three-dimensional surfaces. *BIAS*, **10**, 2.

Morgan, B. J. T., Chambers, S. M., and Morton, J. (1973) Acoustic confusion of digits in memory and recognition. *Perception and Psychophysics*, **14** (2), 375–383.

Morgan, B. J. T., and Leventhal, B. (1977) A model for blue-green algae and gorillas. *J. Appl. Prob.*, **14**, 675–688.

Morgan, B. J. T., and North, P. M. (Eds.) (1984) *Statistics in Ornithology*. Springer-Verlag (to appear).

Morgan, B. J. T., and Robertson, C. (1980) Short-term memory models for choice behaviour. *J. Math. Psychol.*, **21** (1), 30–52.

Morgan, B. J. T. and Watts, S. A. (1980) On modelling microbial infections. *Biometrics*, **36**, 317–321.

Morgan, B. J. T., Woodhead, M. M., and Webster, J. C. (1976) On the recovery of physical dimensions of stimuli, using multidimensional scaling. *J. Acoust. Soc. Am.*, **60** (1), 186–189.

Morgan, R., and Hirsch, W. (1976) Stretch a point and clear the way. *Times Higher Education Supplement*, 23 July.

Morrison, D. F. (1976) *Multivariate Statistical Methods* (2nd edn). McGraw-Hill Kogakusha Ltd., Tokyo.

Morton, K. W. (1957) A generalisation of the antithetic variate technique for evaluating integrals. *J. Math. Phys.*, **36** (3), 289–293.

Moses, L. E. and Oakford, R. F. (1963) *Tables of Random Permutations*. Allen & Unwin, London.

Mosteller, F. (1965) *Fifty Challenging Problems in Probability with Solutions*. Addison-Wesley Publishing Co. Inc., Reading, Massachusetts.

Mountford, M. D. (1982) Estimation of population fluctuations with application to the Common Bird Census. *Appl. Stats.*, **31**, 135–143.

Mudholkar, G. S., and George, E. O. (1978) A remark on the shape of the logistic distribution. *Biometrika*, **65** (3), 667–668.

Murdoch, J., and Barnes, J. A. (1974) *Statistical Tables for Science, Engineering, Management and Business Studies* (2nd edn). Macmillan Press Ltd., London.

Myers, R. H. (1971) *Response Surface Methodology*. Allyn and Bacon, Boston.

Nance, R. E. (1971) On time flow mechanisms for discrete system simulation. *Management Sciences*, **18** (1), 59–73.

Nance, R. E., and Overstreet, C. J., Jr. (1972) A bibliography on random number generation. *Comp. Rev.*, **13**, 495–508.

Nance, R. E., and Overstreet, C., Jr., (1975) Implementation of Fortran random number generators on computers with One's Complement arithmetic. *J. Statist. Comput. Simul.*, **4**, 235–243.

Nance, R. E., and Overstreet, C., Jr. (1978) Some experimental observations on the behaviour of composite random number generators. *Oper. Res.*, **26**, 915–935.

Naylor, T. H. (1971) *Computer Simulation Experiments with Models of Economic Systems*. Wiley, New York.

Naylor, T. H., Burdick, D. S., and Sasser, W. E. (1967) Computer simulation experiments with economic systems—the problem of experimental design. *J. Amer. Statist. Assoc.*, **62**, 1315–1337.

Naylor, T. H. and Finger, J. M. (1967) Verification of computer simulation models. *Management Sciences*, **14**, 92–101.

Naylor, T. H., Wallace, W. H., and Sasser, W. E. (1967) A computer simulation model of the textile industry. *J. Amer. Statist. Assoc.*, **62**, 1338–1364.

Neave, H. (1972) Random number package. Computer applications in the natural and social sciences, No. 14. Department of Geography, University of Nottingham.

Neave, H. R. (1973) On using the Box–Müller transformation with multiplicative congruential pseudo-random number generators. *Appl. Stats.*, **22**, 92–97.

Neave, H. R. (1978) *Statistics Tables*. George Allen & Unwin, London.

Neave, H. R. (1981) *Elementary Statistics Tables*. Alden Press, Oxford.

Nelsen, R. B., and Williams, T. (1968) Randomly delayed appointment streams. *Nature*, **219**, 573–574.

Newman, T. G., and Odell, P. L. (1971) *The Generation of Random Variates*. Griffin, London.

Niederreiter, H. (1978) Quasi-Monte Carlo methods and pseudo-random numbers. *Bull. Amer. Math. Soc.*, **84**, 957–1041.

Norman, J. E., and Cannon, L. E. (1972) A computer programme for the generation of random variables from any discrete distribution. *J. Statist. Comput. Simul.*, **1**, 331–348.

Oakenfull, E. (1979) Uniform random number generators and the spectral test. In: *Interactive Statistics*. Ed. D. McNeil, North-Holland, pp. 17–37.

Odeh, R. E., and Evans, J. O. (1974) Algorithm AS70. The percentage points of the normal distribution. *Appl. Stats.*, **23**, 96–97.

Odeh, R. E., Owen, D. B., Birnbaum, Z. W., and Fisher, L. (1977) *Pocket Book of Statistical tables*. Marcel Dekker, New York & Basel.

Odell, P. L., and Feireson, A. H. (1966) A numerical procedure to generate a sample covariance matrix. *J. Amer. Statist. Assoc.*, **61**, 199–203.

O'Donovan, T. M. (1979) *GPSS Simulation Made Simple*. Wiley, Chichester.

Odoroff, C. L. (1970) A comparison of minimum logit chi-square estimation and maximum likelihood estimation in $2 \times 2 \times 2$ and $3 \times 2 \times 2$ contingency tables: tests for interaction. *J. Amer. Statist. Assoc.*, **65**, 1617–1631.

Ord, J. K. (1972) *Families of Frequency Distributions*. Griffin, London.

Page, E. S. (1959) Pseudo-random elements for computers. *Appl. Stats.*, **8**, 124–131.

Page, E. S. (1965) On Monte Carlo methods in congestion problems: II Simulation of queueing systems. *Oper. Res.*, **13**, 300–305.

Page, E. S. (1967) A note on generating random permutations. *Appl. Stats.*, **16**, 273–274.

Page, E. S. (1977) Approximations to the cumulative normal function and its inverse for use on a pocket calculator. *Appl. Stats.*, **26**, 75–76.

Pangratz, H., and Weinrichter, H. (1979) Pseudo-random number generator based on binary and quinary maximal-length sequences. *IEEE Transactions on Computers*.

Parker, R. A. (1968) Simulation of an aquatic ecosystem. *Biometrics*, **24**, 803–821.

Parzen, E. (1960) *Modern Probability Theory and its Applications*. Wiley, Tokyo.

Patefield, W. M. (1981) An efficient method of generating random r × c tables with given row and column totals. Algorithm AS159. *Appl. Stats.*, **30** (1), 91–97.

Payne, J. A. (1982) *Introduction to Simulation: Programming Techniques and Methods of Analysis*. McGraw-Hill, New York.

Peach, P. (1961) Bias in pseudo-random numbers. *J. Amer. Statist. Assoc.*, **56**, 610–618.

Pearson, E. S., D'Agostino, R. B., and Bowman, K. O. (1977) Tests for departure from normality: comparison of powers. *Biometrika*, **64** (2), 231–246.

Pearson, E. S. and Hartley, H. O. (1970) *Biometrika Tables for Statisticians*, I. C.U.P.

Pearson, E. S., and Hartley, H. O. (1972) *Biometrika Tables for Statisticians*, II. C.U.P.

Pearson, E. S., and Wishart, J. (Eds.) (1958) *'Student's' Collected Papers*. C.U.P. (2nd reprinting; 1st issued 1942).

Pennycuick, L. (1969) A computer model of the Oxford Great Tit population. *J. Theor. Biol.*, **22**, 381–400.

Pennycuick, C. J., Compton, R. M., and Beckingham, L. (1968) A computer model for simulating the growth of a population, or of two interacting populations. *J. Theor. Biol.* **18**, 316–329.

Perlman, M. D. and Wichura, M. J. (1975) Sharpening Buffon's needle. *Amer. Statist.*, **29**, 157–163.

Peskun, P. H. (1973) Optimal Monte-Carlo sampling using Markov chains. *Biometrika*, **60**, 607–612.

Peskun, P. H. (1980) Theoretical tests for choosing the parameters of the general mixed linear congruential pseudorandom number generator. *J. Stat. Comp. Simul.*, **11**, 281–305.

Petersen, G. G. J. (1896) The yearly immigration of young plaice into the Liemfjord from the German sea, etc. *Rept. Danish. Biol. Sta. for 1895*, **6**, 1–48.

Peterson, A. V., Jr., and Kronmal, R. A. (1980) A representation for discrete distributions by equiprobable mixtures. *J. Appl. Prob.*, **17**, 102–111.

Peterson, A. V., Jr., and Kronmal, R. A. (1982) On mixture methods for the computer generation of random variables. *Amer. Statist.*, **36** (3), 1, 184–191.

Pike, M. C., and Hill, I. D. (1965) Algorithm 266. Pseudo-random numbers (G5). *Comm. A.C.M.*, **8** (10), 605–606.

Plackett, R. L. (1965) A class of bivariate distributions. *J. Amer. Statist. Assoc.*, **60**, 516–522.

Poynter, D. J. (1979) The techniques of randomized response. Unpublished undergraduate dissertation: University of Kent at Canterbury.

Press, S. J. (1972) *Applied Multivariate Analysis.* Holt, Rinehart and Winston, Inc., New York.

Rabinowitz, M., and Berenson, M. L. (1974) A comparison of various methods of obtaining random order statistics for Monte Carlo computation. *Amer. Statist.*, **28** (1), 27–29.

Ralph, C. J. and Scott, J. M. (1981) *Estimating Numbers of Terrestrial Birds.* Allen Press Inc., Kansas.

Ramaley, J. F. (1969) Buffon's needle problem. *American Mathematical Monthly*, **76**, 916–918.

Ramberg, J. S., and Schmeiser, B. W. (1972) An approximate method for generating symmetric random variables. *Comm. Assoc. Comput. Mach.*, **15**, 987–990.

Ramberg, J. S., and Schmeiser, B. W. (1974) An approximate method for generating asymmetric random variables. *Commun. Ass. Comput. Mach.*, **17**, 78–82.

Rao, C. R. (1961) Generation of random permutations of given number of elements using sampling numbers. *Sankhya*, **23**, 305–307.

Read, K. L. Q., and Ashford, J. R. (1968) A system of models for the life cycle of a biological organism. *Biometrika*, **55**, 211–221.

Reid, N. (1981) Estimating the median survival time. *Biometrika*, **68** (3), 601–608.

Relles, D. A. (1970) Variance reduction techniques for Monte Carlo sampling from Student distributions. *Technometrics*, **12**, 499–515.

Relles, D. A. (1972) A simple algorithm for generating binomial random variables when N is large. *J. Amer. Statist. Assoc.*, **67**, 612–613.

Ripley, B. D. (1977) Modelling spatial patterns. *J. Roy. Statist. Soc.*, B, **39** (2), 172–212.

Ripley, B. D. (1979) Simulating spatial patterns: Dependent samples from a multivariate density. *Appl. Stats.*, **28** (1), 109–112.

Ripley, B. D. (1981) *Spatial Statistics*. Wiley, New York.

Ripley, B. D. (1983a) On lattices of pseudo-random numbers. *J. Stat. Comp. Siml.* (Submitted for publication).

Ripley, B. D. (1983b) Computer generation of random variables—a tutorial. *Int. Stat. Rev.*, **51**, 301–319.

Roberts, F. D. K. (1967) A Monte Carlo solution of a two-dimensional unstructured cluster problem. *Biometrika*, **54**, 625–628.

Roberts, C. S. (1982) Implementing and testing new versions of a good, 48-bit, pseudo-random number generator. *The Bell System Technical J.*, **61** (8), 2053–2063.

Ronning, G. (1977) A simple scheme for generating multivariate gamma distributions with non-negative covariance matrix. *Technometrics*, **19**, 179–183.

Rosenhead, J. V. (1968) Experimental simulation of a social system. *Ope. Res. Quart.*, **19**, 289–298.

Rotenberg, A. (1960) A new pseudo-random number generator. *J. Ass. Comp. Mach.*, **7**, 75–77.

Rothery, P. (1982) The use of control variates in Monte Carlo estimation of power. *Appl. Stats.*, **31** (2), 125–129.

Royston, J. P. (1982a) An extension of Shapiro and Wilk's W test for normality to large samples. *Appl. Stats.*, **31** (2), 115–124.

Royston, J. P. (1982b) Algorithm AS177. Expected normal order statistics (exact and approximate). *Appl. Stats.*, **31** (2), 161–165.

Royston, J. P. (1982c) Algorithm AS181: The W test for normality. *Appl. Stats.*, **31** (2), 176–180.

Royston, J. P. (1983) Some techniques for assessing multivariate normality based on the Shapiro-Wilk W. *Applied Statistics*, **32**, 2, 121–133.

Rubinstein, R. Y. (1981) *Simulation and the Monte Carlo Method*. Wiley, New York.

Ryan, T. A., Jr., Joiner, B. L., and Ryan, B. F. (1976) *MINITAB Student Handbook*. Duxbury Press, North Scituate, Massachusetts.

Sahai, H. (1979) A supplement to Sowey's bibliography on random number generation and related topics. *J. Stat. Comp. Siml.*, **10**, 31–52.

Saunders, I. W., and Tweedie, R. L. (1976) The settlement of Polynesia by CYBER 76. *Math. Scientist*, **1**, 15–25.

Scheuer, E. M., and Stoller, D. S. (1962) On the generation of normal random vectors. *Technometrics*, **4**, 278–281.

Schmeiser, B. W. (1979) Approximations to the inverse cumulative normal function for use on hand calculators. *Appl. Stats.*, **28** (2), 175–176.

Schmeiser, B. W. (1980) Generation of variates from distribution tails. *Oper. Res.* **28**, 1012–1017.

Schmeiser, B. W., and Baliu, A. J. G. (1980) Beta variate generation via exponential majorizing functions. *Oper. Res.*, 917–926.

Schmeiser, B. W., and Lal, R. (1979) Squeeze methods for generating gamma variates. OREM 78009, Southern Methodist University, Dallas, Texas.

Schruben, L. W. (1981) Control of initialization bias in multivariate simulation response. *Comm. Assoc. Comp. Mach.*, **24**, 4, 246–252.

Schruben, L. W., and Margolin, B. H. (1978) Pseudorandom number assignment in statistically designed simulation and distribution sampling experiments. *J. Amer. Stat. Assoc.*, **73**, 504–525.

Schuh, H.-J., and Tweedie, R. L. (1979) Parameter estimation using transform estimation in time-evolving models. *Math. Biosci.*, **45**, 37–67.

Shanks, D., and Wrench, J. W. (1962) Calculation of π to 100,000 decimals. *Math. Comput.*, **16**, 76–99.

Shannon, C. E., and Weaver, W. (1964) *The Mathematical Theory of Communication*. The University of Illinois Press, Urbana.

Shapiro, S. S., and Wilk, M. B. (1965) An analysis-of-variance test for normality (complete samples). *Biometrika*, **52**, 591–611.

Shreider, Y. A. (1964) *Method of Statistical Testing: Monte Carlo Method*. Elsevier Pub. Co., Amsterdam.

Shubik, M. (1960) Bibliography on simulation, gaming, artificial intelligence and allied topics. *J. Amer. Statist. Assoc.*, **55**, 736–751.

Simon, G. (1976) Computer simulation swindles, with applications to estimates of location and dispersion. *Appl. Stats.*, **25** (3), 266–274.

Simulation I (1976) Unit 13 (Numerical Computation) of M351 (Mathematics). The Open University Press, Milton Keynes.

Simulation II (1976) Unit 14 (Numerical Computation) of M351 (Mathematics) The Open University Press, Milton Keynes.

Smith, J. (1968) *Computer Simulation Models*. Hafner, New York.

Smith, R. H., and Mead, R. (1974) Age structure and stability in models of prey-predator systems. *Theor. Pop. Biol.*, **6**, 308–322.

Smith, W. B., and Hocking, R. R. (1972) Wishart variate generator. Algorithm AS53. *Appl. Stats.*, **21**, 341–345.

Sobol, I. M., and Seckler, B. (1964) On periods of pseudo-random sequences. *Theory of Probability and its Applications*, **9**, 333–338.

Sowey, E. R. (1972) A chronological and classified bibliography on random number generation and testing. *Internat. Stat. Rev.*, **40**, 355–371.

Sowey, E. R. (1978) A second classified bibliography on random number generation and testing. *Internat. Stat. Rev.*, **46** (1), 89–102.

Sparks, D. N. (1973) Euclidean cluster analysis: algorithm AS58. *Appl. Stats.*, **22**, 126–130.

Sparks, D. N. (1975) A remark on algorithm AS58. *Appl. Stats.*, **24**, 160–161.

Spence, I. (1972) A Monte Carlo evaluation of 3 non-metric multidimensional scaling algorithms. *Psychometrika*, **37**, 461–486.

Stenson, H. H., and Knoll, R. L. (1969) Goodness of fit for random rankings in Kruskal's nonmetric scaling procedure. *Psychol. Bull.*, **71**, 122–126.

Stoneham, R. G. (1965) A study of 60 000 digits of the transcendental 'e'. *Amer. Math. Monthly*, **72**, 483–500.

Student (1908a) The probable error of a mean. *Biometrika* VI, p. 1.

Student (1908b) Probable error of a correlation coefficient. *Biometrika* VI, p. 302.

Student (1920) An experimental determination of the probable error of Dr. Spearman's correlation coefficients. *Biometrika* XIII, p. 263.

Sylwestrowicz, J. D. (1982) Parallel processing in statistics. *Compstat Proceedings*, Part I, 131–136. Physica-Verlag, Vienna.

Taussky, O., and Todd, J. (1956) Generation and testing of pseudo-random numbers. In: *Symposium on Monte Carlo Methods*. Ed. H. A. Meyer, pp. 15–28.

Tausworthe, R. C. (1965) Random numbers generated by linear recurrence modulo two. *Math. Comput.*, **19**, 201–209.

Taylor, S. J. L. (1954) *Good General Practice*. Oxford University Press for Nuffield Provincial Hospitals Trust, London.

Teichroew, D. (1965) A history of distribution sampling prior to the era of the computer and its relevance to simulation. *J. Amer. Statist. Assoc.*, **60**, 26–49.

Thompson, W. E. (1959) ERNIE—A mathematical and statistical analysis. *J. Roy. Statist. Soc.*, A, **122**, 301–333.

Tippett, L. H. C. (1925) On the extreme individuals and the range of samples taken from a normal population. *Biometrika*, **17**, 364–387.

Tippett, L. H. C. (1927) *Random Sampling Numbers*. Tracts for Computers, XV, C.U.P.

Titterington, D. M. (1978) Estimation of correlation coefficients by ellipsoidal trimming. *Appl. Stats.*, **27**, (3), 227–234.

Titterington, D. M., Murray, G. D., Murray, L. S., Spiegelhalter, D. J., Skene, A. M., Habbema, J. D. F., and Gelpke, G. J. (1981) Comparison of discrimination techniques applied to a complex data set of head injured patients. *J. Roy. Statist. Soc.*, A, **144**, 145–175.

Tocher, K. D. (1965) A review of simulation languages. *Op. Res. Quart.*, **16** (2), 189–217.

Tocher, K. D. (1975) *The Art of Simulation*. Hodder and Stoughton, London. (First printed 1963).

Tunnicliffe-Wilson, G. (1979) Some efficient computational methods for high order ARMA models. *J. Statist. Comput. and Simulation*, **8**, 301–309.

Uhler, H. S. (1951a) Approximations exceeding 1300 decimals for $\sqrt{3}, 1/\sqrt{3}$ and $\sin(\pi/3)$ and distribution of digits in them. *Proc. Nat. Acad. Sci. USA*, **37**, 443–447.

Uhler, H. S. (1951b) Many-figure approximations to $\sqrt{2}$, and distribution of digits in $\sqrt{2}$ and $1/\sqrt{2}$. *Proc. Nat. Acad. Sci. USA*, **37**, 63–67.

Upton, G. J. G., and Lampitt, G. A. (1981) A model for interyear change in the size of bird populations. *Biometrics*, **37**, 113–127.

Van Gelder, A. (1967) Some new results in pseudo-random number generation. *J. Assoc. Comput. Mach.*, **14**, 785–792.

von Neumann, J. (1951) Various techniques used in connection with random digits, 'Monte Carlo Method', *US Nat. Bur. Stand. Appl. Math. Ser.*, No. 12, 36–38.

Wagenaar, W. A., and Padmos, P. (1971) Quantitative interpretation of stress in Kruskal's multidimensional scaling technique. *Br. J. Math. & Statist. Psychol.*, **24**, 101–110.

Walker, A. J. (1977) An efficient method for generating discrete random variables with general distributions. *Assoc. Comput. Mach. Trans. Math. Soft.*, **3**, 253–256.

Wall, D. D. (1960) Fibonacci series modulo *m*. *Amer. Math. Monthly*, **67**, 525–532.

Wallace, N. D. (1974) Computer generation of gamma random variates with non-integral shape parameters. *Commun. Ass. Comput. Mach.*, **17**, 691–695.

Warner, S. L. (1965) Randomized response: a survey technique for eliminating evasive answer bias. *J. Amer. Statist. Assoc.*, **60**, 63–69.

Warner, S. L. (1971) Linear randomized response model. *J. Amer. Statist. Assoc.*, **66**, 884–888.

Wedderburn, R. W. M. (1976) Remark ASR16. A remark on Algorithm AS29. The runs up and down test. *Appl. Stats.*, **25**, 193.

West, J. H. (1955) An analysis of 162332 lottery numbers. *J. Roy. Statist. Soc.*, **118**, 417–426.

Western, A. E. and Miller, J. C. P. (1968) *Tables of Indices and Primitive Roots.* (Royal Society Mathematical Tables, Vol. 9). Cambridge University Press.

Westlake, W. J. (1967) A uniform random number generator based on the combination of two congruential generators. *J. Ass. Comput. Mach.*, **14**, 337–340.

Wetherill, G. B. (1965) An approximation to the inverse normal function suitable for the generation of random normal deviates on electronic computers. *Appl. Stats.*, **14**, 201–205.

Wetherill, G. B. (1982) *Elementary Statistical Methods* (3rd edn). Chapman and Hall, London.

Wetherill, G. B. *et al.* (1984) *Advanced Regression Analysis.* (In preparation).

White, G. C., Anderson, D. R., Burnham, K. P., and Otis, D. L. (1982) Capture–recapture and removal methods for sampling closed populations. Available from: National Technical Information Service, U.S. Department of Commerce, 5285, Port Royal Road, Springfield, VA 22161.

White, G. C. Burnham, K. P., Otis, D. L., and Anderson, D. R. (1978) *User's Manual for Program CAPTURE.* Utah State University Press, 40 pp.

Whittaker, J. (1974) Generating gamma and beta random variables with non-integral shape parameters. *Appl. Stats.*, **23**, 210–214.

Wichern, D. W., Miller, R. B., and Der-Ann Hsu (1976) Changes of variance in first-order autoregressive time series models, with an application. *Appl. Stats.*, **25**, 3, 248–256.

Wichmann, B. A., and Hill, I. D. (1982a) A pseudo-random number generator. *NPL Report DITC 6/82.*

Wichmann, B. A., and Hill, I. D. (1982b) Algorithm AS183: An efficient and portable pseudo-random number generator. *Appl. Stats.*, **31** (2), 188–190.

Wilk, M. B., Gnanadesikan, R., and Huyett, M. J. (1962) Probability plots for the gamma distribution. *Technometrics*, **4**, 1–20.

Williamson, E., and Bretherton, M. H. (1963) *Tables of the Negative Binomial Probability Distribution.* Wiley, Birmingham, England.

Wold, H. (1954) *Random Normal Deviates.* Tracts for computers XXV. C.U.P.

Worsdale, G. J. (1975) Tables of cumulative distribution functions for symmetric stable distributions. *Appl. Stats.*, **24** (1), 123–131.

Wright, K. G. (1978) Output measurement in practice. In: *Economic Aspects of Health Services* (A. J. Culyer and K. G. Wright, Eds). Martin Robertson, pp. 46–64.

Yakowitz, S. J. (1977) *Computational Probability and Simulation.* Addison-Wesley Pub. Co., Reading, Massachusetts.

Yarnold, J. K. (1970) The minimum expectation of χ^2 goodness-of-fit tests and the accuracy of approximation for the null distribution. *J. Amer. Statist. Assoc.*, **65**, 864–886.

Yuen, K. K. (1971) A note on Winsorized t. *Appl. Stats.*, **20**, 297–304.

Zeigler, B. P. (1976) *Theory of Modelling and Simulation.* Wiley, New York and London.

Zeigler, B. P. (1979) A cautionary word about antithetic variates. *Simulation Newsletter*, No. 3.

Zelen, M., and Severo, N. C. (1966) Probability functions. In: *Handbook of Mathematical Functions.* (M. Abramowitz and I. A. Stegun, Eds) Department of Commerce of the U.S. Government, Washington, D. C.

AUTHOR INDEX

SUBJECT INDEX